Measuring the New World

Measuring the New World

Enlightenment Science and South America

NEIL SAFIER

The University of Chicago Press   Chicago and London

The University of Chicago Press, Chicago 60637
The University of Chicago Press, Ltd., London
© 2008 by The University of Chicago
All rights reserved. Published 2008.
Paperback edition 2012
Printed in the United States of America

21 20 19 18 17 16 15 14 13 12   2 3 4 5 6
ISBN-13: 978-0-226-73355-5 (cloth)
ISBN-13: 978-0-226-73362-3 (paper)
ISBN-10: 0-226-73355-6 (cloth)
ISBN-10: 0-226-73362-9 (paper)

Library of Congress Cataloging-in-Publication Data

Safier, Neil.
    Measuring the new world : enlightenment science and South America /
Neil Safier.
        p.   cm.
    Includes bibliographical references and index.
    ISBN-13: 978-0-226-73355-5 (cloth : alk. paper)
    ISBN-10: 0-226-73355-6 (cloth : alk. paper)
    1. Scientific expeditions—Ecuador—Quito—History—18th century.
2. Ecuador—Discovery and exploration—French.   3. Ecuador—Discovery
and exploration—Spanish.   4. Science—Europe—History—18th century.
5. Mission géodésique (France)   6. Communication in
science—Europe—History—18th century.   7. South America—Historical
geography.   I. Title.
    Q115.S236   2008
    509′.033—dc22

                                                            2008018142

♾ This paper meets the requirements of ANSI/NISO Z39.48–1992
(Permanence of Paper).

For my parents—teachers and travelers

My Muse calls you to America. I had grown weary of the monotonous ideas of our theater, and I needed a new world . . . Now, all of the arts are to be found in Peru: there are some who measure it; I put it to song.

VOLTAIRE (1694–1778)

Contents

The Ascent of Francesurcu

In a long and narrow valley in southern Ecuador, only a few miles from the stately plazas and cobblestone streets of colonial Cuenca, sits the unassuming town of Tarqui. Inhabited centuries ago by a population that produced exquisite ceramic bowls and bulbous vases reminiscent of human forms, today's Tarqui appears to bear little witness to its colorful indigenous past. I had come to this valley in the central Andes to examine the material remains of an expedition that chose Tarqui as the site to resolve one of the greatest scientific debates of the eighteenth century. The dispute over the Earth's true shape, whether it was flattened or elongated near its poles, had been hatched in the halls of European academies. But it was to be resolved through observations carried out on South American soil, near the equator, thanks to ten Frenchmen, two Spaniards, and an endlessly shifting entourage of Amerindians, Africans, Creoles, and mestizos who transformed this tranquil valley into an early modern laboratory of instrumental science.

Before visiting Tarqui, I had spent a few hours poking around one of Cuenca's provincial archives. While examining a modern topographical map of the region, a certain place struck me as especially peculiar, even among names that to my eye appeared equally foreign, such as Bubugloma, Uchucloma, and Pucashpa. It was a small mountain named Cerro Francesurcu. From my reading of early Spanish travel narratives, I knew that *urcu* meant "mountain" in Quichua, the region's most prevalent indigenous tongue. Could Frances-*urcu* mean the mountain of the French, a fusion of Spanish and Quichua in a single word? Since the mountain lay just west of Tarqui, I surmised that the eighteenth-century savants whose trail I had been following may have carried out some of their measurements at that spot. What traces might they have left of their presence? What commemorative markers would I find at the summit? These questions engaged my imagination as I hailed a taxi on Cuenca's bustling Calle Simón Bolivar. Within twenty minutes, I was climbing my way toward an answer, past modest cornfields and Andean scrub brush to the summit of Francesurcu itself.

Despite a spectacular panoramic view with a bird's-eye perspective on Tarqui below, I was largely disappointed with what I found atop the 2824-meter peak. There was a small brick hut open to the east with a large wooden cross affixed to an inner wall, scrawled with illegible graffiti of local confection. Just beyond the shelter stood a plain white obelisk, with two thin wooden crosses placed beside it. There was also a stunted geodetic marker put there by the Military Geographical Institute in Quito. I later learned from a local man that this mountain was called "Ouaoua-tarqui," one of the sites I knew had been used by the Europeans to establish a baseline for their land surveying. But as I sat in the still and silent air at the summit of Francesurcu, I asked myself how such a place might have functioned as a laboratory for instrumental experiment two and a half centuries ago: how the various individuals might have worked together to transform an Andean hilltop into a site to resolve one of eighteenth-century Europe's most burning scientific questions. I tried to imagine young mestizos porting sticks and boulders along the same path I had just taken, while others set up European-made quadrants and telescopes in the valley below. The peak of Francesurcu seemed to retain only the faintest memory of these activities. And I began to suspect that something more than tectonic tremors and volcanoes had complicated the process of commemorating experimental science at this site.

Ascending mountains and waxing philosophical about monuments to human striving have a long and symbolic history in Western culture, and they have a particularly strong resonance in South America. Two centuries before my ascent of Francesurcu, a Prussian mineralogist named Alexander von Humboldt assailed an even higher summit in the equatorial Andes: the snow-capped peak of Chimborazo, then thought to be the world's tallest mountain. There, he battled ice and fog to surpass the record of another explorer, Charles-Marie de La Condamine, a member of the Paris Academy of Sciences who had preceded Humboldt to Chimborazo by more than half a century. Following in their footsteps, the great liberator of South America Simón Bolivar yearned to surpass the achievements of both Humboldt and La Condamine as he ascended Chimborazo's icy peaks in 1822. For Bolivar, no activity was more pregnant with dramatic symbolism than scientific exploration. On one occasion, delirious and possessed by what he called a "strange and other-worldly fire," Bolivar wrote that he had gone beyond where the "the footsteps of La Condamine and Humboldt" left off in Chimborazo's glacial snows. Perched high above what Humboldt had famously described as an "Avenue of the Volcanoes," Bolivar sought to emulate the high-reaching achievements of his European predecessors and thereby exalt his own political accomplishments. But the only testimony to these perilous ascents were the heroic narratives written by the climbers themselves: prose shrines recalling exhilarating explorations to frigid summits, but uncorroborated by observers who might have told a different story.[1]

My admittedly more modest ascent led me toward a set of questions rather than a grandiose narrative, prompting me to consider the perspective from which histories of scientific exploration should be recounted in the first place. What would the story look like, for instance, if we were to take in the view from the *foot* of Francesurcu as well as from its summit? What might we encounter if we were to examine the practices of scientific exploration not only from the soaring perspective of a Humboldt or a Bolivar but also from a vantage that was closer to the ground? How might the traces of these triumphant tales look, in other words, if we were to peer back at colonial science through a more expansive lens?

———

Assailing peaks is not a typical component of historical research. Most historians would rather spend their time burrowing into archives than climbing mountains. But from the very outset of this project, as I made my way from the dramatic landscapes of South America to equally picturesque European cities on a journey that comprised more than a decade of writing and research, I came to rely on far more than books and archival papers to pursue the historical roots of this extraordinary expedition. Throughout my travels, I was aided and encouraged by friends, family members, colleagues, and mentors who deserve high praise for their participatory role. I also accumulated a number of more specific debts to librarians, archivists, and institutions in Europe and the Americas, and it is my pleasure to acknowledge them here.

This book has expanded considerably from a doctoral dissertation defended at the Johns Hopkins University in September 2003, a project that saw its earliest incarnation as a seminar paper on La Condamine's journey down the Amazon during my first year of graduate school. I would like to thank my advisors, David A. Bell and Anthony Pagden, for their guidance, support, inspiration, and wise counsel during these many years. A special note of appreciation goes to Richard L. Kagan, who first suggested I study an eccentric French explorer in South America and who served as a model advisor in everything but name. Other members of the Hopkins faculty deserve my gratitude as well: sincere thanks to Philip Morgan, A.J.R. Russell-Wood, Orest Ranum, Gabrielle Spiegel, Wilda Anderson, Deborah Poole, and Mary Fissell for their advice and encouragement. In addition, the friends and fellow graduate students I met while in Baltimore always provided both camaraderie and good cheer: I would like to thank François Furstenberg, Sonja and Craig Hamilton, Anna Krylova, Matt Lauzon, Guy Lazure, Vaishali Patel, Anoush Terjanian, Paul Tonks, Craig Yirush, and Giovanni Zanalda for their friendship then and now.

As I crossed the Atlantic to begin my research, I was fortunate to have perceptive interlocutors who from the very outset took a sustained interest in my project. In Paris, I benefited greatly from seminars and conversations with Luiz Felipe de Alencastro, Francisco Bethencourt, Marie-Noëlle Bourguet, Roger Chartier, Christiane Demeulenaere-Douyère, Philippe Descola, Jean-Marc Drouin, Pierre-Antoine Fabre, Serge Gruzinski, Jean Hébrard, Yves Laissus, Kapil Raj, Carmen Salazar-Soler, Sanjay Subrahmanyam, and Nathan Wachtel, generous scholars many of whom I now count as colleagues and friends. Thanks also go to Caroline Comola, Cláudia Damasceno Fonseca, Elizabeth Lewin, Emmanuel Lézy, Yasmine Marcil, Stéphane Martin, James E. McClellan III, Ilda Mendes dos Santos, Raffaele Moro, Jacques and Héloïse Neefs, Lucette and Jean-Jacques Petit, Barbara Revelli, François Regourd, Fernando ("Pacho") Roca Alcázar, Alessandra Russo, and Ernesto and Cristina Santos for serious and not-so-serious conversations while residing in the French capital. During my research trips to Madrid, I was fortunate to be welcomed at the Consejo Superior de Investigaciones Científicas by Fermín Del Pino, Mercedes García Arenal, Leoncio López-Ocón, Juan Pimentel, and Miguel-Ángel Puig-Samper, while María Luisa Martín-Merás offered a similarly gracious reception at the Museo Naval and in her home. In Lisbon, special thanks go to Tiago Baptista, Miguel Cardoso, José Felipe Costa, Elisa Lopes da Silva, Bruno Peixe Dias, and Nuno Senos for their company and conversations in the Bairro Alto and beyond. I owe a special debt of gratitude to Sergio Miguel Huarcaya for joining me on a trip to Cochasquí and Yaruquí (Ecuador) in March 2006, to Costanza di Capua for her stimulating conversations in the quaint Quito neighborhood of Mariscal, and to Susan Webster for a memorable sunset in her apartment overlooking majestic Cotopaxi's snow-capped cone. And, finally, my most sincere appreciation to friends and colleagues in Brazil: Maria Fernanda Bicalho, Valéria Gauz, Guillermo Giucci, Beatriz Jaguaribe, Lorelai Kury, Ronald Raminelli, Manoel Salgado, and Antonio Carlos de Souza Lima all made me feel like an honorary *carioca* during my visits to Rio de Janeiro; in São Paulo, Laura de Mello e Souza took me on unforgettable visits to the library of José Mindlin and to the home of Antonio Candido, John Monteiro and Maria Helena Machado offered spirited conversation on the rural outskirts of the capital, and Íris Kantor and other members of the Cátedra Jaime Cortesão at the University of São Paulo introduced me to the culinary and cultural delights of the *vida paulistana*, for which I am ever grateful. But my true spiritual home in Brazil is Minas Gerais, and especially Belo Horizonte, where friends too numerous to list here have consistently welcomed me and encouraged my South American pursuits. Our many conversations, from the Café com Letras to Bar Social, are inscribed throughout these pages. Of these friends, I would especially like to thank Júnia Ferreira Furtado (and Lucas, Clara, and Alice Lins) and Maurício Meirelles (along with Patrícia Tavares) for their ever-gracious hospitality and friendship, in Brazil and elsewhere.

Archival collections and libraries on three continents formed a central component of my scholarly itinerary as well. Support for this travel and research was made possible through fellowships and grants from the Fulbright Foundation, the Ministry of Foreign Affairs of the French Embassy in the United States, the Luso-American Development Foundation, the J.B. Harley Research Fellowships, the Centre Culturel Portugais Calouste Gulbenkian (Paris), the Program for Cultural Cooperation of the Spanish Ministry of Education and Culture, the Horace W. Rackham Graduate School at the University of Michigan, and the American Philosophical Society. I would also like to acknowledge the directors and staff of the many libraries and archives in which I had the privilege to work: the Archives de l'Académie des Sciences (Paris); the Archives de l'Académie des Inscriptions et Belles Lettres (Paris); the Archives of the Royal Society (London); the Archivo General de Simancas (Spain); the Arquivo Histórico Ultramarino (Lisbon); the Bibliothèque Centrale du Muséum National d'Histoire Naturelle (Paris); the Bancroft Library at the University of California, Berkeley (with special thanks to Susan Snyder); the Biblioteca Nacional de España; the Biblioteca Nacional de Lisboa (and especially Luís Farinha Franco); the Bibliothèque Nationale de France (with special appreciation to Catherine Hofmann of the Département des Cartes et Plans for helping me obtain details of several maps in record time); the British Library (especially Peter Barber of the Map Library); the John Carter Brown Library (with a gracious nod to its former director,

Norman Fiering, and maps and prints curator Susan Danforth); the Clements Library of the University of Michigan (especially Barbara De-Wolfe, Brian Leigh Dunnigan, Clayton Lewis, Don Wilcox, and its former director, John Dann); the Dibner Library for the History of Science and Technology and the Cullman Library of Natural History at the Smithsonian in Washington, D.C. (including Ron Brashear, Leslie Overstreet, Kirsten van der Veen, and Daria Wingreen-Mason); the Geography and Map Division of the Library of Congress (especially its director, John Hébert, and Ron Grim); the Oliveira Lima Library at Catholic University (including Tom Cohen and his assistant curator, Maria Leal); the Museo Naval (Madrid); the National Archives (London); the Rare Book and Manuscript Library of the University of Pennsylvania (with special gratitude to Lynne Farrington and John Pollack); and the Westminster Archives (UK).

Back in the United States, I was offered residential research fellowships at the John Carter Brown Library, the Dibner Library for the History of Science and Technology of the Smithsonian Institution, and the Institute for Research in the Humanities at the University of Wisconsin. Special thanks to Loretta Freiling, Arthur and Janet Holzheimer, Jude Leimer, Neil Whitehead, and Rosalind Woodward for their warm welcome in Madison, to the late David Woodward for his perpetual words of inspiration and encouragement, and to my fellow fellows at the John Carter Brown Library—Jack Crowley, Matthew Edney, Hal Langfur, Matt Pursell, Matthew Restall, and Elizabeth Wright—for a grand time in Providence in the spring of 2002. In addition, I had the fortune of two post-doctoral fellowships that allowed me to complete this project and begin work on several others. To James Boyd White and Donald Lopez, former and current chairs of the Michigan Society of Fellows, respectively, I offer my sincere appreciation for their support. Likewise, I would like to thank the successive chairs of the Department of History at the University of Michigan, Sonya Rose and Mary Kelley, as well as my former colleagues in that department: John Carson, David William Cohen, Fernando Coronil, Sueann Caulfield, Dena Goodman, David Hancock, Martha Jones, Sue Juster, Farina Mir, Rebecca Scott, Julie Skurski, and Richard Turits, among others. I also extend a collective thank you to my fellow post-docs in the Michigan Society of Fellows, as well as individual thanks for their much needed and much appreciated friendship while in Ann Arbor to Enrique García Santo-Tomás, Karen Hebert, Matthew Hull, Katherine Ibbett, Paul Johnson, Valerie Kivelson, Eduardo Kohn, Stella Nair, Ziv Neeman, Mary and John Pedley, Shanan Peters, Jonathan Sheehan, Louise Stein, Daniel Stolzenberg, Miriam Ticktin, Geneviève Zubrzycki, and Jonathan

Zwicker. The final corrections to this book were completed while an Andrew W. Mellon Fellow in the Humanities at the Penn Humanities Forum (PHF) of the University of Pennsylvania. It is my pleasure to thank the forum's director, Wendy Steiner, its associate director, Jennifer Conway, and Sarah Sherger for welcoming me to Philadelphia. I would also like to thank my fellow PHF companions, and especially Edlie Wong, for their support and encouragement in the final stages of a very long process.

Other debts to friends and colleagues are more difficult to place into geographically bounded categories. Valéria Tavares has accompanied this project from its earliest stages, offering frequent and spirited encouragement from Baltimore to Belo Horizonte; I am immensely grateful for her enthusiasm, caring, and support over these many years. Jorge Cañizares-Esguerra read an early draft of this manuscript, suggested several primary sources, and provided crucial advice for turning a dissertation into a book; Tom Conley was always available with perspicacious comments and his ever magnanimous spirit, for which I am truly grateful; James Delbourgo shared his wisdom and humor during many conversations and provided useful commentary on several chapters; and Maya Jasanoff read portions of my manuscript at various junctures and talked me through difficult moments of many shapes and sizes. I also had the good fortune to share conversations with Nicholas Dew, Lissa Roberts, and Simon Schaffer on several occasions, from Halifax to Oxford and Haarlem, and always benefited greatly from our discussions. My friends *de longue date* Simon and Elizabeth Beaven, Andrew Dubrock, Lisa Kaufman, Michael Miller, Carolyn and Ruby Namdar-Cohen, Raphael Rothstein, Rashmi Sadana, Rudy and Susan Saltzer, Elizabeth Scott, and Joel Swerdlow have been unswervingly supportive, and I thank them here as elsewhere for being such loyal cheerleaders. Tom and Lisa Cohen-Fuentes, the Cohen-Fuentes clan, and Amy Scheuer Cohen know how much their encouragement and friendship have meant to me; I am grateful for the closeness we have maintained over these many years. The anonymous readers for the University of Chicago Press, who soon enough cast aside their veils of anonymity, provided important advice for improving this manuscript; I of course remain entirely responsible for any errors or omissions. An earlier version of chapter 6 was originally published as "'. . . To Collect and Abridge . . . Without Changing Anything Essential': Rewriting Incan History at the Parisian Jardin du Roi," *Book History*, vol. 7 (2004): 63–96; I thank the Pennsylvania State University Press for permission to reproduce portions of that article here. And, at the University of Chicago Press, Christie Henry deserves the highest praise for her professionalism and encouragement throughout the editorial process.

The final word of thanks goes to my family. My brothers and their respective families have provided an unending supply of emotional and material sustenance during my frequent visits to northern California. Deepest thanks go to Dan, Jackie, Joshua, Lauren, and Mike, Maurina, and Emma for their extraordinary patience and generosity through a process that was equally extraordinary in its length. Although she left us before I was able to share with her these first fruits, my grandmother Minnie Safier was and always will be a fount of inspiration for my study of history. But most of all I would like to thank my parents, Estelle and Robert Safier, to whom I have chosen to dedicate this book. They first encouraged me to travel the world and to learn about places and cultures beyond the confines of our own backyard, whether we were driving a motor home through the Pacific Northwest, hiking through Chilean Patagonia, or taking a whirlwind tour through Europe. Their spirit of curiosity impelled me toward journeys great and small, and I can only hope that this book will encourage them to continue exploring new worlds, just as those earliest travels guided and inspired me to do the same.

Introduction: New Worlds to Measure and Mime

An October day in 1739 brought commotion and revelry to the highland village of Tarqui. Every year, this small hamlet in the Spanish American province of Quito sponsored an extraordinary festival, one that commemorated historical events that had taken place centuries before in a land that few of Tarqui's inhabitants, if any, had ever seen. In preparation for these festivities, the local residents took out sets of colorful banners, outfitted themselves with mock weapons, and dressed in elegant tunics designed expressly for the occasion. Suitably attired, they mounted specially chosen steeds and trotted out to the open fields, where the day was spent racing, jousting, and parading before one another with their faithful squires (read: their costumed wives) alongside. Framed by the lofty peaks of the Andean cordillera and the church towers of colonial Cuenca in the distance, the locals galloped and whirled and strutted in a florid equestrian ballet. With imagination and verve, the residents of Tarqui merged drama and defiance as they reenacted in a picturesque valley of the central Andes one of Christian Europe's greatest territorial triumphs: the mythic reconquest of the Iberian Peninsula.[1]

Witnessing the spectacle that day in Tarqui were several even more oddly dressed visitors. Dubbed by one European journalist as "a New Kind of Argonauts, who had gone to the Land of Gold to search for truth," Louis Godin, Pierre Bouguer, and the most outspoken of the trio, Charles-Marie de La Condamine, had been sent to South America by one of Europe's most prestigious scientific academies to resolve what may have seemed a child's riddle: what was the true shape of the terrestrial globe? This group of French academicians was well aware, however, that the question of the Earth's form was causing quite a stir back in Europe. Eighteenth-century astronomers knew that the Earth was not a perfect sphere, but they had been unable to determine empirically whether it bulged vertically toward the poles or horizontally at the equator: in more colloquial terms, whether it was long like a lemon or squat like a carnival squash. In order to assess the Earth's curvature, these "geometers"—schooled in the practice of trigonometric surveys and astronomical observation—had come to the most inhospitable of scientific laboratories, spending the better part of their days scaling volcanic peaks, traversing roaring rivers, and shielding themselves from frosty winds, all to transform the jagged Andean landscape into a crisscrossing series of angular shapes on paper. After years of triangulating with telescopes, quadrants, and compasses, they were nearly ready to proclaim the definitive resolution of this fiercely disputed contest of European science. Their scientific spoils would owe much to this Andean village, thousands of miles and worlds away from the drawing rooms, academies, and powdered wigs of their Old World colleagues.[2]

Unbeknownst to the Europeans sitting in the audience that October day, however, the local villagers had long been planning a special performance for their foreign guests. At an opportune pause in the equestrian races, a rapt crowd watched as several youths took to the stage. This group of young mestizos, children of mixed European and indigenous descent, wheeled out several large wooden devices, constructed and painted especially for the occasion, and began to fiddle with the virtual parts just as a Parisian astronomer might calibrate a quadrant or a zenith telescope. With theatrical aplomb, they turned their heads upside down and peered into imaginary eyepieces as they adjusted the angles of the oddly shaped shafts. Using improvised writing implements, they scribbled data onto small sheets of invisible paper, scurrying back and forth between the large devices and conferring amongst themselves as to their results. And finally, when the audience might have thought their performance at an end, they began the whole process again, acting out the repetitive gestures they had witnessed so many times while observing the instrumental rituals of their European visitors.

We owe the description of this extraordinary Andean pantomime to Charles-Marie de La Condamine's *Journal du voyage fait par ordre du roy* (1751), a narrative account of the expedition sent in 1735 to determine the length of a three-degree arc of the meridian near the equator. Like the eleven other Europeans who accompanied him, La Condamine had gone to Quito as part of a new wave of traveling empiricists, hoping to trans-

Figure 1. This engraving shows two European savants taking celestial measurements with a zenith sector in a candlelit observatory. Visible through the open door, a volcano spews smoke and flames skyward while a young child, gazing intently at the European astronomers, takes careful note of the foreigners' peculiar costumes and rituals. La Condamine, *Mesure des trois premiers degrés* (1751). Courtesy of the Harlan Hatcher Graduate Library, University of Michigan.

form instrumental measurements in places like Tarqui into authoritative texts that garnered praise and prestige from admirers back home (fig. 1). They referred to the transfer of scientific knowledge from Europe to the Andes as a "sacred fire," passing mysteriously through European astronomical instruments to curious observers in South America. To encompass the globe with their instrumental observations, however, they had to rely on the active participation of resident guides, laborers, and knowledge brokers, locals who considered interlopers like La Condamine either highly peculiar or hopelessly mad—or both. The French academicians certainly never expected the local populations to subject their behavior to such dogged scrutiny. Nevertheless, as one of the objects of this staged ridicule, La Condamine recounted each and every detail of their pantomimic sketch, marveling at the young mestizos' keen ability "to imitate perfectly everything they see." In awe and admiration, he was astonished to report that "none of our movements [had] escaped [their] curious gaze."[3] And with a sense of fascination mixed with narcissistic self-recognition, La Condamine hardly contained his laughter as he dutifully described the Tarqui pantomime in print:

[T]hese buffoons [had] mimicked us with such accuracy that it was impossible for us not to recognize ourselves . . . And I was taken with such a strong desire to laugh that for a few moments, I forgot my more serious occupations.[4]

Through the mimicry of these mestizos, a French astronomer in a foreign land who had spent three years peering through eyepieces and narrow tubes was able to observe with his own eyes the strangest sight of all: himself. But from our own perspective, albeit with considerable historical and ethnographic distance, it is difficult to decide for sure who was the playwright, who were the actors, and which direction the stage in this Andean village was actually facing.[5]

What did this pantomime mean, in other words, to its performers and its audience? The question is as intriguing as it is inscrutable. If we were to ask ethnohistorians of the colonial Andes, they would tell us that mimicry was a common practice of indigenous cultures living under colonial Spanish rule and that ritual enactments and nonverbal performances like these constituted an unwritten archive of social memory.[6] Even in the eighteenth century, European visitors to South America frequently acknowledged that the native populations "have a genius for Arts, and are good at imitating what they see."[7] Following postcolonial critics, we might acknowledge that local populations the world over used colonial mimicry as a form of active resistance to European attempts to control them, and that imitation and mimesis undermined Europe's "panoptical vision of domination" by "[mocking] its power to be a model" at all.[8] Cultural historians in turn might suggest we focus on a Frenchman's laughter, observing that humor masked deeper sociocultural tensions that those wearing the greasepaint on an Andean stage might have sought to bring to light.[9] And anthropologists might explain how this "reverse ethnography" constituted a wholly new form of cultural encounter between Europeans and the non-European world, a "second contact" in which the observer and the observed lose their inherent polarity, blurring the lines between the subject and object of ethnographic description.[10]

Each of the above theoretical explanations has its drawbacks as well as its merits. But I draw attention to this scene because it encapsulates some of the fascinating dynamics that emerged as eighteenth-century science moved away from the laboratories and landscapes it had traditionally inhabited back in Europe. The pantomime forces us to confront the *peculiarities* of the encounters that emerged when Europeans extended their field sites across oceans and into radically different social and cultural spaces. The pantomime reminds us that a group of French natural scientists were engaged not with members of their own national

communities but rather with a Spanish monarch's diverse colonial subjects, who were in turn capable of reacting in a fervent, creative, and
sometimes hostile manner to European observational practices. In short,
the powerful and unexpected juxtaposition of a group of French savants
being observed and mimicked by a group of Andean villagers in colonial
Spanish territory dispels any overly simplistic ideas about how imperial
expansion and instrumental science may have marched in lockstep in
the Americas or elsewhere.

What if we were to follow the Tarquian mimics in looking at the gestures and practices of European science overseas: the inchoate and intermediate stages of scientific observation rather than the ornate title
pages and manicured charts by which natural knowledge was often self-
consciously portrayed by European authors? In emphasizing the physical rituals they had seen rather than the results those operations had
produced, the local performers at Tarqui provided one of the earliest
non-European representations of the gestures associated with European
instrumental practices in the Americas. Whatever their ultimate intent,
they expressed eloquently through pantomime what historians of science
have taken pains to address in scholarship of far more recent confection:
that European empirical practices and instrumental observations were
socially embedded performances comprising repetitive and theatrical
gestures; that the recording of data in material form was a crucial stage
through which instrumentally based observations were transformed into
codified measurements; and, finally, that public display was an inherent and essential element in the collection, analysis, and communication of those measurements to a broader audience. In essence, the pantomime portrayed science as a socially circumscribed and materially
bounded spectacle. For an experiment to prove its authority, it needed
to be physically executed, textually codified, and publicly demonstrated
to its audience—whether through a monumental structure, a theatrical performance, or a textual narrative. These forms of public persuasion
were all the more necessary when what separated the laboratory from the
printing house was not a few city blocks but rather an intercontinental
divide: an expanse that reached over a hemisphere's highest peaks, along
one of the world's most voluminous rivers, and across an ocean sea.

———

So how did empirical practices change when they moved outside the
confines of Europe, and how were they affected by the social and material

landscapes through which they passed? *Measuring the New World* answers these questions through a sustained examination of the joint French and Spanish expedition to colonial Quito, a mission whose goal was to determine definitively the true shape of the Earth. From 1735 to 1744, its twelve core participants crossed the Atlantic, traversed the Caribbean, and made their way to the western coast of South America, where they sought local assistance to establish semipermanent observatories from Quito to Cuenca in order to carry out their measurements (plate 1). Never before had the experimental resolution of such an important scientific debate been attempted on American soil. Never before had two European nations under the auspices of a royal academy cooperated so explicitly for the pursuit of a common, nonmilitary goal. And never before had the results of an expedition been so self-consciously celebrated in maps, journals, and official publications back in Europe. In order to understand such an unprecedented convergence of efforts in the name of natural knowledge, it may be useful to provide some historical background to the expedition itself.

Fierce debates over the Earth's shape began to rage following the publication of Isaac Newton's *Principia Mathematica* (1687), which had predicted the flattening of the Earth near its poles. In one corner stood London's Royal Society, supporting Newton's assertion that "the axes of the planets are smaller than the diameters that are drawn perpendicularly to those axes" due to centrifugal forces inside a fluid Earth. In the other corner was the Paris Academy of Sciences, invested in seeing the Earth as a "prolate spheroid" that bulged at the poles rather than at the equator following the theories of René Descartes. Since earlier measurements carried out in Europe had been inconclusive, two royally sponsored expeditions were dispatched to reevaluate the Earth's shape. One group of scientists traveled with Pierre-Louis Moreau de Maupertuis to Lapland in 1737, while Louis Godin, a senior member of the Academy and a mathematician, led fellow "geometers" Charles-Marie de La Condamine and Pierre Bouguer to the mountains near Quito in the Spanish viceroyalty of Peru. These academicians were accompanied by seven other affiliates of the French Academy: the surgeon Jean Seniergues, later killed by an angry mob in Cuenca during a popular uprising; Théodore Hugo, a watchmaker and guardian of the group's scientific instruments; the young Couplet, an apprentice who was the nephew of the treasurer of the Academy of Sciences, Claude Couplet (few frowned upon nepotism in the Old Regime); Jean Godin des Odonais, Louis Godin's nephew; Jean Louis de Morainville, another assistant who helped construct maps and other monuments to the expedition; Jean-Joseph Verguin, who also composed several important maps; and Joseph de Jussieu, brother to the famed Parisian botanists Antoine and Bernard de Jussieu, who spent thirty-five years as the expedition's botanical specialist traversing the jungles of South America only to arrive disoriented and destitute in France.

Why would a group of Frenchmen be sent to Spanish America in the first place, of all the sites they might have chosen to measure the length of the meridian at the equator? The primary reason was the much-disputed ascension to the Spanish throne of a Bourbon monarch, whose subsequent reign allowed the members of the French academy to carry out their experiments in a region that had been officially off limits to non-Spaniards for centuries. Following the death of Charles II of Spain in 1700 and a war that ensued over who could legitimately succeed him, the victorious monarch Philip V (a grandson of Louis XIV of France, who had grown up at Versailles) effected a grand restructuring of Spanish administration that included taking stock of the political and economic status of its overseas territories. In a period of significant economic pressures on the Crown, Philip believed that sponsoring scientific missions—including the expedition to Quito as well as a host of other expeditions later in the century—would demonstrate the "enlightened" nature of the new Iberian monarchy and augment the prestige of the Spanish nation.[11] To this effect, he named two young Spanish military officers, Antonio de Ulloa and Jorge Juan, to accompany the ten Frenchmen to Quito and to ensure that they did not engage in illicit trade or espionage. In the end, the Spaniards also carried out many of the observations in tandem with their French colleagues and published one of the most important accounts of the expedition, the *Relación histórica del viage a la América meridional* (1748). They took part alongside their French counterparts in constructing ephemeral markers, measuring distances, constructing triangles, and drawing maps. For Spain, the participation of these two young naval officers represented the first official step toward a Spanish scientific presence on the European stage. But for the French, the stakes of these calculations were literally global in their expanse: quoting a contemporary, Jean Le Rond d'Alembert (1717–83) remarked that "it was a question of national honor not to let the Earth have a foreign shape, a figure imagined by an Englishman or a Dutchman."[12] Unfortunately for d'Alembert and the rest of his cohort, the measurements in Lapland and the Andes made utterly clear that the Earth was shaped in the Newtonian (English) manner. Cartesian pride would be slow to recover.

Rather than following the monarchical edicts and macropolitics that made it possible for a group of French savants to determine the Earth's shape in South America, this book redirects our attention toward the local conditions that enabled them to carry out their observations on the spot. Experimental science in a colonial context, as elsewhere, was not practiced in a social vacuum. Instead, encounters unfolded spatially and temporally in unpredictable ways, especially between Europeans and the populations they found in the "New World." The chapters in this book trace these pathways through a series of textual and material "monuments" as the Quito expedition moved between continents and across cultures: from manmade markers on a plain near Quito to a raft on the Amazon River, from copperplates in a cartographic printing house to the hothouses of European gardens. The "measuring" of the title refers not only to the new shape of the Earth predicted by Newton and confirmed by their expedition, but also to the process of understanding a social world that was radically different from the one they had left behind. They were measuring in the traditional sense, but they were also weighing and evaluating in a far more metaphorical vein: by observing indigenous cultures, by assaying plants and minerals, and by assessing historical truths. Previous scholarship on the Quito expedition has tended to emphasize the ways that scientific exploration went arm in arm with colonial domination, arguing for seeing scientific expeditions like the Quito mission as one of "Europe's proudest and most conspicuous instruments of expansion."[13] But if we examine the social and material processes by which experimentation actually took place, as well as the range of spaces in which these activities occurred, we find that the results often left a far more ambiguous legacy than scholars of European and Latin American history have led us to believe. The numerous actors and contingent factors through which measurements and observations came to be discerned and communicated underscore the physical materials used and the resources employed to produce them. The myriad publications produced by the expedition bear witness to how their authors attempted to understand natural phenomena in a social world that was often incommensurable with European norms. And the polemical debates illuminate the conflicts and challenges that arose not only among nations and academies but also between individuals on both sides of an ocean as they fought for their own piece of scientific prestige.[14]

These social conflicts belie the universalizing aspirations of European science and its representational practices. Extra-European scientific practice relied on local forms of knowledge and frequently suppressed the sources from which that knowledge derived. Subsequently, this "other worldly" knowledge was incorporated, transformed, and packaged to satisfy the presentational protocols of European print culture. Through their desire to portray themselves as intrepid heroes fighting single-handedly to overcome the challenges stacked against them, scientific actor-protagonists frequently effaced the processes by which this knowledge was provided to them by other agents—not only Amerindians, Africans, and mestizos but American-born Creoles and other Europeans as well.[15] The many sources of the knowledge European travelers acquired in situ were often disguised and displaced, disenfranchising those who were responsible for its collection, including their assistants as well as their adversaries. Sometimes this was conscious. But more often than not, as the chapters that follow make clear, this hidden knowledge was a result of editorial practices that scraped away the record of how such knowledge was derived. One of the goals of this book, then, is to show how social and material practices served to efface a far wider range of actors than has previously been recognized, including individuals who were poorly placed in material as well as geopolitical hierarchies of scientific knowledge production. Since printed texts tend to paper over the processes by which they came into being, the voices of those whose traces do not appear emblazoned on the title pages of European texts, such as those of African-descended peoples and South America's native populations, need to be reinscribed into a broader social history of knowledge production. At the same time, tracking the intermediate stages of textual elaboration provides an important corrective to a kind of intellectual history that has paid scant attention to the material realities of knowledge production, and that often imagines the formulation and circulation of ideas as a disembodied enterprise. Only by reconstituting the material practices, the social relationships of the actors involved, and the successive stages of textual elaboration and production can we understand the broad contours and unintended consequences of science and its representations during this period. By recovering what was invisible in the final versions of European accounts, we recognize Enlightenment science in an age of imperial expansion for what it was: not an omniscient, universal knowledge of the natural world but rather a partial and contingent knowledge, one that silenced and suppressed its sources just as often as it acknowledged and represented them.

Three centuries ago, this limited knowledge produced by European travelers to the New World created an epistemological paradox for Enlightenment authors. Throughout the eighteenth century, travelers and their accounts were both revered and reviled in the same breath, praised and lauded but also frequently criticized. As the French academician and natural philosopher Condorcet wrote in his posthumously published *Outlines of an Historical View of the Progress of the Human Mind* (1795), "travelers are almost always imprecise observers; they examine objects all too quickly, with all the prejudice of their own country."[16] His sentiment was not an uncommon one. Two-and-a-half decades earlier, the Dutch philosophe Cornelius de Pauw had written that "as a general rule, . . . out of a hundred travelers, there are sixty who lie for no reason but their own stupidity, thirty who lie for some gain, or if you will, out of malice, and lastly ten who tell the truth, and act as men."[17] But travelers also served as the portable eyes and ears of more sedentary philosophers, those who sought knowledge beyond European shores but were unwilling or unable to travel themselves. Travelers represented the physical manifestation of Europe's thirst for and fantasies about an omniscient understanding of the outside world. Expeditions to Asia and the Americas in the seventeenth century were undertaken by individuals who either already were or soon would become members of Europe's most prestigious scientific academies: Willem Piso and Georg Marggraf's expedition to Dutch Brazil (1636–44), Jean Richer's voyage to Cayenne (1671–73), Hans Sloane's journey to Jamaica (1687–89), and the French Jesuit mission to China (1685) are only a sampling of the voyages that served as precursors to later expeditions in the eighteenth century, most notably the Pacific voyages of Malaspina, Lapérouse, Bougainville, and Cook. Exotic evidence was an especially coveted commodity in the scientific communities of the Old Regime. Ideas and material objects brought in through intercontinental, as opposed to domestic, networks and connections came to hold a special place in the knowledge economy of early modern Europe. It is no coincidence that writers such as Diderot and Herder took up the subject of non-European peoples and places in their writings on empire. And while the accounts penned by eighteenth-century travelers frequently attained the status of best sellers (Bougainville's and Cook's Pacific voyages being the best examples), they were also increasingly subject to the vituperative attacks and harsh critiques of erudite readers. Anticipating Condorcet, Diderot wrote in a letter to Madame Necker that "one must have a long sojourn in order to understand the most common phenomena with even a small degree of accuracy; the traveler who takes notes on his writing tablets as the wheels [beneath his carriage] are spinning is quite aware that what he composes is [nothing but] lies."[18] To err and to be errant were thus seen as two sides of the same coin: travelers' observations were necessary for the accumulation of eighteenth-century knowledge, but the foundation of that knowledge was treated with suspicion if not outright contempt by many of the writers who relied most heavily upon it.[19]

The figure of the "philosophical" traveler emerged as a conscious antidote to these critiques and offered the greatest hope for satisfying

Europe's craving for knowledge of the outside world. Many eighteenth-century authors came to believe that such traveling sages might ultimately bring order to the moral chaos that had emerged through faulty observations. While Rousseau in his *Discourse on the Origins of Inequality* famously indicted those traveling dilettantes who described only what they might have already observed "without leaving their own street," he also wrote encouragingly of a new wave of travelers who would someday be capable of providing legitimate knowledge about many new parts of the world. He praised the "magnificence" of those curious sorts who traveled to the Orient in search of ruins and inscriptions, but he also insisted that what was truly needed was a new kind of traveler that would not consider "stones and plants" but human customs. Rousseau called this figure a "philosophical" traveler, "traveling to instruct [his] compatriots," and reflected optimistically on the kinds of texts that would emerge if the likes of Montesquieu, Diderot, or d'Alembert were to take to the road in search of human nature:

Let us suppose that upon returning from their memorable travels, these new Hercules were to write at their leisure the natural, moral, and political history of what they had seen: then we would see a new world issue from their pens, and would thus learn to know our own.[20]

Rousseau's hope that a new world would magically emerge from philosophers' pens may have been little more than eighteenth-century reverie. But scholars since then have largely followed Rousseau in ignoring the social and material conditions under which knowledge flowed from these "memorable travels" into travelers' memoirs. Students of print culture know that intellectual history cannot be untied from its real-world moorings: philosophical travelers were not lone truth-tellers, they did not produce unmediated accounts of their journeys, and they were not sealed off from the social world that surrounded them. The elaboration of travel narratives describing scientific voyages overseas occurred in specific scribal and editorial contexts and was dependent as much on material processes as it was on an author's philosophical predispositions. What is more, as Roger Chartier has observed, the production of texts in early modern Europe ultimately led to such an "uncontrollable textual proliferation" that new strategies were developed in order to erase what had been written previously, a practice that was "as essential for [the act of] remembering as [that of] forgetting." By focusing on the multiple sites of collection and textual production through which a single eighteenth-century expedition passed from Europe to South America and back again, I show not only that these practices of effacement were prevalent *outside* of Europe as well, but that extra-European expeditions were one of the main reasons such practices of erasure existed in the first place. As narratives of overseas exploration proliferated, an impressive array of seeds, saplings, and other specimens flooded into European repositories, which included the burgeoning public spaces of museums, curiosity cabinets, and botanical gardens. Naturalists and collectors had to filter, separate, and package these materials; compilers and editors in turn had to "reduce" the descriptions to their utmost "simplicity" by trimming what was seen as extraneous or useless information for the printed books and dictionaries of nature they were producing. These two impulses of accumulation and abridgement reflected two extremes of Enlightenment knowledge production. Exploration and observation beyond European shores allowed for the "emergence" of a "new world" from the pens of heroic travelers, in the words of Rousseau, while editorial practices made possible the "erasure" or diminution of that same knowledge in order to control what Chartier calls "the accumulated discourses" of European philosophy.[21]

Measuring the New World attempts to situate the production of Enlightenment knowledge about South America between these two poles. In order to do this, each chapter focuses on the social and material practices that comprise what I have called transatlantic scientific commemoration. This term brings together under one umbrella the broad set of activities through which empirical observations were transformed into tangible, memorable products in the emerging space of a noninstitutionally bound transatlantic public sphere. Whether through print, monumental architecture, collected specimens, or manuscript narratives, the expedition to Quito was one of the first officially sponsored expeditions in which academicians appealed through commemorative gestures and objects to a new arbiter of scientific utility: the European reading public. The Quito expedition garnered unprecedented attention in the popular press, and the wave of "official" publications that recounted the mission's exploits would have provided ample intellectual sustenance for even the most voracious eighteenth-century reader. As the academicians were departing from Europe, publications such as the abbé Prévost's *Le Pour et Contre* were already remarking on the monuments and inscriptions that the academicians planned to construct at the very site of their operations. Even prior to their departure, memory and publicity seemed to accompany one another hand in hand.

But while commemoration in a modern sense typically refers to the ritualized, temporally fixed celebration of a particular event, I use "scientific commemoration" in the sense that the Académie française implied in its 1694 dictionary, where it defined *commémoration* as "that which is done, or that which is said, to renew (*renouveller*) the memory of someone, or something."[22] In the context of science moving across transatlantic spaces, the idea of renewing or reviving the memory of an experimental act is particularly germane. Given the distance that separated the European scientific community from sites of empirical observation in the Americas, the repetition of these experiments would have been impossible. Thus, "scientific commemoration" could connote both the rehearsal ("that which is done") and the relation ("that which is said") of a particular event or activity, without necessarily being bound to a particular date or anniversary. Most of the time, scientific observation across the Atlantic was ephemeral and sui generis: the scientific actors would not be returning to the sites at which these experiments initially took place. So in order for the results to be remembered on a more permanent basis, they needed to be inscribed materially and narrated textually; as they reached a wider audience, they were read, corrected, and criticized; and for them to become subsumed within the realm of universal knowledge, they had to be translated into the appropriate language and abridged to fit within a particular physical or generic format. These are some of the components of scientific commemoration as conceived and carried out in a public sphere, and this book will examine each of them—from inscription through abridgement—in the chapters that follow.

These voyages of scientific exploration took place at a time when natural philosophical practice was undergoing fundamental change as well, especially with regard to the public status of the early modern laboratory. Historians of science have identified the public verification of experimental results as one of the main criteria by which early modern theories and hypotheses became transformed into experimental facts, and they have demonstrated the fallacy of seeing the experimental laboratory as a secluded chamber isolated from the broader social sphere. Scientific instruments became the sine qua non of experimental facticity, and natural philosophers subsequently learned to manipulate instruments as well as contemporary social codes to validate their knowledge claims. But historians have been less intrepid about studying what happened when these sites of knowledge production and accumulation moved to other parts of the world. They have been even more reluctant still to traverse the national and linguistic boundaries that have traditionally defined their fields, a methodological impulse that is essential for moving toward a pan-Atlantic or even global understanding of early modern scientific cultures.[23]

In addition to taking a social and material approach to the history of scientific exploration, then, this book places these practices into a context that is consciously transnational and transimperial. Historical studies of the European appropriation of non-European spaces have traditionally focused on institutions sponsored by individual states that collected and codified new information brought in from overseas. Under the aegis of colonial administrators, these institutions often managed political and territorial dominion as European empires expanded. The Casa de la Contratación in Spain, the British East India Company, the Dutch East India Company (VOC), and the French Compagnie des Indes became privileged loci for analyses that placed a near-omnipotent empire at the forefront of colonial knowledge collection, and they have been studied for their important role in transforming emerging colonial societies into economically viable arms of the larger imperial body.[24] My approach is different. While not denying the important and often central role of these institutions, I emphasize instead the shifting and transnational platforms on which scientific knowledge was produced. My analysis reconfigures exploration as a series of progressive spaces or moving "laboratories," in which experiments are performed across traditional geographical and geopolitical divides. The model for how these portable laboratories transported knowledge is the itinerary itself—defined in the late-seventeenth century as an "account [*mémoire*] of all the places one passes going from one country to another, and sometimes also the things that happen to those who have taken the route."[25] The focus on itineraries deemphasizes a static center-periphery scheme and demonstrates instead the interconnecting and contingent pathways that link individuals, objects, and impulses not only from one country to the next but also *between* institutions and *across* empires. Viewing these encounters along strictly national or imperial lines imposes excessively narrow geographical and linguistic parameters. It also misrepresents the complexity of these interactions as cultures mixed, peoples interacted, allegiances changed, and identities shifted. These methodological limitations ultimately lead to a monolithic and somewhat stagnant picture of how colonial science functioned: an image that may adequately reveal the powerful institutional resources of the imperial state but that fails to recognize the inconsistencies, unexpected itineraries, and occasional colossal failures in the accumulation and deployment of natural knowledge during an age of imperial expansion.[26]

In short, colonial science was not merely produced in a narrow "contact zone" where "agents of empire" and their "imperial subjects" faced off in a rote and predetermined fashion. Instead, knowledge emerged from a broad narrative interaction involving multiple sites of collection and codification. While the Quito expedition did of course take advantage of preexisting imperial pathways, and while some of the knowledge derived from the mission was eventually folded into imperial networks, the expedition itself was neither a colonial project to reconnoiter terra incognita nor did it make any explicit territorial claims on behalf of any European power. The specific context of the Quito mission defies conventional understanding of the institutional, national, and imperial boundaries of the eighteenth century, and it challenges the reader to ask which individual nation's "Atlantic world" would be capacious enough to contain the expedition's bifurcating paths. Indeed, many scholars still insist on speaking of British, French, and Iberian "Atlantics." But none of these is a separate ocean unto itself; only an approach that integrates local, national, and imperial perspectives rather than merely comparing across geographic lines can begin to account for the borrowings, interactions, and conflicts through which the Atlantic became a stage for empiricism and exchange in the early modern period.[27]

————

The book begins by examining a set of pyramidal monuments conceived in Europe and erected on site in Quito to commemorate the results of the academicians' measurements. This elaborate if ultimately failed project was devised by Charles-Marie de La Condamine, arguably the most effective communicator of the expedition's achievements, and it exemplifies how La Condamine used the resources available to him in situ to increase his own prestige and augment his right to speak on behalf of the Quito mission as a whole. Although he did not explicitly portray his actions in this way, the pyramids came to serve La Condamine as symbolic representations of scientific truth, embodying in architectural form a larger philosophical struggle over who had the right to proclaim the results of experimental inquiry on foreign soil. Chapter 2, which treats La Condamine's return journey down the Amazon River, shows how La Condamine transformed the raw materials of the explorer—including letters, maps, manuscript treatises, and scientific instruments—into a potent arsenal of tools by which he transferred intellectual possession of Amazonian history and geography from previous explorers to his own text. In the process, he suppressed the contributions of Jesuits, Creoles, and Amerindians alike while portraying many of their local observations as his own. In the guise of "autopsy" and eyewitness observation, he substituted knowledge he had culled and collected from other sources throughout his journey, never revealing that many of his most bold and infamous assertions about Amerindian culture and Amazonian myth were not rooted in on-the-spot observation but rather were based upon bookish and epistolary consultation both before and after he descended the river.

Chapters 3–5 begin to focus on actors that were more peripheral to the actual expedition but who nonetheless played a direct role in the ultimate shape of its reception in Europe and the Americas. In the months, years, and decades that followed La Condamine's return, a host of readers responded to his account in manuscript and in print; the reactions elicited by the *Relation abrégée d'un voyage fait dans l'intérieur de l'Amérique méridionale* (1745) in the periodical press and in private correspondence, from Amsterdam to Amazonia, are examined in chapter 3. What is striking about these reactions from both sides of the Atlantic are the similarities rather than the differences in their responses. Independent of any particular national or religious tradition, these readers saw La Condamine's account as one that placed "metaphysical" arguments above on-site observations, and they criticized an epistemology that allowed a transitory explorer to make broad assertions about the cultures through which he had passed without adequately understanding the linguistic and historical parameters of the communities in question. Nevertheless, northern European authors did respond differently from several Iberian agents, who perhaps not surprisingly remarked above all on the political and territorial implications of La Condamine's account. Chapter 4 brings us into the cartographic atelier of Jean-Baptiste Bourguignon d'Anville, as we trace the communication between the printers and engravers who constructed the *Carta de la Provincia de Quito* (1750) based on Pedro Vicente Maldonado's manuscript annotations. Maldonado descended the Amazon with La Condamine, crossing the Atlantic soon thereafter and working alongside his French friend to publish his map in Europe. The plurality of authorial and editorial voices in the cartographic workshop nevertheless calls into question the claims for a specifically Creole vision of Spanish American territory and complicates the relationship between cartographic representation and imperial domination in a colonial context. The dialogue between French and Spanish diplomats, the political stakes, and the reappropriation of Maldonado's work after his death speak to the multilayered web of negotiations through which this highly praised map of Quito emerged. In chapter 5, we follow the arguments of an anonymous critic as he attempted to defend the Indians of Quito against the observations of the Spanish naval officers who participated in the Quito expedition, Jorge Juan and Antonio de Ulloa. The critic used the *Relación histórica del viage a la América meridional* (1748), and especially Ulloa's chapter on the Indians of Quito, to denigrate the status of the transient observer and emphasized instead the importance of linguistic and ethnographic standards. This dialogue between an anonymous critic and a printed text both enabled and inhibited certain kinds of criticisms in the intellectual culture of eighteenth-century Spain, despite the tremendous efforts and expenditures of the Crown to publish the *Relación histórica* in the first place. In the process, the representation of the Indians of Quito became a weapon wielded by both sides: a malleable substrate that could be used to determine the standards by which eighteenth-century ethnography was and should be written.

Chapters 6–7 concern the uses to which the expedition's findings were put. Chapter 6 focuses on a 1744 French translation of a classic account of Incan history that was compiled and reorganized based on natural specimens that had been sent from South America to the Parisian Jardin du Roi. Transformed into both a political tract aimed at relieving social distress in France and a museum in miniature, the *Histoire des Incas* is an unexpected context to find the material traces of the Quito expedition, revealing the physical world of natural specimens and the printed world of New World histories. The text also points to the tendency toward abridgement as new specimens crowded out old information, similar to the way that an Aristotelian worldview was slowly displaced by new empirical methods and specimens following the arrival of Europeans in the Americas. Along these lines, the final chapter looks at the manner in which abridgement functioned to dissect parts of the expedition and redistribute them in the Enlightenment's most iconic text: the *Encyclopédie, ou Dictionnaire raisonné*. Merging bibliographic techniques of abridgement with the material processes of collecting, packaging, and transporting natural historical specimens, this chapter reveals the complexities of European collection and collation practices as editors diced and dissected the expedition's contributions to the Enlightenment's greatest compendium of purportedly universal knowledge.

This structure functions to show the trajectories of knowledge acquired through extra-European travel, like a cinematographic close-up that slowly widens its field of vision as a story unfolds. Each chapter focuses on a particular text or set of texts but also emphasizes a particular activity whereby scientific memory was transmuted from one state to another: inscription, narration, reading, correction, criticism, translation, and abridgement. These are ephemeral and interconnected states that should not be seen as chronological or progressive; sometimes these stages took place consecutively, at other times they operated concurrently. Taken as a whole, however, the singularities related to the Quito mission underscore the contingent factors and multiple actors that coalesced around the production of particular "monuments," and they argue for a multidisciplinary and methodologically adaptive approach to the history of travel and scientific exploration in the transatlantic context.

―――――

Two-and-a-half centuries ago, on the plain of Tarqui, the mestizo residents of an Andean village put on a pantomime that emphasized the bizarre and idiosyncratic practices of European scientific experiment. By mimicking the gestures of instrumentally based empiricism, they reminded their own audience in October 1739 that science, like all creative endeavors, is a branch of human culture; that it is a reflection of the people that carry it out and ultimately depends for its meaning on the audiences to which it is addressed. This understanding of science as theater was not unfamiliar to eighteenth-century natural scientists as they cast themselves as intrepid heroes in the public eye. La Condamine certainly portrayed himself in this way, and his image as a cartographic conquistador calmly taking measurements amid snakes, animals, and fire-belching volcanoes was projected to audiences throughout Europe in several narrative accounts (as shown in the frontispiece to the *Geschichte der zehenjährigen Reisen der Mitglieder der Akademie der Wissenschaften zu Paris*, a German edition of La Condamine's voyages published in 1763 [fig. 2]). Nearly two decades after his return from South America, theatrical overtones abounded in La Condamine's reception at the Académie française, where the *voyageur-philosophe* was welcomed into its esteemed ranks as a tireless interrogator of the secrets of nature, and whose sole recompense was his society's broad acclaim:

Having traveled both hemispheres, traversed the continents and the seas, overcome the supercilious summits of flaming mountains, where perpetual glaciers brave both the subterranean fires and the harsh conditions of the South; having been exposed to the precipitous descent of frothy waterfalls . . . ; having penetrated vast deserts, the immense wilderness (*solitudes immenses*) where few traces of humanity can be found, where Nature, accustomed to the most profound silence, must have been stunned to hear itself questioned for the first time; having done more, in a word, for the sole

Figure 2. Following his return from South America, La Condamine became a pan-European symbol of the intrepid and unflappable savant, subduing the New World not with a sword but with a scaled quadrant, like the one he displays prominently in this image. His scientific feats were often dramatized in print alongside astronomical instruments, exotic creatures, and fire-belching volcanoes. When admitted into the prestigious Académie française in 1760, La Condamine was canonized as a peripatetic philosophe who had conquered "the supercilious summits of flaming mountains" and the "subterranean fires" of the torrid zone. Frontispiece to *Geschichte der zehenjährigen Reisen der Mitglieder der Akademie der Wissenschaften zu Paris* (Erfurt, 1763). Courtesy of the Northwestern University Library.

benefit of literary glory than was ever done for the desire for gold: this is what Europe knows [of your exploits], and how you will be remembered by posterity.[28]

The fiery cordillera of the Andes and the vast solitude of the Amazon served as the dramatic backdrop for La Condamine's apotheosis as an Argonaut in a new age of scientific exploration. The mountains and rivers along the equatorial line became tablets and platforms upon which science and publicity came together. And South America became the new Rhodes of classical antiquity, where geometric figures sketched ephemerally in sand turned into stone pyramids and measured triangles, durable symbols of a mobile scientific knowledge that in the eyes of many had arrived serendipitously on American shores.

In the early modern period, the texts and material objects scientific actors used to portray their exploits frequently resembled theatrical scenarios such as these.[29] Several graphic moments explored in this book are powerful examples of the artful use of chisels and sketch tools to depict the continent as a theater for European scientific endeavors. They include the inscription marking the intersection of the equator and the coastline at Palmar, examined in chapter 1 (fig. 6 below); the thick curtain in the cartographic cartouche showing Maldonado's life coming to a premature close in chapter 4 (plate 10); a set of puttis on a pedestal conducting scientific experiments (fig. 58); and in chapter 2, La Condamine's self-portrait in the Pongo de Manseriche at one of the more dramatic moments of his Amazonian journey (plate 8). The travel narrative was itself a conspicuously theatrical text, following a diachronic style and portraying its author-protagonist in a series of challenging scenes through which, by innovation and skill, he was able to triumph over danger and adversity. Meanwhile, back in Europe, South America was on display in many dramatic representations, ranging from Voltaire's *Alzire, ou les Américains* (1736) to Boissi's *La Péruvienne* (1748). During this period, the European theatergoing public was treated to spectacular displays of temples and tropical utopias on stage. The culture out of which these savants emerged already valued representations linking drama with social and moral concerns; authors and barristers alike saw the theatrical genre as an effective means to negotiate the boundaries between public and private spheres. Success at the theater was seen as the surest way to gain honor and prestige in a broader social world. And in the scientific realm, displays of scientific showmanship such as the abbé Nollet's electricity experiments or Mesmer's *séances* blurred the lines between the private laboratory and the public stage. Condorcet's *éloges* at the Academy of Sciences, published both in the scientific periodicals and in the popular press, even read like dramatic monologues, seeking to bring the individual exploits of savants into the public arena through heroic displays of virtue and achievement.[30]

Tracing the effects of these commemorative practices in Europe is important for understanding South American history as well. The French and Spanish mission to Quito marked a critical moment in South America's long path toward becoming—for better or for worse—a public theater for experimental science. For anthropologists and historians seduced by the Romantic image of exploration, South America's scientific "discovery" began in earnest only after Latin American independence in the 1820s. But it was substantially earlier than this nineteenth-century heyday that the mythical "Land of the Amazons," with its bronze-skinned warriors and cities of gold, began its inexorable transformation into "Amazonia," connoting a tropical storehouse that teemed with exotic treasures. Long before Carl von Martius, Charles Darwin, Henry Walter Bates, and Claude Lévi-Strauss turned South America and its native inhabitants into a laboratory for studying human, plant, and animal diversity, naturalists, missionaries, and imperial agents could be found combing the forests and descending the riverways of Spanish and Portuguese America, seeking to master swaths of uncultivated territory by assessing the region's tremendous resources. Already in the eighteenth century, they were influenced in no small measure by a group of French and Spanish astronomers that had preceded them along the equatorial line. And it was these dramatic gestures and monuments to scientific achievement that most captured the imagination of explorers, mercenaries, and statesmen as they ascended the highest peaks and traversed the broad and inhospitable expanse of the continent's interior.

To answer the question of how cultures of empirical observation made their way from the Andes down the Amazon and across the Atlantic, we need to know how instrumental measurements appeared both from the vantage point of a plain near Quito and from the salons near the Seine. We need to address the material and rhetorical means by which the frontiers of these new scientific "laboratories" were construed and legitimated. And we must also understand the editorial practices that allowed these experiments to be collected and communicated from the channels of the Amazon River to the canals of Amsterdam. In an age of expanding geographic frontiers, the accumulation and inscription of knowledge about the natural world became an Enlightenment obsession. The "performance" of experiments that captured this knowledge was equally essential for its public reception. And paradoxical though it may seem, the erasure and abridgement of that knowledge were also ubiquitous and necessary practices: ones that were motivated not only by an impulse to display exotic specimens from distant shores but also by a desire to portray those spoils as having been wrested single-handedly from the veiled portions of America's most mysterious lands.

Let us move on, then, to South America, as the actors, mimes, and supernumeraries prepare to take the stage.

ONE

The Ruined Pyramids
of Yaruquí

The works of Bouguer and La Condamine have had an extraordinary (*singulière*) influence on the Americans from Quito to Popayan. The ground of this land has become classic . . . The Audiencia of Quito was able to destroy the Pyramids, but they were incapable of drowning the spark of genius that rises up from time to time in this land, and that shines brightly along the trail that Bouguer and La Condamine have blazed.   ALEXANDER VON HUMBOLDT (1802)

I will dwell in solitude amidst the ruins of cities; I will enquire of the monuments of antiquity, what was the wisdom of former ages: I will ask . . . what causes have erected and overthrown empires.

CONSTANTIN FRANÇOIS CHASSEBOEUF DE VOLNEY (1791)

All they could discuss were Pyramids: the word alone awakens great ideas.

LA CONDAMINE, *HISTOIRE DES PYRAMIDES DE QUITO* (1751)

The luminous constellations of the Southern Hemisphere shone brightly overhead as the Prussian naturalist Alexander von Humboldt stood in silent reverie on a highland plain near Quito. As he eagerly scoured the heavens for signs of a passing comet on this March night in 1802, he recalled the stunning panoramas through which he had passed earlier that day. Images of the majestic snow-capped volcanoes of the Andean cordillera—Cotopaxi, Illiniza, Corazón, Pichincha—glowed in his memory like fading beacons against the evening sky. Turning his gaze to the Southern Cross and finding its nebulous core darker and more beautiful than he had ever observed, Humboldt used the opportunity to reflect on the inscrutability and opacity of human knowledge. The visual senses could only perceive a mere fraction of the universe's boundless matter, and Humboldt seemed to throw up his hands at the enormity of the task before him: "[H]ow many things exist," he wrote in his personal journal, "that man cannot see."[1]

As if humbled by the profundity of his own reflections, Humboldt turned the next morning to less celestial concerns: visiting the destroyed remnants of a pyramid that had been erected nearby as a monument to European science. Trampled fragments of broken marble tablets, shards of a broken fleur-de-lis made of ash-green porphyry stone, illegible inscriptions formed with Latin characters, and discarded cube-shaped stones employed as pedestals were the vestiges he found scattered about at the farm of Oyambaro. Like the French traveler Constantin de Volney who sojourned among the decaying monuments of Egypt, Humboldt saw in these Andean ruins evidence for the rise and fall of civilizations. But where Volney saw only the downfall of territorial empires, Humboldt saw the demise of the human spirit as well. As he sifted through the rubble picking up various "astronomical relics" as souvenirs, he marveled at history's destructive hand and shuddered at the forces that ravaged such objects of human ingenuity. "We might imagine ourselves among the Turks," he wrote, "when we find the most important monument to the progress of the human spirit ransacked before us."[2] To Humboldt, the state into which this ruined pyramid had fallen reflected a mysterious and unpredictable encounter between the aspirations of human culture, the destructive power of nature, and the fleeting character of historical memory, like a brilliant meteor that burns its path across the darkest sky and then just as quickly disappears from view.

These decrepit structures on the plain of Yaruquí had been built to function as permanent markers of an astronomically determined baseline, but instead they stood out rather ironically as symbols of decay. Half a century before a broken fleur-de-lis caused Humboldt to reflect on the "vicissitudes" of scientific memory, an intrepid and inventive astronomer from France had managed to persuade others that pyramids would be suitable monuments to scientific achievements. Charles-Marie de La Condamine had arrived in Yaruquí with an ingenious plan, having left behind a culture in which monuments were performative and theatrical gestures were part and parcel of intellectual life.[3] He and the rest of his colleagues had come to South America not to read demurely from the book of nature; they wanted to be the scribes that wrote and performed their own entrance onto the scientific stage. In the words of one of La Condamine's colleagues, South America would be "a vast theater . . . which few or none duly qualified to make observations have as yet seen, or travel'd over."[4] The French savants wanted this South American laboratory for geometric triangulation to be their own performative space, where they would serve as the playwrights, actors, designers, and directors of a scientific drama unveiled in print.

In order to build this performative theater of science at the very site of their operations, La Condamine worked assiduously to transform the raw materials of the Andean landscape into monumental "texts" inscribed with his own heroic and empirical accomplishments. And of the many sites onto which he inscribed his achievements, none was more symbolic than the pyramids at Yaruquí. Affixed permanently at either end of the academicians' northern baseline, these monuments were to serve as material markers of transcendental truths (plates 2 and 3). Each tetrahedral pyramid would sit upon a square base of between five and six feet in height, constructed from rocks taken from nearby ravines and covered in a brick exterior, also of local confection. An inscription carved in marble stone would then be affixed to one of the four faces, and a 4-by-6-inch inscribed silver plate inside a sealed box was to be lowered down into each pyramid to protect it from eventual mutilation or theft. What La Condamine was unable to foresee, however, was that the pyramids would eventually be destroyed, not only by those whom he disparagingly referred to as "people from the neighborhood" (*gens du voisinage*) but also by those who had authorized their construction in the first place. In the wake of their destruction, La Condamine would nonetheless profit from their demise by publishing an account where his own authorial talents came to the fore, and in which he would claim proprietary rights over the expedition's achievements by suppressing the accomplishments of others who had assisted and supported his work. The *Histoire des pyramides de Quito* (1751) was the result (fig. 3).

La Condamine's narrative account of the construction of the Quito pyramids emerges out of a context in which members of eighteenth-century scientific academies began to see themselves as authors as well as academicians, seeking to lay claim to their own intellectual rights by publishing accounts in which their actions and ideas took center stage.[5] To recount the history of the pyramids, La Condamine used symbolic, material, and linguistic strategies, fashioning himself as the conceptual and practical foreman of a project in which physical monuments functioned as markers of intellectual, if not territorial, hegemony. The pyramids were not only a literal imposition of European architectural and scientific codes onto the Latin American landscape. They also effectively erased the contributions of those who had assisted in their construction. The attempt to commemorate the results of the Quito expedition

Figure 3. La Condamine's *Histoire des pyramides de Quito* (1751) recounted a dispute over the construction of two pyramidal monuments at the base of the academicians' measurements near Quito. The citation from Lucanus's *Pharsalia*, "Even the ruins perished," is ironic, since the account for which it serves as epigraph guaranteed that the controversy lived on well after the physical monuments themselves had been destroyed. Courtesy of the Special Collections Library, University of Michigan.

thus brought into sharp relief the ways in which Europeans accommodated themselves and their projects to local conditions; in this case, they sought to make a plain near Quito conform to the universal aspirations of Enlightenment science.[6]

This moment also signaled an important transition between a period in which great achievements were commemorated on the spot through construction and inscription—a carryover from earlier traditions that saw the construction of monuments as an homage to individual monarchs or deities—and the more modern procedure of setting down the results of scientific experiment through published works. Humboldt believed that mountains would make more lasting monuments than pyramids. The volcano of Cayambé that straddled the equator near Quito was in his words "one of those eternal monuments by which nature has marked the great divisions of the terrestrial globe."[7] For Humboldt, empirical science should follow suit in using these sites as markers of scientific truth. But La Condamine knew that monuments to individual achievement had symbolic value as well, especially since they could be forged out of local materials and captured iconographically in print. The inscription of scientific results *onto* the American landscape would thus provide a visual referent that could accompany published texts and treatises. And as a historical symbol of permanence and stability, the pyramid was the ideal choice to commemorate the mission's activities on a highland Andean plain.

Sturdy Symbols

As a young man, La Condamine had front-row seats for learning about science as a performative art. While in Paris in the late 1720s, he regularly attended the informal salon at the Café Procope, which was presided over at the time by the flamboyant dramaturge Antoine Houdart de La Motte. Later, Pierre-Louis Moreau de Maupertuis took the reins, an astronomer with whom La Condamine would collaborate throughout his career and whose encouragement and support may have facilitated La Condamine's entrance to the Academy of Sciences in 1730. His close friendship with Voltaire may have been a factor encouraging his theatricality as well, since this was one of Voltaire's most active periods as an author of drama, much of which seemed redolent of themes and locations dear to La Condamine. But perhaps improvisational drama was merely in La Condamine's blood. Voltaire told of an evening at the home of Charles-François Cisternai du Fay in which a young La Condamine, recently returned from an expedition to the Levant, had disguised himself as a Turk. He then proceeded to beguile those in attendance about his "Oriental" origins. In a letter to Maupertuis, Voltaire commented that "whenever [Maupertuis] wished to have another dinner at the home of M. du Fay with the honest Muslim who speaks such excellent French," he would be happy to comply.[8] Following Voltaire's return from exile in England, during which time he had become a vocal advocate for Newtonian science, the two friends conceived and carried out a brilliant scheme to defraud the French treasury of large sums of money. For anyone willing to make the massive initial investment, La Condamine had realized with devious mathematical prowess that the poorly designed lottery would pay out more money than it actually took in. They bought up as many tickets as they could, and enriched themselves greatly as a result.[9]

This image of La Condamine as a trickster and deceiver did not always go over well at the Academy, however. Indeed, following his death, Condorcet wrote that La Condamine's curiosity and need to be constantly active "made all prolonged meditation impossible, preventing him from going deeply enough into any scientific area to arrive at any new discoveries."[10] Likewise, Jacques Delille conceded that La Condamine's knowledge was perhaps "more extended than profound." But both were quick to acknowledge the merit of La Condamine's career within the context of eighteenth-century culture: "[I]f others have made more sublime discoveries in philosophy," Delille wrote, "none has left a greater example for the *philosophe*."[11] And in what was perhaps the most perspicacious assessment of La Condamine's theatrical and literary talents, Condorcet provided an explanation for what had allowed the errant astronomer to so successfully mold opinions to his favor:

[La Condamine was] widely known in every society, possessing the art of persuading the ignorant people to whom he had listened, bringing back singular observations to pique the frivolous curiosity of the *gens du monde*, writing with enough charm (*agrément*) to have people read his work, with enough neglect and too simple a tone to foster envy or threaten the self-esteem of others, interesting for his bravery and piquant for his faults.[12]

The "art of persuasion" and ability to "pique the frivolous curiosity" of his European colleagues were skills that were honed throughout La Condamine's career.[13]

The Turkish disguise he had used for the practical joke on Voltaire and Maupertuis was likely a prize from his first voyage as a member of the Academy. La Condamine established his reputation by accompanying an expedition to the Mediterranean and the Levant only a year after being offered membership. This early experience melding travel writing and scientific observation provided a basis for his later writings in South America and undoubtedly influenced the Academy of Sciences to choose La Condamine to participate in the Quito expedition. But it also gave him an opportunity to showcase the broad range of his analytical skills:

the relatively brief, twenty-seven-page report he wrote on the heels of this maiden voyage to the Levant in 1732 featured extensive sections on astronomy, navigation, geography, mechanics, anatomy, chemistry, botany, physics, and the natural history of the Mediterranean world. He also included a newly "corrected" map of the Golfe de Contesse in Macedonia, ensuring that his voyage would be recorded graphically as well.

But when it came time for La Condamine to depart from Europe to South America, his previous experiences abroad gave him no immunity from criticism. Just weeks before he and his colleagues crossed the Atlantic, the Parisian periodical *Le Pour et Contre* published a cautionary tale of love and literacy in colonial Peru that was directly linked to La Condamine's project. The abbé Prévost's message functioned as a preemptive plea on behalf of a people he felt might be unfairly excluded from the fruits of their labors: the native inhabitants of colonial Quito. His Lima love story, which described a forlorn Spanish viceroy who threatened to destroy a culture's patrimony because of his unrequited love, functioned as a dramatic foil to criticize the unilateral imposition of European scientific memory on American soil. According to the article's author, the project La Condamine had orchestrated to commemorate the Quito expedition's measurements in situ was ill conceived because it linguistically disenfranchised the local populations, "whose rights," according to Prévost, "are apparently not judged to be worth much." By insisting that the pyramids' inscriptions be composed exclusively in Latin, La Condamine had excluded the natives from sharing in the spoils of a project executed on their own territory. Prévost's modified version of the project envisaged a text written in four separate languages: French, Spanish, Latin, and "Peruvian," providing the indigenous populations of Quito with a translation in their own language. At the end of his article, the abbé Prévost explained that "our enlightened Travelers (*sçavans Voyageurs*) will be interacting with men who are of good sense and capable of reason . . . [T]hey should perhaps translate the Inscription into the Peruvian language on behalf of the Natives (*Naturels*) alone."[14]

Prévost's article thus raised an important series of questions as to how the results of experimental practices executed *outside* of Europe were to be publicized, to whom recognition should extend, and in what form such recognition should be inscribed. According to Prévost, the inscription of scientific results onto structures built in colonial territory brought with it the obligation to communicate to those who collected the raw materials for the experiment as well as those who constructed the physical structures that contained the inscription. By raising this issue, Prévost foreshadowed the disenfranchisement of the native populations from the fruits of the entire expedition. And he provided a moral argument for the recognition of an entire nation that, in his words, had been "forced by their Masters into slavery and ignorance."[15]

The ornate Latin inscription which La Condamine intended to affix on the side of the pyramids had been devised prior to his departure in consultation with the Academy of Inscriptions and Belles Lettres, a royal academy that was established informally in 1663 to study the monuments, texts, and cultures of ancient civilizations. The original function of this academy was linked to a controversy during the reign of Louis XIV over the language in which monumental inscriptions should be written.[16] La Condamine's text proclaimed in florid Latin the glorious achievements of the French king's scientific minions before they had measured a single angle, and the Academy saw itself as being "uniquely qualified and worthy" to compose this kind of public statement on behalf of the French nation. On May 13, 1735, the day following the departure of the academicians from La Rochelle, the Academy made reference to its role by proudly acknowledging "the journey of Messieurs Godin, Bougher [sic] & la Condamine to Peru, a voyage about which the *Mercures* and other Journals have also spoken, and which we ourselves mentioned in our Minutes, when the Academy [of Inscriptions and Belles Lettres] composed the Inscription meant to mark the Center of the Operations they will undertake at the Equator."[17] With the imprimatur of one of the Old Regime's most venerable institutions, the inscription La Condamine carried across the seas was meant to acknowledge publicly the instrumental measurements on American soil and to recognize auspiciously the glorious accomplishments of King Louis XV and his academies.

La Condamine did not deny certain ceremonial aspects of the pyramids and their inscriptions. But he emphasized that these markers were part of a larger empirical project, "specially suited to fix the endpoints of the measurement that had been the foundation of all our geographical and astronomical operations." By constructing permanent and immoveable physical structures, the scientists would be able to protect their work from "the fate of all experiments (*travaux*) that the Ancients carried out related to the measurement of terrestrial degrees . . . whose fruit has been lost to posterity for not having taken similar precautions."[18] Since geodetic measurements relied on geometric triangulation, the exact measurements could not be reproduced without marking their precise location on the ground.[19] La Condamine argued that the site first chosen for the geodetic measurements in France should be recuperated, and he made a plea for a similar set of monuments at the site of his expedition's observations at the equator:

[I]n order to prevent similar difficulties in the measurement that we were to undertake, I judged that we should fix the two endpoints of the fundamental base of our operations with two lasting monuments, such as two columns, obelisks, or pyramids, whose purpose would be explained by means of an inscription.[20]

Like Lapland, Quito would serve to extend the set of observations that had already been carried out in Paris early in the eighteenth century. But unlike La Condamine, Maupertuis did not attempt to construct pyramids near the Arctic Circle as monuments to his own expedition's measurements.[21]

The origins of La Condamine's decision to construct stone columns, obelisks, or pyramids in a land of steep ravines and earthquakes can be traced to his eighteen-month excursion to the Levant aboard a ship captained by M. le Chevalier de Camilly. This vessel was part of the squadron of M. Duguay-Trouin, a redoubtable sea captain then in the later years of a prominent career.[22] During a journey dedicated primarily to improving scientific knowledge in the fields of navigation, geography, and natural history, La Condamine had the opportunity to visit Alexandria, where he witnessed with his own eyes Egypt's most ancient monuments. While in Egypt, he was able to see the elaborately inscribed ornamental columns and obelisks lined with exotic hieroglyphs dotting the horizon. Although his formal observations dealt primarily with topics of interest to the Academy of Sciences, La Condamine did note that "the Inscriptions and other ancient monuments provided me with an abundant harvest":

One can see along the seashore, among the ruins of Alexandria, Egypt, two obelisks composed of this granite, or Theban stone . . . As far as the obelisk is concerned, the side exposed to the northwest, which faces the sea, and the side facing the new city, are those that are best preserved, and one can clearly make out the hieroglyphic figures that are engraved there, and which I have sketched.[23]

Hieroglyphs delighted the young academician on his maiden voyage and made a powerful visual impression that would have been amplified through the drawings he produced. But La Condamine was also conscious of the material state of these inscriptions. The hieroglyphs on "Cleopatra's Needle," for instance, had eroded to such an extent that "one could hardly make out any of its characters." These physical observations made

La Condamine aware of the damage that exposure to natural causes such as wind or rain could cause. So when he suggested that tetrahedral pyramids upon a cubed pedestal were most appropriate for the Quito monuments since they would not be exposed to the "harmful effects (*injures*) of the air," it is possible that he had in mind the eroded inscriptions he observed in Alexandria.

To justify his choice, La Condamine repeatedly emphasized the simplicity of the pyramidal design, careful to downplay any notion of grandeur, glory, and pomposity. He explained deferentially that such a marker "was neither large enough nor magnificent enough to serve as the forum for a ceremonial monument [*éloge*] to the two most powerful Monarchs in Europe."[24] For La Condamine, these were structures that would support the interests of science rather than serving to bolster national or imperial glory. And in the *Histoire des pyramides*, La Condamine explained that the design of the pyramids had been based almost entirely on pragmatic concerns:

There was never any discussion of constructing a fancy edifice, but rather a simple and durable monument appropriate for showing clearly the two endpoints of our base. As far as their form was concerned, the most convenient to accomplish our purposes was the pyramid, and the simplest of all pyramids was the tetrahedron*; but since we wanted to orient the structure in relation to the regions of the world, I decided to give our pyramids four sides, not including the base, [a choice] which also happened to facilitate their construction.[25]

The symbolic aspects of the pyramid found little place in La Condamine's practically minded description. And yet, the emphasis on orienting the structure so that the faces of the base corresponded with the "regions of the World" reveals an interest in the pyramids' reaching out beyond the plain of Yaruquí. While the durability and stability of the tetrahedron vis-à-vis other architectural forms may have been factors, the impulse of making a four-sided monument radiate to the four corners of the globe— embodying the universal aspirations of European science—seems to have been an equally significant symbolic motive.

La Condamine may have emphasized utilitarian factors such as durability and ease of construction in his structural description of the pyramids, but the discussion at the Academy of Inscriptions and Belles Lettres had revolved around more symbolic questions. Prior to the expedition's departure for Quito, the Academy devoted three sessions to questions over the monuments' form and significance. One proposal came from Scipione Maffei (1675–1755), an Italian marquis and collector of antiquities

Figure 4. Initially presented before the Académie des Inscriptions et Belles Lettres and later inserted into La Condamine's *Histoire des pyramides*, Scipione Maffei's sonnet glorified scientific knowledge as a form of possession, and likened the members of the Quito expedition to the heroes of classical antiquity. La Condamine, *Histoire des pyramides de Quito* (1751). Courtesy of the Rare Book and Manuscript Library, University of Pennsylvania.

who was then living in Paris. Maffei, who was particularly interested in the expedition's potential for making observations regarding pre-Columbian history, proposed the idea of an inscribed marble column. Along with this proposal, he penned a poetic inscription to commemorate the savants' work that La Condamine honored by reproducing in the *Histoire des pyramides* (fig. 4). The sonnet provided flattering comparisons between the academicians and the heroes of classical antiquity, giving a sense of the grandiosity with which the expedition and the monument's inscription were perceived at the Academy. As a paean to the role of knowledge in the construction of empires, Maffei's sonnet gave poetic voice to the notion that science could be understood as a form of intellectual possession: "And this you shall also know, he who discovers and understands / the Form, extension and measure [of the terrestrial sphere] . . . will possess and comprehend [the Earth]."[26] According to Maffei, those who discovered the Earth's shape would come to dominate it, a powerful articulation of the linkage between scientific knowledge, power, and possession. It was also, ironically, a view that belied La Condamine's seemingly innocent portrayal of the monuments as mere placeholders for future surveying experiments. Maffei's poetry and the discussion at the Academy emphasized the symbolic aspects of obelisks and pyramids, as well as the texts to be inscribed upon them.

While pyramids had many symbolic functions in the eighteenth century—from mausoleum architecture and Masonic iconography to informational markers in "geographical gardens"—over time they had come to be associated most directly with the articulation of lasting and verifiable scientific truths. This connection between science and hieroglyphic writing can be followed all the way back to classical antiquity, but there was a conscious revival of interest during the seventeenth century, especially through the writings of the humanist philosopher and polymath Jesuit Athanasius Kircher. Kircher made his way to Rome in 1633 in order to deepen his already burgeoning knowledge of the hieroglyphs and while in residence under the protection of Cardinal Barberini became one of the leading advocates for hieroglyphs' importance in deciphering and interpreting Egyptian scientific knowledge. Indeed, he believed that he had been able to interpret Egyptian hieroglyphs because of his own knowledge of Near Eastern languages.[27] Other humanists and philosophers in the seventeenth and eighteenth centuries followed Kircher in seeing the hieroglyphs as containers in which the doctrinal teachings of the Egyptian sages had been preserved and as symbols that could only be decoded through careful study of their arcane teachings.[28]

The text that most amply reflected the pan-European enthusiasm for Egyptian hieroglyphs was the Benedictine Bernard de Montfaucon's ten-volume *L'antiquité expliquée et représentée en figures*, published between 1719 and 1724 in Paris. Montfaucon never visited Egypt, but he did explore some of Europe's most comprehensive Egyptian collections, including Kircher's museum in Rome and Scipione Maffei's personal holdings in Verona.[29] His volumes comprised an encyclopedic gathering of materials related to diverse aspects of many ancient cultures, including, but by no means limited to, the architectural structures of ancient Egypt. Five years after he completed these earlier volumes, Montfaucon published the first tome of a five-volume work entitled *Les monumens de la monarchie françoise, qui comprennent l'histoire de France, avec les figures de chaque règne que l'injure des tems à épargnées.* Within a decade, then, Montfaucon published two significant treatises on monuments and monumentality, the one cataloging with sweeping range the visual and material customs of ancient cultures, the other focused specifically on France. Both were widely available, and both made explicit the connection between monumentality and permanence in the study of ancient structures.

While La Condamine likely had exposure to Montfaucon's treatise prior to his expedition to the Levant, he most certainly had read Paul Lucas's *Voyage d'Egypte*, an Egyptian travel narrative he cited in the *mémoire académique* composed following his journey.[30] Lucas's three-volume account of his travels to Turkey, Asia Minor, and Egypt ranged from geographical observations to discussions of architecture and culture. Like some earlier Egyptologists, Lucas believed that the Alexandrians learned about mathematics, astrology, and philosophy through the hieroglyphs, which they had invented because "they did not wish for the Sciences to be shared by everyone."[31] Obelisks, temples, and labyrinths adorn the pages of Lucas's text; an image of "Cleopatra's Needle," the elegant obelisk referred to above, appears in volume 2. And several plates highlight pyramids, including one image showing an eclectic assortment of bulbous, egg-shaped structures found in Upper Egypt.[32] Traveling along the western shore of the Nile, Lucas explained that many cities along the great river contained "Pyramids to serve as Tombs to the Kings of this land" and that it was possible to tell many things about the makers of the pyramids from their size and grandeur: "[I]f some are higher and more magnificent than others, it is due to either the power or the vanity of those who had them made."[33] The character of the pyramid, to Lucas's eye, was a direct reflection of the personality of its creator.

During the decade between Montfaucon's treatise and the departure of the academicians to Quito, French publications tended to confirm the status of Egyptian inscriptions as symbols of continuity and truth rather than occult monuments to arcane knowledge. In 1731, the abbé Terrasson sought to uphold pyramidal monuments as secure repositories for human knowledge as well. In his *Sethos,* he wrote about the king of Thebes's prescient strategy to protect his civilization from the damaging impact of a terrible flood. Within the subterranean alleys of the city, the king ordered them to construct

square or pyramidal columns whose sides were laden with the principles of all kinds of doctrine, in hieroglyphic symbols, so that in case the art of writing were lost, [the images] could be explained through conjecture, and if [their meaning] escaped some men, at least they would have an advantage and would not be reduced . . . to the tedious work that requires them to invent everything from scratch.[34]

The hieroglyph came to serve not merely as a substitute for writing but also as superior to it for its indelibility and hence its permanence, inscribed upon these freshly placed vertical panels to protect Theban teachings and send knowledge to future generations. Indistinguishable from the pyramidal columns upon which they were inscribed, these strands of encoded understanding served as symbols of strength and permanence rather than ephemeral and mystic knowledge.

So when La Condamine wrote in his *Histoire des pyramides* that "pyramids . . . awaken great ideas," he was responding to the eighteenth-century symbolism that saw pyramids as the embodiment of universal scientific truths.[35] When reinscribed into the cultural and literary universe of eighteenth-century France, it makes perfect sense that the Academy of Inscriptions and Belles Lettres was abuzz with the talk of pyramids. Inscribing characters onto architectural symbols drawn from ancient Egypt had been an effective means of preserving scientific ideas. La Condamine submitted his proposal to the Academy not so much to hear their suggestions for architectural structures as to "consult [with them] on the most appropriate manner to communicate meaning on a lapidary stone."[36] Regardless of the language that would eventually be used for the inscriptions, however, La Condamine knew that pyramids would effectively symbolize the monumental grandeur of European science on the plain of Yaruquí. Whether in the Andean highlands or along the Nile, the pyramidal shape radiated authority, strength, and permanence.

"I have always been impressed," wrote Voltaire in his *Lettres anglaises*, "by the fact that we have discovered such sublime truths with the help of a quadrant and a little arithmetic."[37] Instruments, geometry, and arithmetic—the building blocks of geometrical triangulation—provided the source materials not only for accurate measurements but also for sublime truths that would emerge through experimental science. The pyramid, a geometrical edifice with overtones of monumentality and myth, was the natural symbol to commemorate a process that would yield an accurate understanding of the Earth's shape and that used the triangle as its fundamental tool. La Condamine chiseled his message of scientific discovery onto stone tablets and attached them to the symbolic form of the pyramid so that his monumental efforts would pass on to posterity. The only challenge that remained for him was finding the raw materials with which to build these sturdy symbols on the spot.

Matters of Inscription

Transporting empiricism from Europe to field sites overseas created a host of practical difficulties. Back in France, experimental observers could conduct geometric observations by using any of a number of fixed, preexisting locations as triangulation landmarks: clock-towers, castles, windmills, and isolated trees were among those locations that La Condamine described as commonly employed by European astronomers carrying out geodetic measurements. Once the landmarks were chosen, those doing the measuring would use these points to construct a series of triangles, measuring in succession the angles between these various points. After determining the angles between the markers, which represented the endpoints of each triangle, a single measured baseline—La Condamine's baseline was the one he hoped to commemorate in Yaruquí—was all that was needed to enable the astronomers to measure the distance on the ground between the northernmost point and the southernmost point of the extended array. At the conclusion of this process, astronomical observations at the endpoints—using zenith telescopes and watching some celestial event, usually the occultation of one of the moons of Jupiter—would determine the difference in latitude between the two stations, allowing the team to measure the arc distance of the entire set of triangles. This would provide both a terrestrial measurement of distance between the two points and an astronomical measurement of the angle that separated the two locations. The resultant calculations would enable them to determine the length of an arc of the terrestrial spheroid and thereby to assess the curvature of the Earth, the ultimate goal of these operations. But the process required at the outset choosing solid and durable stations on the ground.

By contrast from what was the case in Europe, in the Andes the astronomers had to create what La Condamine called "artificial signals" made according to "the nature of the terrain, either with pieces of wood shaped into pyramids and covered with straw or [with] masses of cylindrical or conic stones." Where wood was scarce, these markers were often stolen, which required that they be rebuilt with considerable effort and expense. The raw materials for these signals occasionally came from the Indians' own ancestral structures: in one instance, La Condamine described "an ancient fortification of the country's native populations" that they used for extra stones and boulders. In turn, La Condamine made the site "respectable" by placing an eighteen-foot-tall cross above it, a way of symbolizing the emergence of European science from the literal rubble of an ancient Amerindian civilization. These artificial signals, including

PREMIÈRE PARTIE.

MESURE GÉOMÉTRIQUE

DE L'ARC DU MÉRIDIEN,

O U

OPÉRATIONS SUR LE TERREIN,

*Pour fixer la position & déterminer la longueur de
la Ligne Méridienne.*

J'AI cru que le meilleur moyen de préfenter au Lecteur
avec clarté & précifion le détail d'un grand nombre de
différentes opérations, étoit de former une Table qui raffem-

Figure 5. In order to transform the province of Quito into a series of interconnected triangles, whose lengths could then be measured to determine the curvature of the Earth's surface, La Condamine and his fellow academicians used quadrants and other instruments to calculate the angles between fixed stations that were placed on hillsides and volcanic peaks. La Condamine, *Mesure des trois premiers degrés* (1751). Courtesy of the Rare Book and Manuscript Library, University of Pennsylvania.

wooden pyramids, cylindrical stones, and this cross, were examples of how La Condamine used the material fabric of the New World to fashion physical monuments for symbolic and scientific purposes (fig. 5).[38]

Once the practical difficulties were overcome, these artificial locations needed to be incorporated symbolically into a broader network of extra-European field sites. Repeating previously performed measurements was one way of embracing these new locations as part of a global empirical enterprise. But experiments needed to be observed and verified to be credible, and European academies—the traditional arbiters for these kinds of credibility tests—were far away, on the other side of a vast ocean. Credible witnesses were also in short supply in a land where common languages of facticity and trust were either incommensurable with European standards or nonexistent. Natural philosophers conducting empirical observations far from the European metropole thus seemed to require something that was performative as well as graphic.[39]

Making the results of experiments visible through physical inscription and reproducing these inscriptions in print were strategies used to effect this scientific commemoration at a distance. La Condamine's *Journal du voyage fait par ordre du Roi*, which chronicled the history of the expedition as well as the history of the pyramids, was replete with images showing the inscriptive measures used to make monuments out of empirical observations. Prior to erecting the pyramids, La Condamine's first "public" act in this sense was to chisel an inscription into a rock at Palmar, the symbolic site where the equator crossed the South American coastline. This was where La Condamine would make his first mark on the land, claiming South America not with a flag or a wooden cross but with a measurement. The hallmark for incorporating a new location as part of the dominion of European science overseas was to take an instrumental reading, jot it down in a notebook for future reference, and capture the event with a commemorative inscription that was later reproduced in print:

I determined the point where the coast is cut by the equator; it is a place called *Palmar*, where I engraved upon the most visible rock an inscription* useful to Mariners.[40]

The rhetorical importance of these inscriptive gestures should not be dismissed. While the act of proclaiming the monument's utility to mariners reaffirms the prominent role of the maritime sciences at the eighteenth-century Academy of Sciences, the Palmar rock would have been far more visible to the reading public in Paris than it ever was to sea captains off the Pacific coast.[41] Not coincidentally, La Condamine chose this inscriptive moment at the Palmar rock as the plate to adorn the opening page of his *Journal du voyage*. The symbolic power of the American landscape and the equally potent force of the European scientist, chiseling his own authority onto the land, suffuse this commemorative graphic allegory (fig. 6). The modern Prometheus is an academician, engaged intently by his scriptural task and seated beside a protruding rock with a sheer face that is perfectly sculpted to fit the inscription being chiseled into it. Three rowers, presumably indigenous inhabitants of the region, appear in the image but are nowhere indicated within

Figure 6. The inscription at Palmar, where the equator crosses the Pacific coast of South America. Plate 4 shows an alternate view of this stone. From La Condamine, *Journal du voyage fait par ordre du Roi* (1751). Courtesy of the Special Collections Library, University of Michigan.

the body of the text.[42] Conversely, the mosquitoes that pursued and pestered La Condamine during the day and kept him from sleeping for five nights straight were included in the text but not in the engraving. La Condamine even quipped that perhaps he should have included information about the mosquitos and flies as part of the inscription on the rock, warning mariners not to stop in the region around Palmar precisely because of these pests. Quotidian details such as these, however, were effaced from the graphic representation that adorned the text's title page, an image that was meant to serve as a symbolic justification for this European presence in a foreign, often inhospitable, land.

The extent to which this scene conforms to what Stephen Greenblatt has called an "anthology of legitimating gestures" is striking. In the engraving, a small cabana provides shelter and support for the astronomer and his assistants and anchors their presence. Just up the coast stand a pendulum and a quadrant, instruments that enable La Condamine to determine the precise location of the site he enshrines in the engraving. Once inscribed, the large boulder no longer functions as a neutral feature of the natural landscape but instead becomes a sign within a much larger symbolic system, a site that could be located on the small globe crisscrossed by lines of longitude and latitude that is located at the bottom of the boulder. The scantily clad rowers either serve as unwitting witnesses to this legitimating act or are meant to function as a bridge between the uncivilized nature of the rugged seashore and the forward march of European science, represented by the neatly inscribed veneer of the rock face. And science did march forward alongside this depiction: a similar representation of the Palmar stone was sent by La Condamine to the Academy in 1736, included as part of the cartouche of an unpublished map entitled "Carte de la Côte du Pérou" that was based on La Condamine and Bouguer's earliest reconnaissance of the South American coast (plate 4). In the coastal map, however, no artist was inscribed within the confines of the graphic representation. But in the frontispiece to the *Journal du voyage*, this scene reveals the observer transforming scientific empiricism into an inscription, and that inscription in turn is transformed into a graphic trope. These features legitimate the act of scientific as opposed to territorial possession. Only by closely examining the engraving do two barely discernible details show the entire scene as a fictive trompe-l'oeil, produced in a Parisian atelier: the name of the artist and the name of the engraver are included in miniscule script just outside the borders of the plate. The faux chiseled script on stone is nothing more than a technical artifice to foster verisimilitude in this dramatic representation of science as both material inscription and geographic spectacle.[43]

As La Condamine made his way inland, he placed other legitimating scientific texts along his route as well. He arranged for one such public inscription to be hung on the wall of the Jesuits' College in Quito, alongside an elaborately crafted bronze rule that displayed the calculation they had made of the length of the pendulum (in seconds) taken at the equator. Always attentive to material details, La Condamine also inserted a silver nail and golden pin into the rule to ensure that when the inscribed lines wore down, the length of the measurement would still be visible. This data connected La Condamine's observations with those pendulum experiments that Jean-Jacques Dortous de Mairan and Father Richer had had carried out in France and Cayenne, respectively, in the previous century. These experiments studied the changing length of the distance between two points traced by a pendulum's motion, which could then be used to determine the relative distance of any point on the Earth's surface from the Earth's core. By extension, this experiment would demonstrate the approximate curvature of the terrestrial surface. In 1672, Father Richer had discovered that the pendulum beat more

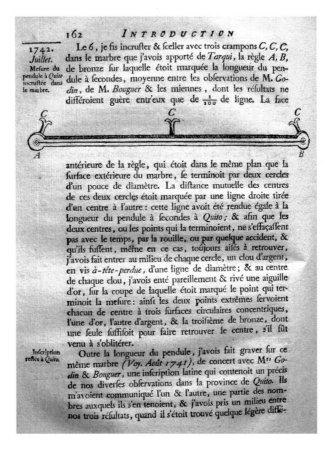

Figure 7. The bronze rule and accompanying inscription on the wall of the Jesuit college were meant to link Quito symbolically to other sites in Europe where measurements were taken. La Condamine, *Journal du voyage fait par ordre du Roi* (1751). Courtesy of the Special Collections Library, University of Michigan.

slowly in Cayenne than in Paris, meaning that Cayenne was likely farther from the center of the Earth than the French capital. This result lent further credence to the Newtonian hypothesis and inspired La Condamine to repeat Father Richer's experiments in French Guyana after completing his journey down the Amazon River. The bronze rule and accompanying inscription that La Condamine mounted on the wall of the Jesuit College were thus meant to embody physically and symbolically this additional extra-European field site, adding credibility to the surveys they had carried out during their years in Quito (fig. 7).

Unlike the Palmar inscription, however, which La Condamine presented as having engraved himself, the inscription in the Jesuit College in Quito was executed by a native engraver, an expert technician in the carving of wood:

The engraver who had come recommended as the best for this kind of work was an Indian, a wood sculptor by trade. He was illiterate, so I was obliged not only to trace the lines and the spaces for him but also to sketch with the utmost precision all the letters, periods, and commas, so that all he needed to do was to follow the outlines with the burin. He worked under my supervision, and if I left for a moment, I was not sure that I would find him upon my return, unless I had locked him up with a key. Many days often passed without his showing up. Normally, he only engraved one line per day: his work took six weeks to complete.[44]

The result of this apparently coerced labor was an artifact of European design and indigenous technical prowess, a product of artistic hybridity in the colonial world. The scene in the artistic atelier reveals La Condamine not so much in the role of pedagogue, instructing a novice in the material grammar of scientific commemoration, but rather as a master with his servant, ordering the engraver to "follow the outlines" of a textual language not his own. Only through what La Condamine represented as his persistent surveillance was the inscription completed. And despite his description of the indigenous artist, La Condamine left no doubt as to who deserved full authorial credit for the inscription's execution and who had maintained exclusive control over the final product. Just as he had with the graphic representation of the inscription at Palmar, La Condamine emphasized the material strategies he employed to affix a commemorative text to the wall of this Jesuit college. Underlining the rhetorical place of visual representation in validating empirical observations, La Condamine legitimated his on-the-spot observations for his audience back in Europe by reproducing this graphic display in print, acknowledging only obliquely the assistance of an anonymous native engraver.

Like the *Journal du voyage* to which it was appended, the *Histoire des pyramides* emphasized materiality as well; the text La Condamine penned in response to the destroyed pyramids was as much an architectural history as it was a tale of erudite dispute. The *Histoire* placed its readers on the slightly uneven plain of Yaruquí alongside the academicians, guiding them step-by-step through the drawn-out process by which two pyramids were raised from rubble into structures meant to embody scientific truths. In the text, La Condamine detailed the many concurrent processes through which raw materials from the Peruvian highlands—dirt, trees, stone, slate, and limestone—were transformed into commemorative markers encapsulating the fruits of measurement and calculation. He emphasized the digging of holes, the building of piles, and the construction of devices to raise stones out of quarries.

These material details played a significant role both in justifying the elaboration of the *Histoire des pyramides* and in bestowing La Condamine with authorial credit for the project overall. References to rocks, quarries, and limestone may seem like colorful sideshows to entertain an eighteenth-century readership. But by providing a painstaking reconstruction of the material circumstances under which the pyramids came to be erected, La Condamine used notions of labor and property to credit himself with the project's conception, elaboration, and execution. Part of this process was also meant to exclude the Spaniards from receiving authorial credit by arguing that nonconceptual labor (i.e., work that did not directly relate to the initial formulation of a particular plan or project) did not entitle an individual to public recognition. The entire project to construct the pyramids was founded upon the idea that the exclusive right to intellectual and authorial credit should stem from the initial conception, as opposed to the actual execution of measurements and calculations.

But the Spaniards were not the only individuals from whom La Condamine withheld credit. Skilled and unskilled manual laborers also received little recognition from the author of the account, even though they were ever-present as part of the project's execution. These faceless workers functioned as intermediaries between the stark landscape of the central Andean cordillera and the sites of the pyramids' construction. They forged a link between the region's material conditions and the intellectual aspirations of La Condamine and his fellow academicians. La Condamine could not have brought his project to fruition without the well-orchestrated physical performance of these native and mestizo laborers, individuals who appeared intermittently throughout the *Histoire des pyramides* as disembodied figures moving contrapuntally against a backdrop of legal proceedings, instrumental observations, and philosophical musings. They ranged from brick makers to day laborers, from Indians to "people from the neighborhood" (*gens du voisinage*).[45] They appeared in the text when material conditions became difficult, often saving La Condamine from his own despair.[46] But rather than praising their assistance, the French academician frequently complained both about their work habits and their propensity to disappear. And watching over them became part of his managerial responsibilities:

It was only in the month of May, 1740, after we had made our observations at Cotchesqui, that I was able to observe the construction of the Pyramids up close and in person. I was assisted by M. de Morainville, who was responsible for executing and supervising the [project], and he instructed the workers, who could not be left alone.[47]

If one of La Condamine's primary purposes was to watch over the native laborers, the sole function of the "workers" whom de Morainville commanded—aside from carrying rocks and other menial tasks—seems to have been to escape from the watchful eye of their European master.

Indigenous laborers functioned as essential figures in the construction project but were shown devoid of any ability to perform independent, creative work. They were represented as fleeing from the worksite at every opportunity with any materials or tools that happened to be at hand. In order to discourage the "locals" from running off with the stones and using them for their own purposes, La Condamine carefully defaced each millstone in the pyramids with a small hole or notch in its circumference. He even molded his bricks differently "so that no one would be tempted to debase the Pyramids in order to use the bricks in another location."[48] Such devious acts of preemptive security show the antagonistic attitude that the expedition's members had developed toward Amerindians, mestizos, and many Spaniards as well, sentiments with which La Condamine's texts were redolent. European observers treated local populations as suspicious, mischievous, and entirely untrustworthy.[49] They not only disenfranchised local populations from any recognition for their labors, but they relegated them to silent observers of the process by which instrumental observations came to be inscribed and commemorated in their own land. For La Condamine, the natives' attitudes were explained by the cultural decline of Peruvian civilization, despite the architectural glories of the Inca past and the artistry with which they were once able to manipulate stone without the use of iron: "What history teaches us about the ancient buildings built by the Peruvians during the time of the Incas could lead some in Europe to think that the construction of these new Pyramids would be nothing more than a game for such an industrious people; but things have changed dramatically during the last two hundred years in Peru."[50] Like the pyramids whose destruction La Condamine described, the culture of ancient Peru had undergone a demise as well, leaving only ruins in its wake. For La Condamine, the epigraph on the title page of his *Histoire des pyramides*, "Etiam perière ruinae," could be just as applicable to the descendants of the Incas as it was to the pyramids of Quito.[51]

Destruction and decay, of course, were what motivated La Condamine to pen his history of the pyramids in the first place: "In every place and especially in Quito," he wrote, "it is easier to destroy than to build."[52] Following his return to Europe, and after learning that his pyramids had been destroyed at the hands of the Spanish authorities, La Condamine imagined a scenario in which his carefully wrought monuments were pillaged down to the two silver plates that contained a copy in miniature of the inscription, each of which had been carefully inserted into the center of the respective pyramids. Construing the natives as willing accomplices to this act of demolition, La Condamine was certain that "the materials out of which the ancient Pyramids were made had been dispersed at the moment of their destruction... [and] that the locals (*gens du voisinage*) took them and used them elsewhere."[53] After narrating the series of events that led to the razing of the markers at the end of the *Histoire des pyramides*, La Condamine once again repeated the litany of material efforts undertaken to bring the project to fruition:

We have seen that in order to construct the pyramids that were demolished, it was necessary to dig five hundred feet deep through twelve to thirteen-thousand quintals [a weight measurement equal to 100 kilograms] of rock; to search out two stone tablets large enough for the inscription...; to build machines and cables to raise them up, and instruments to work them; to place one of the two pyramids on piloting; to find wood appropriate for this use in a region where there was none at all; to bring water from two spots by a channel made expressly for this purpose. I do not even speak of the difficulty of deciding upon and transporting the materials, and the paucity and rudeness of the workers.[54]

An encore performance in the final section with recourse to familiar themes, this monologue by La Condamine emphasized that anyone who might wish to undertake the reconstruction of the pyramids would have "neither the same motives... nor the same resources in a country where one can safely say the arts are still in their infancy."[55] According to La Condamine, the opportunity to commemorate this unique experiment on South American soil had been lost and could only be recovered partially through the narration of the controversy in print. But in the end, La Condamine profited from the demolition of the pyramids far more than the "local thieves" he imagined to have stolen fragments of his work. It was he who picked up the damaged and dispersed materials and used them for his own benefit across the Atlantic. By conspicuously inscribing the results of his activities on the very theater of his observations, he built commemorative monuments to his own achievements. And by diminishing the creative role of those toiling alongside him, he ensured that he alone would benefit in print from the controversial construction—and subsequent destruction—of the pyramids.

Juridical Spaces of Extra-European Science

As indigenous laborers were laying the foundations for the pyramids under La Condamine's watchful eye, Anglo-Spanish rivalry arrived unexpectedly off the coast of South America in the form of British warships. Tensions had been on the rise since 1731, when the Spanish coastguard boarded a British merchant ship near Havana and mutilated its captain, Robert Jenkins, as they scoured the vessel for contraband. Armed conflict erupted eight years later in a series of military engagements that came to be known as the War of Jenkins' Ear (1739–42). In September 1740, two separate British fleets were speeding toward South America and the Caribbean to attack Spanish positions. Because of their familiarity with the waters of the South Seas as well as their proximity, the Spanish naval officers, Antonio de Ulloa and Jorge Juan, were dispatched to protect South America's Pacific flank. Traveling first to Lima and later down the coast to Chile to fend off a group of ships commandeered by Captain George Anson, the two returned to Quito only the following year. By that time, a dispute between them and La Condamine over the pyramid's inscription was to demand the lion's share of their attention as they argued their respective causes before the Audiencia of Quito.[56]

Anglo-Spanish conflict quickly segued into Franco-Hispanic strife as the two Spanish naval officers reacted critically to the pyramids and their proposed inscriptions. They argued that La Condamine had failed to secure the "required permission" of the Audiencia to place commemorative markers on the plain of Yaruquí and they rejected the status they were accorded on the stone tablets. They were also enraged that the participation of the Spanish Crown was symbolized by a fleur-de-lis, age-old symbol of the Bourbon monarchy and an icon that only in the most oblique sense could be associated with the new Bourbon king installed on the Spanish throne. Although the Spanish participants had initially acquiesced to the project, they ultimately came to believe that La Condamine had not only used inappropriate words to characterize King Philip's patronage but had also arrogated a role that belonged to the Spanish king himself.

The contours of the struggle that ensued between the French and Spanish plaintiffs before the Audiencia reveal the nature of the extra-European juridical space in which their experimental observations took place. To defend himself and his project, La Condamine claimed to be operating in a jurisdictional vacuum. While he made every effort to abide by the *ceremony* of the Audiencia, he saw himself as inhabiting a juridical sphere to which the Audiencia's temporal powers did not extend. What is more, he accused the Spaniards of condemning him in a forum within whose jurisdiction they too saw themselves as exempt: "In a Court by whose authority they do not consider themselves bound," La Condamine wrote, "[Juan and Ulloa] denounce a foreigner who by his very status is exempt from its jurisdiction."[57] By claiming this immunity, La Condamine gave himself the legal right to carry out his observations without the need to seek the approval of the society or administration that surrounded him. He referred to Philip V as the "sovereign of the most extensive monarchy of the globe" but donned for himself a similar mantle of sovereignty when conducting his own experimental research in a foreign land. By virtue of his status as a member of one of the Old Regime's most venerable institutions, La Condamine asserted his special rights throughout the pyramid controversy and sought to profit from the privileges such a position accorded him. Building upon a narrow interpretation of authorship, La Condamine was able to constuct his own identity free from legal proscription of any sort. It was in this way that he was able to inscribe his own glory onto permanent monuments and exclude others from those rights he claimed as his own.[58]

Juan and Ulloa also asserted their right to receive public acknowledgement for observations and measurements. But their claims extended beyond semantic questions of honorific recognition. In a joint statement to the Audiencia, they not only disputed the content of the textual inscription but also took issue with the process by which the Audiencia had approved the project in the first place and the pretense under which it had been executed. This was more than a contest between nations: it was an indictment against a colonial institution's ability to prescribe rules of conduct and ceremony for instrumental observers operating under the auspices of monarchical authority.[59] But the Spaniards' fiercest criticism was reserved for their French colleague. It was entirely unjustified, they claimed, for someone to build a monument that criticized the Spanish king when laws that expressly forbade "similar inscriptions" already existed in Quito. The Spaniards argued that it was unconscionable to claim credit for work carried out by someone else. And they perceived it as the Audiencia's duty to see that the pyramids and their inscriptions were either transformed or destroyed.[60]

Using a narrow interpretation of honorific language and patronage, La Condamine turned to linguistic evidence to ground his arguments. He explained that he and the Academy of Inscriptions and Belles Lettres had quite consciously chosen "auspiciis" (as in "under the auspices of") to describe the patronage offered by Philip V, a term that Juan and Ulloa had argued gave the impression that the king of Spain had no greater authority to sponsor a mission in Spanish territory than did the ministers of France. La Condamine contended that the Spaniards had misinterpreted the meaning of the term and accused them of "not wanting to discover the etymology and true meaning of the term *auspiciis.*"[61] This argument was typical of a broader anti-Spanish prejudice in the eighteenth century, a northern European response to an intellectual tradition often perceived as backward and ignorant.[62] From La Condamine's perspective, the application of the term *auspiciis* to describe the king's sponsorship of the expedition was a generous and fitting tribute to the Spanish monarch:

[T]he word *auspiciis* was chosen because it was the best and most comprehensive of any of the other terms, and was considered appropriate for His Royal Majesty just as it was in ancient times . . . [A]ll of this is clear from the ancient inscriptions, in which you will find no term that is more honorific to describe the Roman adulation and adoration of their emperors than *auspiciis*, [and] to dedicate and consecrate columns, coliseums, triumphal arches, or any other type of monuments.[63]

La Condamine's references to dedicatory customs and inscriptions drawn from classical antiquity comprised one aspect of his strategy for legitimating his project. But the citation also reveals La Condamine's implicit recognition of the link between the pyramid's inscription and the honorific meanings inherent in monuments of classical and preclassical provenance. His acute attention to linguistic details and the fluid penumbra of meaning represented by words such as "auspiciis" enabled him to wrest proprietary control from his competitors in the juridical sphere.

But La Condamine also used jurisprudential language to articulate notions of intellectual property in the context of empirical observation. He rejected outright the notion that merely to labor on behalf of a project was to retain any proprietary stake in that project's completion. This argument extended implicitly to the native and mestizo populations, but La Condamine explicitly deployed it to strip and dispossess the Spaniards of any claim to public recognition as well. Responding to Juan and Ulloa's assertion that he had "unjustly usurped [the Spaniards'] right... to receive credit for [their] work," La Condamine declared that he and his colleagues had an implied privilege to employ the data deriving from their geodetic measurements. What is more, this prerogative carried with it the privilege of publicizing the results as if they were personal possessions:

No one will dispute the right of the academicians sent for said measurement to declare that they have carried it out; and, as such, having placed our names [on the pyramids], we are employing our right and have usurped nothing and no one.[64]

La Condamine created a jurisdictional sphere around this right to publicity. While acknowledging that physical as well as intellectual labors were necessary for the execution of commemorative projects, he lambasted the Spaniards for thinking that "the greater or lesser toil corresponds to a greater or lesser right to honorific prizes":

Until the world is reformed, there will be no place for this kind of jurisprudence. If it were so, the general would have less glory than his soldiers, the architect less [glory] than the bricklayers, the author less [glory] than the printer of his works.[65]

La Condamine's employment of the term "jurisprudence" indicates that the framework in which he was thinking was a legal one, one that articulated concentric layers of protected rights to authorship and that included protection for those who conceived and executed projects with or without the assistance of others. These ideas about the protection of authorial privilege prefigured later disputes over intellectual property rights in France and throughout Europe, especially noteworthy in the conflict between Diderot and the Parisian booksellers in the 1760s.[66] The entitled manner in which La Condamine spoke of deserved glory and the possession of rights suggests that these ideas were percolating in the scientific realm long before ideas of authorial property circulated more broadly. La Condamine sniped that until the "world is reformed," glory would remain in the possession of those who by virtue of their status and position were entitled to it. Accordingly, only the author of a project deserved credit for the achievements carried out in his name.

While these categories bespeak a typically hierarchical notion of Old Regime society, the functions of architect and author noted above were also highly particular to La Condamine's commemorative project. Architect and author, after all, were occupations that La Condamine fused together by constructing pyramids as monuments to instrumental science and subsequently writing of their demise. These examples helped to bolster La Condamine's standing and may have convinced the Audiencia in the end that the author and architect of these monuments deserved greater credit than the bricklayers, printers, or Spanish officers who toiled to erect them. By tightening and loosening the penumbra of meaning around certain words or phrases and couching his arguments in juridical terms, La Condamine may have successfully excluded others from the fruits of their collective scientific labors. The social, cultural, and intellectual stakes of the pyramids offer a clear example of the scientific space European natural scientists began to claim during this period. They sought increased independence and autonomy from colonial institutions as well as the European monarchs who ruled them. In turn, these monarchical structures and their representatives overseas attempted to restrict the status of the very agents of instrumental empiricism who had proclaimed immunity from colonial rule.

Conclusion

Even though La Condamine denied their symbolic overtones and explicitly rejected any association they may have had with ancient monuments in his 1751 *Histoire des pyramides*, the pyramids of Yaruquí were nonetheless understood some years later as having been inspired by the monumental vestiges of Egyptian civilization. In his 1769 treatise on French rivers entitled *Canaux navigables,* the barrister and journalist Simon-Nicolas-Henri Linguet explained that La Condamine was a gifted writer and admirable traveler whose inspiration had arisen from pharaonic models: "Imitating the Egyptians, he attempted to preserve the memory of his discoveries by erecting a pyramid on the very theater of his work."[67] The *Histoire des pyramides* appears to have convinced Linguet that La Condamine was commemorating *his* discoveries and that he alone was responsible for erecting the monumental structures that recorded them. What is more, Linguet lauded La Condamine as being superior to the "savants of Memphis" in that he was able to communicate his discoveries in a "clear and intelligible language":

What places [La Condamine] above the savants of Memphis is that he did not envelop [the pyramids] with impenetrable hieroglyphs . . . He placed his research within the reach of his readers for the purpose of their instruction, and I join them in giving him the recognition that he deserves.[68]

Linguet recognized the importance of transparency in language and seems to have been impressed by the articulation of scientific results represented on the faces of the pyramids. He referred to La Condamine as having "enlightened an infinite number of countries," thereby vindicating La Condamine's commemorative activities and justifying the broader project of European scientific discovery as a whole.[69]

Not all reactions to La Condamine's *Histoire des pyramides* were as positive as that of Linguet, however. The anonymous Spanish author of the "Historia de las Pyramides de Quito" presented La Condamine's project as both sneaky and motivated by a desire for intellectual profit.[70] He claimed that La Condamine had "disfigured" the truth in his published account and had "dissimulated" with "his artifices." The ire toward what this author saw as La Condamine's having slighted the efforts of the two Spanish officers (and by extension the honor of the Spanish Crown) expressed itself as irony. The author used the term "Historian" to describe La Condamine, sardonically undermining La Condamine's efforts toward writing a true "History" of the pyramids by showcasing his partiality and biases at every turn: "Thus, the Pyramids and the Quito Inscriptions were not necessary, but were rather part of the gallant fantasy of Monsieur de la Condamine, and the fruit of efforts by which he ripped off his two companions."[71] The author implied that La Condamine had not only invented the "necessity" of this commemorative monument but that he had robbed his fellow academicians of the "fruits" of this project. In the end, the attack against La Condamine turned virulent and personal, attempting to expose him as a traitor to the expedition and to portray his activities as inhumane examples of greed and manipulation:

Monsieur de la Condamine has his head and his heart in a different place than other men. It seems unbelievable that he did not foresee these and other less significant difficulties; but above all he hoped to triumph by means of sophisticated maneuvers of his fertile genius, or he wanted to expose himself purposely to great risk so that he could receive the short-lived satisfaction of the worst and most shameful of all the passions of Man.[72]

The identity of the Spanish author who composed what he called a "reformed" history of the Quito pyramids remains a mystery, although there is circumstantial evidence that it may have been Andrés Marcos Burriel (1719–62), the Jesuit polymath who was primarily responsible for organizing the publication of Juan and Ulloa's *Relación histórica del viage a la América meridional* (1748).[73] Burriel was concerned that there be some kind of response to what he called La Condamine's "distur-bance" (*bullicio*) but hesitated as to whether or not a description of the episode should be provided in the *Relación histórica*. In the end, when Juan and Ulloa published their much acclaimed *Relación histórica*, they passed over the pyramid controversy without any significant explana-tion of the drawn-out legal battle that had embroiled them and their monarch for years, both in Spain and in the Americas. Only a single paragraph referred to the conflict over the pyramids in Juan and Ulloa's text, likely an attempt to smooth over in print what had in fact been a turbulent and contentious process before the Audiencia.[74]

But well before the publication of the *Histoire des pyramides* or the revised "Historia de las Pyramides de Quito," Juan and Ulloa were already conscious of their powerlessness to defend the king and themselves from the permanent injuries the French academicians could inflict with a pen rather than a chisel. Their fears were confirmed. The *Histoire des pyramides* prolonged the life of the pyramid controversy far longer than the inscriptions on the sides of the pyramids:

[E]ven if they were to take down the tablets in which the inscription is to be found after this excess is recognized, ... they will not be able to avoid the insult of what those French academicians will put into the books they are likely to print, the case of having been allowed in the fields of Y.[our] R.[oyal] Person to engrave and calculate arms and writings against his honor.[75]

The permanency of the "tablets" upon which the inscriptions were en-graved was far less significant than the injurious comments about the Spanish king in print. For a European readership, the texts written and published by the French academicians forged the central image of how the expedition was executed, commemorated, and disputed in the mountains near Quito. Once back in Europe, La Condamine must have surmised that his best revenge against the destruction of the pyramids was to rewrite the history of the pyramids with the tools, instruments, and priorities of a plaintiff in the court of public opinion. The *Histoire des pyramides* was to unveil in print what the destroyed pyramids of Quito had been meant to proclaim in stone.

The debates over scientific language in the eighteenth century re-flected an increasing anxiety over the legitimate format in which to pro-claim scientific achievement and an even greater concern with what the broad circulation of instrumentally derived knowledge would eventu-ally achieve. The philosophes were conscious of "the power of language to shape destiny ... [and] ... effectively linked intellectual and social progress to linguistic advance."[76] No place is this tension expressed more fundamentally or resolved as elegantly as in the tenth chapter of Condorcet's *Esquisse d'un tableau historique des progrès de l'esprit humain*, entitled "Des progrès futurs de l'esprit humain." In the first few lines, Condorcet asserted the predictability of human events based on an understanding of universal laws. In affirming the permanence and fixity of the natural laws of the universe, Condorcet extended this natural code to the moral sphere, insisting that human societies would little by little diminish the vast social, cultural, and intellectual distances between them. Eventually, these advances would pave the way for the liberation of the human spirit and the "real perfection of intellectual, moral, and physical faculties."[77] One of the most effective ways toward achieving this equality would be through what Condorcet called "the art of instruction," buoyed by the progress of scientific thought.[78] But scientific instruction could not proceed without the institution of a "universal language," one that was characterized by clarity and simplicity. Older forms of scientific communication served to divide society into "two unequal classes: the one composed of individuals who know this language and have the key to all the sciences; the other that is unable to learn [the language and] finds itself incapable of attaining Enlightenment (*acquérir des lumières*)."[79] In Condorcet's vision, the acquisition of language would lead to the understanding of science, which in turn would conduct those impoverished or "savage" nations to participate like other nations in the collective march toward universal civilization. The tablet inscribed in Latin on the pyramids of Yaruquí would have failed such a test of universal suffrage, since it failed to speak to the very nations upon whose territory its ornate and elite language was affixed.

In addition to raising questions of enfranchisement and education for less enlightened nations, these arguments over the form that scientific language should take represented a breach that emerged during this period between collective and individual authorship: between the publication efforts made by members of an academic corps and the same individuals' efforts to seek prestige outside of an institutional context. Prior to 1777, the individual scientific author, like authors in other fields, had no right to official protection under law for published ideas or experimental results.[80] But the inscription of scientific results onto physical markers beyond the traditional jurisdiction of European bodies operated outside these legal frameworks and functioned within an alternative economy of intellectual prestige. These presumably unconscious allusions to ideas of property and labor were not unique to La Condamine or his entourage. In fact, such ideas were already percolating during the pre-Revolutionary period and ultimately received explicit expression in debates over literary property in the late 1770s in France. Their origins were to be found in arguments meant to seal and justify the royal practice of giving literary privileges to certain Parisian publishers. Louis d'Héricourt, who was retained by the Publishers' Guild to construct an argument about literary property in 1726, wrote that "[property in ideas] is the fruits of one's own labor, which one should have the freedom to dispose of at one's will."[81] La Condamine's language is ultimately different but seems to express a similar sentiment: that an academician has the right to proclaim empirical truths wherever in the world an experiment may take him, be it on a rock at the edge of the Pacific Ocean or upon a pyramidal monument in the mountains near Quito.

———

During his five-year expedition to South America from 1799 to 1804, Alexander von Humboldt came to see the exuberant landscapes he traversed as monuments to the underlying unity of the natural world, majestic symbols that man-made objects could never equal. After his encounter with the pyramids of Yaruquí, he concluded that it had been "foolish" to attempt to build physical monuments in a region where jungles grow incessantly and earthquakes shake the foundations of even the highest mountains. When Humboldt stood near the Hacienda of Oyambaro (fig. 8) and surveyed the ruined outlines of La Condamine's pyramids, he saw in the scattered shards and illegible tablets proof that earlier ideas about commemorating scientific activities had allowed feeble monuments to be built on shaky ground. Exactitude was "an immaterial and disembodied achievement," and La Condamine and his cohort had attempted to use stones and mortar to transform their scientific results into solid objects.[82] Even if these markers were intended to allow future generations to determine the accuracy of the original measurements, Humboldt suspected that the region's native populations would take anything of value from these pyramids, and the incessant rains of the tropics would finish off the job soon thereafter. "In the land of the Indians," he concluded, "far greater monuments [would have been] required for [the pyramids] to remain intact."[83]

But while Humboldt may have felt that the pyramids were useless for scientific purposes, they could still stand as monuments for the three scientists—Godin, Bouguer, and La Condamine—sent by the French king to measure the shape of the Earth. Indeed, Humboldt hoped to encourage a local colonial administrator in Quito to rebuild these monuments: "I

Figure 8. The pyramid at Oyambaro, letter "B," is framed by the majestic volcanic peaks near Quito, including a smoking Cotopaxi ("2"). The farm where Humboldt reflected on the broken shards of the pyramid's plaque is visible at letter "I". Detail from La Condamine, *Journal du voyage fait par ordre du Roi* (1751; plate 3). Courtesy of the Rare Book and Manuscript Library, University of Pennsylvania.

would consider myself very happy if I contributed with my trip to the efforts to enshrine the names of Bouguer, La Condamine, and Godin in the great valley of the Equator."[84] Underlining the ephemeral nature of all physical monuments, Humboldt asked:

Had La Condamine known, when he was crowning his pyramid with a fleur-de-lis, that these flowers would disappear sixty years later because they were seen as symbols of tyranny, and that France [itself] would be annihilated?[85]

With hindsight, Humboldt could recognize the patent instability of a physical structure designed to encapsulate universal scientific truths. The construction of the pyramids was, in the end, an ephemeral spectacle. But as monuments immortalized in print, the pyramids were also part of a more extended history: a history of science as physical inscription. As such, they were bound as symbols not only to a broader history of the social and material cultures that produced them, but also to the history of language, politics, and polemics that emerged in the New World and traversed the Atlantic in both manuscript and print.

TWO

An Enlightened Amazon, with Fables and a Fold-Out Map

Arriving in Borja, I found myself in a new world, far from human commerce, upon a sea of fresh water, amid a labyrinth of lakes, rivers, and channels that penetrate the immense forest in every direction . . . I encountered new plants, new animals, [and] new men. LA CONDAMINE, *RELATION ABRÉGÉE D'UN VOYAGE FAIT DANS L'INTÉRIEUR DE L'AMÉRIQUE MÉRIDIONALE* (1745)

An affinity for marvels and the desire to adorn the descriptions of the New Continent with certain characteristics drawn from classical antiquity have no doubt contributed to the great importance [given] to these early accounts.
HUMBOLDT, *RELATION HISTORIQUE DU VOYAGE AUX RÉGIONS ÉQUINOXIALES DU NOUVEAU CONTINENT* (1814)

A Long-Expected Return

Can a river be enlightened? The Spanish Jesuit José Gumilla certainly thought so. In 1741, when he published *El Orinoco ilustrado*, he introduced his account by making specific reference to the illuminating power of the past. In his description of the Jesuit missions in the Orinoco basin, he explained that history was not only a "witness to the [present] times" but that it could be a "light" for all posterity as well, as long as it was written with "clarity, distinction, and method." Without these elements, he proclaimed, even "the most fascinating history" would become the "origin of many doubts" and a "chaos of confusion." A thirty-year resident of the South American region he described in his text, Gumilla organized his account of the Orinoco with these principles in mind, creating a natural, civil, and geographical history that echoed the classic New World chronicles of his sixteenth-century forebears, especially the Jesuit José de Acosta. Gumilla believed that his descriptions of the singularities found near and along the river, including the peculiar animals, insects, trees, resins, and roots that the "temperament of those climes" yield, would cause "the Orinoco . . . to be reborn . . . luminous." For Gumilla, an enlightened Orinoco would be a Christian utopia as well as a Spanish colonial territory; his treatise was both a religious history and a pedagogical handbook for missionaries whom "God Our Lord calls . . . to the cultivation of his American vine, [a section] abounding in fruit [yet] retaining much evil to trim away."[1]

Four years later, on April 28, 1745, a very different kind of South American chronicler stood before the Paris Academy of Sciences. Unlike Gumilla, La Condamine chose to present the Amazon River by narrating his journey downstream, entertaining his audience with eyewitness description rather than instructing the faithful through the details of history. Instead of dividing his text into chapters, La Condamine packaged his account as a first-person *relation*, hearkening back to a narrative form mastered by early seventeenth-century French explorers and missionaries and eschewing the methodological principles that would have parsed or dissected his account arbitrarily into separate pieces.[2] Without clear chapter divisions, the *Relation abrégée d'un voyage fait dans l'intérieur de l'Amérique méridionale* (1745) was a text that flowed as smoothly as the river it described, with no systematic demarcation between La Condamine's loosely chronological retelling of his journey and discussions of the region's geography, natural history, and indigenous customs and beliefs. Astronomical measurements stood comfortably alongside mythical accounts of mist-laden forests, "enchanted" palaces of gold, and bare-breasted woman warriors. The *Relation abrégée*'s colorful descriptions of Amazonian flora and fauna were interspersed with notations of latitude and longitude, prefiguring a new formula for attracting a broad audience while conforming to the more rigid expectations of the academic *mémoire*. The narrative unity of the text thus provided its readers with an uninterrupted account of La Condamine's expedition across the full extent of the South American continent, from the Andes to the Atlantic. His was a strategy that showcased the Amazon using the instrumental standards of Enlightenment empiricism but in a format that would not tax those who preferred *Robinson Crusoe* to the *Principia Mathematica*.[3]

Piquant and pleasurable though it may have been, La Condamine's narrative strategy nevertheless concealed from its readers as much as it revealed. Over the course of his expedition, he carefully collected maps, manuscripts, epistolary correspondence, and missionary histories from individuals he met along his route. He also relied upon local knowledge and material support from Creoles, Jesuits, Amerindian informants, and enslaved peoples of African descent. But in order to provide the coherence and first-person authority with which he hoped to impress his superiors at the Academy of Sciences, he suppressed the sources of much of this information and hid much of the assistance he received. To present the Amazon in print, La Condamine highlighted his own eyewitness observations and criticized those texts that might have been seen as competing narratives. He lifted ethnographic observations from manuscript accounts without citing his sources and used epistolary correspondence by learned Creoles as the blueprint for many of his geographical conclusions. Interactions with native populations enabled him to make broad conclusions about Amazonian hydrography while providing grist for his caustic assessments of their customs, language, and character. In short, to assemble the Amazon for an academic audience, La Condamine had to fuse exceptionally diverse materials together and provide the illusion that they were collected and compiled using unified and recognizable standards. The deft deployment and strategic effacement of certain forms of textual knowledge were the glue holding together the disparate building blocks of La Condamine's enlightened Amazon, interlocking pieces that when assembled as a unit afforded a new image of South American geography for one of Europe's premier scientific institutions.

But La Condamine's most effective strategy for depicting the Amazon basin was through cartographic, rather than textual, means (plate 5). The map that accompanied the *Relation abrégée* was the first graphic representation of the entire navigable length of the Amazon based upon measurements drawn from astronomical instruments and celestial observations. According to contemporary reviews and articles, La Condamine's *Carte du cours du Maragnon* (1745) was superior in accuracy to both the Sanson map of 1680 and Father Samuel Fritz's of 1707 (reprinted in the *Lettres édifiantes et curieuses* in 1717), the most recent printed maps of the Amazon basin at the time of La Condamine's descent (figs. 9 and 10).[4] But what is most striking about the *Carte du cours du Maragnon* is how La Condamine arrayed his materials in order to forge a smooth representation out of what was in fact a latticed assortment of diverse sources. In certain places, he insisted on astronomically determined observations, but in others, he relied on maps, travel logs, and oral accounts by missionaries and Amerindians. Certain geographical features were measured meticulously over a period of several days, while others were inscribed onto his map without more than cursory knowledge, often supported by conjectures based on scanty evidence. Alexander von Humboldt was one of many who criticized La Condamine's reliance on earlier textual narratives, stating that he had "multiplied beyond measure" erroneous hypotheses and sown confusion that had lasted until Humboldt's own expedition. "La Condamine's voyage," he wrote, "which has spread so much light on the different parts of America, has muddled everything related to the course of the Caqueta, Orinoco, and Negro [rivers]." For Humboldt, La Condamine had created an "imaginary system" based not on eyewitness observations but on uninformed speculation. And because of La Condamine's successful rhetorical strategies and persuasive graphic representations, those who followed in his footsteps had spread the same mistaken geographical notions.[5]

Recent scholarship in cartographic studies has tended to view mapping as a mediated process imbued with complex layers of social, cultural, and political meaning. Rather than focusing on the content of a map and its purported "accuracy," scholars have begun to treat maps as "artifacts," emphasizing "the way maps store, communicate, and promote spatial understanding."[6] La Condamine's Amazonian map is an important

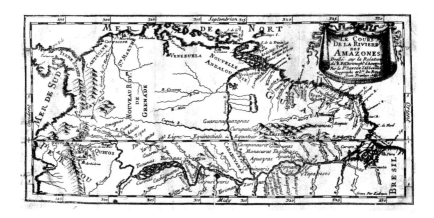

Figure 9. Sanson's "very defective" map of the Amazon River. The *Cours de la Rivière des Amazones* (1680) accompanied Marin Le Roy Gomberville's *Relation de la rivière des Amazones* (1682), a French adaptation of Pedro Texeira's seventeenth-century Amazonian narrative. Textual citations indicate a golden village ("village d'or"), a "river of gold," and "Amazones," hiding among the mountains at center-right just above the equatorial line. Courtesy of the Fundação Biblioteca Nacional, Rio de Janeiro, Brazil.

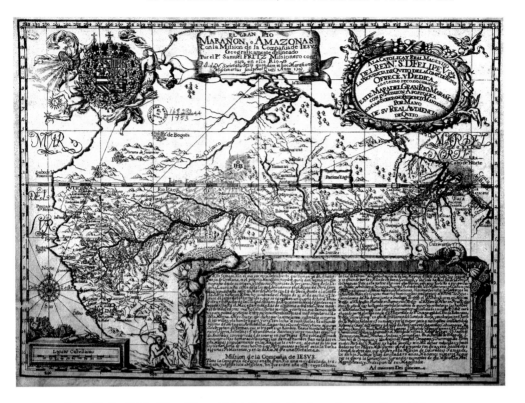

Figure 10. The 1707 edition of Samuel Fritz's *Gran Rio Marañon, o Amazonas*. Samuel Fritz was a Czech Jesuit who had been sent to the Amazonian mission at Omaguas in 1687, and he produced several maps of the Amazon River basin between 1690 and 1707, one of which is shown above. Ten years later, a reduced version of this map was published in volume 12 of the *Lettres édifiantes et curieuses écrites des missions étrangères* (Paris, 1717). The Marqués de Valleumbroso, a resident of Cuzco, did not trust Fritz's observations and, consequently, neither did La Condamine, who justified the need for a new Amazonian survey based on Fritz's purported errors. Courtesy of the Fundação Biblioteca Nacional, Rio de Janeiro, Brazil.

example: like the narrative that accompanied it, the *Carte du cours du Maragnon* reflected the explicit expectations of the European scientific community and the inchoate demands of an emerging public reader-ship just as much as it did the "true" geography of the Amazon basin. La Condamine manipulated the standards by which he assembled his observations in order to conform to the expectations of his audience. He wielded third-party evidence in order to settle longstanding debates whose significance he deemed crucial for satisfying the Academy, es-chewing for those moments the eyewitness observational methods he had previously espoused and upon which the superiority of his own cartographic project depended. He chose to argue in favor of the fabled existence of El Dorado and a gynocratic society of female warriors, the American Amazons, to appeal to a public that desired an affirmation or refutation of one of America's most longstanding and intriguing myths. On the map, he decided to include an unobserved lake to mark the site of El Dorado, and he extended the Rio Negro far beyond the places he himself had observed in order to inscribe a reported connection between the Amazon and Orinoco rivers. Employing both textual and cartographic evidence, La Condamine narrated and remapped the Amazon according to standards forged within the academic culture of eighteenth-century Europe and in line with the expectations of a broader public. His discursive skills helped him to justify and ultimately vindicate his project to a diffuse group of potential critics. Both the *Relation abrégée* and the *Carte du cours du Maragnon* demonstrate how an impulse toward synthesis, unity, and intellectual dominance was justified and executed in an eighteenth-century travelogue doing double duty as exotic narrative and scientific *mémoire*.

Porters and Papers

La Condamine and his fellow travelers departed from La Rochelle on May 16, 1735 for a transatlantic crossing that lasted thirty-seven days. After several weeks at sea, La Condamine was astonished to find a felicitous memento from his former Parisian life upon arriving in Saint Domingue (present-day Haiti) in mid-July. "I never doubted that your name was known everywhere," he wrote to Voltaire in December of that same year, "but I did not expect to find friends of yours in the Torrid Zone and much less people with whom I had dined at your home."[7] It was the poet Sinetti, with whom La Condamine had shared a meal some eighteen months earlier and whose "large cheeks" he still remembered vividly from this earlier encounter. He wrote to Voltaire that the two had spent the length of the journey from Saint Domingue to Portobelo (present-day Panama) discussing their much-beloved mutual friend, and that their sole "pastime" on board had been to cite from memory passages from Voltaire's many publications, including *La Henriade*, *Zaïre*, and *Adelaïde*. This poetic moment during which two Europeans on a Caribbean traverse discussed texts by Voltaire raises a larger issue: how the geographical conceptions of European travelers were fueled by their previous reading and how literary memories intertwined with local observations in the narrative construction of South American geography and mythology. To understand what ancient and modern histories flashed upon their inward eyes as they plowed the seas, scaled mountains, and recorded their observations, and how the content of La Condamine's account may have been shaped by what he read previously, we turn to examine how La Condamine's books and papers traveled alongside him and his companions in the "Torrid Zone."

In order to read at all, European travelers were dependent upon hired laborers to convey their libraries up and down mountains, across rivers, and through the forest. The academicians had traveled to South America with several trunkloads of books, and once arrived in Portobelo, physical laborers transported their baggage for them. These individuals loaded and unloaded trunks but only rarely appeared in the texts whose production they facilitated with their efforts. Like the indigenous workers described in chapter 1, they remained in the interstices of the narrative text. Their only archival appearance was in an inventory written by Don Bernardo Gutierres Vocanegra, the governor of Portobelo, when he wrote that "six blacks (*negros*) [carried] the baggage listed below to the city of Panama."[8] The objects in this meticulous inventory included "twenty-one [trunks] of Books, and one with four thousand pesos in reales."[9] Scattered throughout the rest of the list is evidence of the bibliophilia of the expedition's participants: "One trunk of clothing, Andalusian tobacco, books and other small items . . . One trunk of Books and white clothing . . . [and] one Trunk of dress clothing and Books." Specific book titles were only occasionally mentioned in the expedition's published accounts. But those who toiled to cart and drag these books from one location to another were almost always suppressed in the narratives; only in this one administrative document did the presence of books and their porters explicitly, if briefly, appear.[10]

The appearance of slaves and coerced laborers of African origin was also intermittent in La Condamine's published narratives. On the rare occasions they did appear, their presence was often quickly papered over. Prior to descending the famed Pongo de Manseriche, one of the most narrow and treacherous portions of the Amazon River as it descends from the Andes, La Condamine recounted how he had been left alone on his raft "with a Black slave"; the four Indians he hired in Jaén de Bracamoros had chosen to pursue the journey on foot. As La Condamine and the slave spent the night together on the raft, the water level in the river began to descend rapidly and a peculiar situation ensued that nearly saw

La Condamine lose all his papers to the waters below. As the height of the river decreased, La Condamine became aware of the sound of a large branch rubbing up against the bottom of his raft:

[The branch] had lodged itself between the pieces of wood on my raft, and it continued to penetrate deeper and deeper as the level of the river fell. Had I not been present and awake, I would have ended up with the raft attached to a branch of the tree and suspended in midair. The least that would have happened was to lose my journals and papers, the fruit of eight years of work. Luckily for me, I finally found a way to release the raft and set it afloat once again.

But what happened to La Condamine's nocturnal companion? The "black slave" disappeared entirely from the narrative. La Condamine re-counted in heroic fashion how he had released the raft himself once it became suspended, placing it afloat entirely on his own. But given that the two of them were present on the raft, it is likely that the "black slave" had some part to play in loosing the raft and placing it back in the water as well, which in turn saved all of La Condamine's written work. Yet La Condamine only gave *himself* credit for this miraculous fortune. Why was the slave included in the first place, if his presence was removed only two sentences later? Did he serve as an unwitting witness to La Condamine's heroic efforts, or did he aid him without narrative recompense? Unlike his South American friend Pedro Vicente Maldonado, whom La Condamine described as a "zealous" companion and "a great help for his intelligence and energy," the black slave re-ceived no acknowledgement in a situation that potentially prejudiced all of La Condamine's scientific work. Slaves and coerced laborers may have served as porters and guides, but they received only marginal recog-nition, if any, in the textual accounting of the journey. Instead, their ac-tive presence was effaced through the narrative conventions of the heroic account.[11]

Like the journals and papers he so assiduously protected as he de-scended the river, previous travel accounts were crucial companions for La Condamine during his journey. He had read extensively both in Europe and in the Jesuit libraries at Quito and Maynas, and was well aware that he was operating within a broader literary and ethnographic tradition of European travelers to South America. While he occasion-ally cited the work of French travelers—including Jean Richer, Amédée Frézier, and Louis Feuillée—he made far more frequent references to Amazonian explorers who had traveled under the auspices of the Spanish and Portuguese monarchs, including Pedro de Úrsua, Pedro de Texeira,

Cristobal d'Acuña, and Samuel Fritz. But in order to contextualize La Condamine's knowledge and experience of the Amazon, it is necessary to go back even further than these sixteenth- and seventeenth-century accounts and briefly review the river's initial discovery by Europeans, which was carried out in the early 1540s by Gonzalo Pizarro, governor of the province of Quito, his military captain Francisco de Orellana, and the Dominican friar Gaspar de Carvajal.[12]

The initial exploration of the Amazon basin was linked to two myths that had circulated since the arrival of Europeans in the New World. Early in the sixteenth century, rumors had spread through Central and South America of a "país de la canela," a site of mythical and geographical speculation where gold, spices, and other natural treasures were thought to abound. Europeans were particularly interested in a specimen resembling Indian cinnamon (*Cinnamomum zeylanicum*) that was reputed to exist in the eastern foothills of the recently conquered Andes. Pizarro and Orellana set out to explore this region in 1541. On the same expedition, Pizarro was determined to find the domain of an indigenous warrior-chieftain named "El Dorado," who was said to dip himself each day in a lake of gold to impress his subjects and frighten his enemies. In his attempt to find this city of gold, Pizarro undertook the first European exploration of the Amazon River. During his journey, Orellana also claimed to see a group of bellicose women "who set about fighting in front of [their Indian subjects] as captains, so valiantly that the Indians did not dare to turn their backs."[13] A century later, the Portuguese captain Pedro de Texeira chose the reverse itinerary, leaving from the Portuguese city of Pará (present-day Belém) to explore the river westward. In Quito, he met up with the Jesuit Cristobal d'Acuña, who accompanied him on his return to Pará. Acuña's text was the first full narrative account of the river's descent, published in 1641 as the *Nuevo descubrimiento del gran Rio de las Amazonas*. La Condamine made frequent reference to the 1682 French translation by Gomberville, to which he had regular access as he descended the river.[14]

Amazonian travel accounts were at least partly responsible for La Condamine's having chosen to return to Europe by this route and for determining his itinerary once he had made his decision. But according to La Condamine, it was Louis Godin who initially suggested they cross the continent via the great river for the journey home: "M. Godin had thought that after our mission was completed, all of us could embark on a journey down the river of the Amazons to return to Europe."[15] In fact, La Condamine had initially viewed this idea with some skepticism, since he believed the Amazon River would be far too risky for such uninitiated travelers: "After reading [Gomberville's *Relation de la rivière des Amazones*], I could not consider this route as anything but the longest and most difficult of all, and I was far from enjoying the idea of a undertaking a project that would only delay our eventual return to France."[16] Yet La Condamine's later discussions with Jesuit missionaries in Quito convinced him that an impression based merely on his reading of Gomberville's account was wrong:

My ideas changed after learning more specific details from various missionaries following my stay in Quito. I was convinced that this route was truly impracticable for a large group like ours, since we would need a canoe and a crew of seven or eight rowers for each person, or at least for every two persons, and such a large number [of rowers] might at times be difficult to find. But things were very different for one or two travelers... As far as the discomforts that are part and parcel of this kind of journey, I was certain that they were exaggerated.[17]

While portable libraries may have been important for La Condamine's scientific observations in the Andes, the information provided by Gomberville's text was quickly superceded by conversations with locals on the spot. When La Condamine had to choose between what was presented in Gomberville's 1682 *Relation* and the information he gleaned from discussions with Jesuits, the latter appear to have won out. Even local information was eventually treated with suspicion, however, and subordinated to his own curiosity: "[E]verything that I had heard made me all the more eager to confirm their reality for myself."[18] For La Condamine, only firsthand observations would reveal the truth about the Amazon. He did not wish to portray himself as having been swayed by hearsay, speculation, or legends arising from an out-of-date text or the gossip of Jesuits. In the end, however, written materials did weigh heavily in forming La Condamine's geographical and ethnographic conceptions; they were not limited to Gomberville's account nor were they available in print. Instead, they were locally produced and provided by observers whose texts could be consulted and discarded with little more than a trace.

Tropical Cheat Sheets: Manuscript Instructions for an Amazonian Adventure

Once decided on returning to Europe via the Amazon, La Condamine wrote to Cuzco informing his friend, the Marqués de Valleumbroso, of his intentions. The Marqués de Valleumbroso was Don Josef Pardo de

Figueroa y Acuña, a Creole originally from Lima who was the brother of the bishop of Guatemala and nephew of the Mexican viceroy, the Marqués de Castel-Fuerte. Valleumbroso enthusiastically approved of the academician's planned itinerary. "I am delighted by your determination to undertake a journey down the Marañon River," the marquis wrote in response. As it turned out, Valleumbroso was the great-nephew of Cristobal d'Acuña, the chronicler of Pedro de Texeira's 1637–39 expedition down the river. Not only had Valleumbroso long been curious about Amazonian geography, but he was especially eager for more up-to-date information about the state of the river. "You will no doubt already have seen how strange the mathematical sciences are in these lands," he lamented to La Condamine, "and how little you can trust the common maps, which for the most part are based on uncertain, inaccurate accounts." Valleumbroso had earlier requested accounts from Father Andrés de Zarate, a curate at the Jesuit mission in Quito, who had compiled the journals of Pablo Maroni and in turn had sent some of these notebooks back to Valleumbroso in Cuzco. But his thirst for new information was not quite sated: "I wish to have more detailed and up-to-date news about the Marañon," Valleumbroso continued, "and to know the present state of this great river." As such, La Condamine was able to benefit from some of the most recent accounts and observations taken from the field, folded into a manuscript letter in digest form and sent to the Frenchman a year and a half before he descended the Amazon on his own.[19]

In particular, La Condamine used the information he learned from Valleumbroso's letters to critique earlier explorers. The accounts to which Valleumbroso had access were primarily from Spanish Jesuits, a detail that was not lost on later Portuguese critics of La Condamine's *Relation abrégée*, as we will see in chapter 3. But whatever their provenance, La Condamine was offered a preview of what Valleumbroso perceived as the faults and flaws of explorers that had gone before him, and the *Relation abrégée* by and large reflects the marquis's interpretations. Valleumbroso's letter reads like a litany of errors contained within various maps and narratives of the Amazon, beginning with Orellana, Pedro de Úrsua, passing through to his great-uncle Acuña, Father Manuel Rodriguez, and finally ending with La Condamine's predecessor, the Jesuit Samuel Fritz, whom La Condamine used as a straw man to demonstrate the superiority of his own observations. Valleumbroso's contempt for the methods used by the earliest explorers of the river comes through clearly in his analysis of why so little accurate geographical information had made its way from the Amazon basin to the printed page:

[T]hey [do not] know with mathematical precision the course of that great river, since neither Orellana nor Pedro de Ursua were men who could make mathematical observations, let alone create historical accounts. And so it [happens] that until Father Acuña published his journey, which was merely historical, nothing was known with certainty about this river.[20]

For Valleumbroso, the representation of the Amazon had come to be a site of controversy because the histories and maps produced by these early travelers were flawed on so many levels. "[Modern authors want] to usurp the right of the first discoverers to provide names [for the rivers]," wrote Valleumbroso, "and in this way they arrogantly consider the Marañon to be one and the same as the Amazon River, without having proven this, nor even having seen the waters of this great River, nor having peered into the original histories of the discovery of America; [and] many foreign authors follow them down this path."[21] According to Valleumbroso, previous explorers lacked crucial information, and the timing of La Condamine's planned expedition was especially propitious for a reevaluation of ancient myths and legends. Valleumbroso seemed to place high hopes on the benefits that would accrue from his journey: "It will be a worthy use of your curiosity to examine [the question of the Amazons] with great precision, since today there is greater ease [of access] than in times past . . . Today all of those riverbanks are well traveled . . . and in this way you will be able to write accounts (*tomar memorias*) of things that we still do not know about [the Amazon's] geography, with which you will be able to regale the Public (*enriquecer al Publico*)."[22]

Valleumbroso's invocation of the interests of the "public" foreshadows La Condamine's discussion of the dual nature of his audience, both lay and academic. And while it is impossible to know if La Condamine relied exclusively upon Valleumbroso's observations to formulate his own strategies for denigrating previous knowledge about the river, he does seem to have taken at least some of the flavor of the marquis's letter and adapted it to his own ends. For example, La Condamine referred to Acuña's narrative as a "purely historical account," echoing almost verbatim Valleumbroso's assessment that little was known in detail about the river until Acuña's visit in 1640: "[U]ntil Father Acuña published his journey, which was merely historical," Valleumbroso wrote, "nothing was known with certainty about this river." In a rhetorical twist, La Condamine used the syntactical logic of Valleumbroso's phrase but transformed it to insist that nothing had been written of any value *since* Acuña's account either: "In Europe today," wrote La Condamine in 1745, "all that is known concerning the countries traversed by the Amazon was learned more than a century ago from the account of Father Acuña."[23] In this way, La Condamine paved the way for his own entrance upon the Amazonian stage by mirroring the strategy employed by Valleumbroso and implying that nothing had been learned since Acuña's journey one hundred years earlier.

Borrowing yet again from his friend, La Condamine discussed Father Fritz's inadequate account of the river's twists and turns using terms that are strikingly similar to those employed by Valleumbroso. The marquis had remarked that "as the said Father [Fritz] was sick during his journey, I do not put much trust in his Longitudes and Latitudes, since he would not have been in a condition to make his observations with the precision that is required."[24] La Condamine criticized his Jesuit predecessor for being ill and for lacking the appropriate instruments to reach conclusions that were scientifically verifiable, precisely those aspects of Fritz's journey that Valleumbroso had raised in his letter. "One need only read his manuscript diary," wrote La Condamine in his *Relation abrégée*, "to see that many obstacles, then and upon returning to his mission, prevented him from making the necessary observations to render his map precise . . ."[25]

Finally, Valleumbroso offered explicit instructions to La Condamine as to the kinds of observations he should make along his journey, the source materials he should consult, and the form his narrative should take: "I do not doubt that in Quito you will have been able to acquire [maps], and together with the information that you can get from the Portuguese, [you will be able to] create a very perfect map, in which you could show the errors that up until now have occurred in the maps that have been made of this River and its environs." He suggested where particular accounts were trustworthy (". . . up until the place where you embark, you can count on the Map of Father Fritz") and which accounts should be used in concert with on-the-spot observations to compile a more complete view: "[A]nd in this way try to get the route descriptions (*derroteros*) of these journeys, and if some maps have been made of them, to serve in the Map that you will make of this great River . . . you will shine much light on a Geography that until now has been quite ignored."[26] At certain points, Valleumbroso's suggestions became so forceful in tone that they appeared to command La Condamine to follow his prescribed course. When, for example, he explained that certain authors had wished to portray the Xauxa and Apurimac as the sources of the Amazon River without taking into account their respective courses, Valleumbroso wrote that this misinterpretation would be rectified by La

Condamine's new study: "You will show this very clearly in the Map that you create, and in the precise account you make of your journey."[27]

This written description that Valleumbroso laid out like a bountiful picnic before his friend dovetailed elegantly with La Condamine's own ambitions for his trip down the river. The marquis had provided La Condamine with an arsenal of arguments he could use to establish the preeminence and authority of his own observations, and the echoes of Valleumbroso's passionate rebuke of previous Amazonian explorers reverberated throughout much of La Condamine's *Relation abrégée*. Despite his interest in "assuring himself of the reality" of the river's myriad features, La Condamine relied to a great extent upon a letter written to him by a Cuzco Creole who had spent his life searching for accurate information about the route his great-uncle had taken many years before. As such, the background to La Condamine's geographical conceptions seems to have derived as much from Valleumbroso's close readings of Amazonian travel narratives as from the observations La Condamine made with his own eyes while descending the river.

––––––

Aside from its discussion of previous geographical and mythological debates, however, the feature that most distinguished the *Relation abrégée* from contemporaneous academic treatises was its lengthy discussion of Amerindians, a subject that lay somewhat outside the normal purview of the Academy but one that was certain to capture the enthusiasm of the philosophically minded among its members and the broader public. La Condamine's scathing assessments of indigenous peoples in South America as lazy, unintelligent, gluttonous, and incapable of rational thought formed part of a common vocabulary by which many European authors came to discuss Amerindians in the wake of his journey, especially as a view that laid out multiple stages of human development came to predominate among European naturalists and philosophers from Montesquieu to Hume.[28] It would be hard to overestimate the pejorative effect La Condamine's observations had on the European image of the Amerindian, especially since he was frequently perceived as a credible and trustworthy eyewitness. The testimony of indigenous Amazonians, on the other hand, was regularly treated as fanciful and fallacious in La Condamine's own account. Even as he incorporated a vast storehouse of indigenous knowledge about Amazonian botany, animal life, and geography into his *Relation abrégée*, La Condamine proclaimed without equivocation that "all or most of the American Indians are liars," leaving little doubt to what extent he trusted their knowledge on scientific or geographic matters. He asserted that "insensitivity" formed the base of their character, leaving the reader to decide whether this insensitivity should be "honored with the name of apathy, or vilified with that of stupidity." Meanwhile, he used "the virtues that are attributed to many [plants] by the natives of the country" as the basis for his discussion of the usefulness of the "multitude and diversity of trees and plants that one encounters along the banks of the Amazon River." These same Indians who "go through life without thinking and . . . grow old without leaving their infancy" were also responsible for the rain shelters "made from woven palm leaves [and] artfully prepared."[29] This disjunction between vilifying the character of indigenous populations, on the one hand, and praising their artistic skills and applied knowledge of Amazonian flora and fauna, on the other, exemplifies the conflicting attitude toward native culture within La Condamine's text.

What was the basis of these caustic assertions? While most analyses of La Condamine's attitude toward Amerindians have argued that his pejorative view of native populations arose from cultural prejudice and a sense of European intellectual superiority, the detailed comparison of La Condamine's *Relation abrégée* with a manuscript account he consulted in situ demonstrates that other texts, which he copied nearly verbatim into his own account, constituted the true source of many of his observations. Just as Valleumbroso's letter predisposed the academician to particular geographical conclusions before he had ever seen the river with his own eyes, La Condamine absorbed much of what he knew about native Amazonian cultures from what he had read as opposed to what he had observed himself on the spot. Unable to make methodical observations of Amerindian culture during such a swift journey downstream, La Condamine was forced to rely almost entirely on manuscript treatises given to him by Jesuit missionaries.

One such text was the "Description de la province et des Missions de Maynas au royaume de Quito," an account written by Jean Magnin (1701–53) on the Maynas missions and the indigenous populations that inhabited the region. Magnin was a Jesuit from the canton of Fribourg (Switzerland) whom La Condamine met in Borja following his descent of the Pongo de Manseriche. At the age of twenty-four, Magnin had arrived in Cartagena (Colombia) from Europe and began a slow traverse across rivers and mountains toward Quito, where he would spend nearly three decades until his death at the age of fifty-two. In the intervening years, he carried out an active program of proselytization among the Yameo, Miguiana, Parrano, and Amaona tribes of the upper Amazon. The final stage of his missionary career was spent at the curiate in San Francisco de Borja, where for three days in 1743 he received La Condamine and shared his knowledge of the region's populations with the passing academician. Magnin also provided him with a manuscript account of his observations.[30]

La Condamine's conclusions regarding the native inhabitants of the Maynas region read as if lifted directly from Magnin's account. The observations La Condamine made of Amazonian culture included biting assessments of native customs, their character, and their language, and Magnin's manuscript was crucial in providing La Condamine with evidence for his broad claims. In his "Description de Maynas," Magnin remarked on what became in eighteenth-century Europe some of the most oft-cited and "bizarre" features of Amerindian cultural life, such as the way in which the Omaguas people attempted to manipulate the shape of a child's face at birth. In a section of Magnin's manuscript entitled "Autres coutumes particulieres, leur nudité, leurs habits de ceremonie," Magnin had written that the Omaguas were "persuaded that it is a beautiful thing to have one's face as flat as the moon."[31] But in his own retelling, by using an active voice, La Condamine made it appear as if the Omaguas had communicated their cultural predilection directly to him:

The name of Omaguas in the language of Peru means flat head; in essence, these people have the bizarre custom of pressing the forehead of their infants between two boards as soon as they are born, which gives them a strange form that, they say, makes them resemble the full moon. The language of the Omaguas is as sweet and easy to pronounce as [the language] of the Yameos is harsh and difficult.[32]

La Condamine's manner of actively narrating this encounter conformed to an expectation of personal testimony within the Academy, a language of experience that allowed him to appear as if the original report had been his own. These observations, initially written down in Magnin's manuscript and later transcribed to La Condamine's *Relation abrégée*, traveled subsequently from the mission of Borja all the way to the pages of the *Encyclopédie*, where they were codified under the entry "Aguas," as discussed in chapter 7. The path from the forests of South America to the folio pages of the *Encyclopédie* shaved off the source of the original observation as it passed through the consecutive stages of translation (Magnin's manuscript was originally written in Spanish), transcription, and abridgement. In the process, La Condamine's debt to Magnin's observations was obscured and displaced, if not entirely effaced.

Beyond pointing out what he perceived as the most peculiar of Amerindian customs, La Condamine made broad and nefarious assessments of their character as well. Once again, many of these observations consisted of collated and reorganized material drawn from Magnin's manuscript account. Magnin had written that their character was similar to that of children, and they were "timid, fearful, and lazy drunks [who are] unfaithful, without recognition, and without memory." After spending nearly twenty years among the natives, Magnin remained convinced of the Amerindians' laziness, their lack of intelligence, and their immorality; to him, they were "contented savages" that lived entirely without cares or fears:

They live so happily with what they possess that they wish for nothing more. They care little for what they own, and have no desire for what they do not have. If you give them something, they are ungrateful (*ils n'en sçavent point de gré*). If you refuse them something, they will say to your face that it is because of stinginess. If they lose [an object], they will not be disappointed; if you reprimand them, they will start laughing. Praise them, and they will laugh all the more . . . they live without needs, they sleep without nervousness, they die without fear. One can call them happy, if happiness consists of being insensitive.[33]

Magnin's vocabulary was donned and deployed by La Condamine in his own caustic description of indigenous customs, even though the academician presented these generalized observations as if they were the culmination of his own extended study of native life. Despite his admission that it was not entirely appropriate to rely on a traveler who merely observed native cultures as he moved swiftly past them, he nonetheless claimed that "all the American Indians from the various countries that I have had occasion to see over the course of my voyage seem to me to resemble one another to a great extent."[34] He then offered a critique of native populations that was cited regularly by detractors of Amerindian culture throughout the century, especially those who supported the view that the South American continent was degraded and that the culture of its indigenous populations was decayed:

Insensitivity forms the base [of their character]. I leave it to the reader to decide if one should honor [their character] by calling [the Indians] apathetic, or vilify [their character] by calling them stupid. This [trait] grows out of the small number of their ideas, which do not extend beyond their needs. Voracious gluttons when they have enough to satisfy themselves; sober when necessity requires it of them, they will relinquish everything without seeming to care. [They are] excessively pusillanimous and cowardly, if drunkenness does not help them overcome [their fears]; enemies of work, indifferent toward every form of glory, honor, or recognition; exclusively focused on the object in front of them and always preoccupied by [that object], without preoccupation for the future. [They are] incapable of foresight or reflection; when nothing bothers them, they enjoy themselves childishly by jumping and laughing beyond what is appropriate, without a reason and without a motive; they pass their lives without thinking, [and] they grow old without leaving their infancy but maintain all of [infancy's] flaws.[35]

To summarize his years of observations, La Condamine drew on the spirit, as well as the letter, of Magnin's previous ethnographic descriptions. Magnin had concluded that one of the characteristics of Amerindian happiness was being "insensitive," and La Condamine responded in kind that "insensitivity" formed the base of their character; Magnin wrote that the Amerindians were "so indifferent, or rather so stupid that they would sooner see someone die before their eyes before they had the slightest thought of helping them," and La Condamine in turn hesitated as to whether their character should be honored as "apathetic" or vilified as "stupid." Each of La Condamine's conclusions seems to have been forged in the ethnographic smithy of Magnin's workshop. It is not far from Magnin's description of Amerindian life to La Condamine's philosophical assertion that "man, abandoned to nature and deprived of education and society, differs little from the beast."[36]

Where natives did differ from the animal world, of course, was in their use of language; but here again, La Condamine's descriptions of indigenous languages echoed Magnin's caustic assessments of their customs and character. La Condamine's critics were quick to point to his inadequate knowledge of native tongues when assessing the broader legitimacy of his geographical and historical conclusions. In his portrayal of the poverty of indigenous languages, however, La Condamine yet again relied on the observations of others. In a chapter entitled "Autres sorts d'obstacles et de difficultés pour les missionaires," Magnin outlined several elements of the Yameo language, which he described as being of an "insurmountable difficulty":

For the Yameos, *Poetararorireneroa* means the number three, beyond which they do not know how to count... [One of their words], which is ten syllables long, is pronounced in such a way that it does not seem comprised of more than three-and-a-half [syllables], not to mention the unique and bizarre manner in which they pronounce their words by inhaling forcefully.[37]

Arriving in the land of the Yameo Indians, La Condamine wrote that "their language is of an inexpressible difficulty, and their way of pronouncing is even more extraordinary than their language." He went on to discuss their peculiar habit of breathing in when speaking, without making a single vowel sound, and ridiculed the fact that many of their words have nine or ten syllables. And, finally, in assessing their linguistic system for counting, La Condamine indicated that the word "Poettarraror-incouroac means the number three in their language: luckily for those who have commerce with them, their arithmetic extends no further."[38]

More than his passing off of Magnin's data on Amerindian customs or language as his own, however, what is especially striking is how La Condamine wove the Jesuit's observations into his own narrative account; these details appeared in the *Relation abrégée* as if La Condamine had personally observed these native cultures on the spot. Rather than summarizing Magnin's account and giving credit to the Jesuit for his observations, he provided a smooth narration so that his readers would discover the Amazon and its inhabitants as if they were accompanying La Condamine down the river. Narrative structure thus formed a central part of the strategic integration of these materials and a simultaneous suppression of the origins or sources of this material. The assembling of an enlightened Amazon was as much a process of compiling, editing, and narration as it was the empirical accumulation of eyewitness observations. This strategy not only comprised the incorporation of material without citation of its sources; it included the conscious repression and active erasure of knowledge as well.

These examples show clearly that La Condamine was not only familiar with Magnin's manuscript but that he relied on it and used Magnin's observations while presenting them as his own. According to Jacques-François Artur, a naturalist and medical doctor in Cayenne, La Condamine "revised and corrected" the French translation of Magnin's manuscript "sheet by sheet" during his stay there, and thus had an intimate knowledge of what was contained within.[39] And yet, beyond explaining that Magnin had given him "a map that he had made of the Spanish Missions of Maynas and a description of the habits and customs of the neighboring nations," he did not make much mention of Magnin. Other than remarking that Magnin's manuscript had been translated by Dr. Artur with his assistance and that it was "worthy of the public's interest," La Condamine made no reference to the text that would have revealed it as the true source for many of his observations.[40] What is more, La Condamine took the French translation of Magnin's account back with him to Europe, apparently "with the intension of having it printed, but other occupations and circumstances did not allow him to do so."[41] Not only was the account never published; it was never even acknowledged appropriately within La Condamine's account. Diderot and Vosgien, author of the *Dictionnaire géographique portatif*, did not know that their entries on the "Aguas" nation in their respective encyclopedic and geographical dictionaries derived from the manuscript account of a Swiss Jesuit from Maynas rather than a philosophical traveler, who in actuality had floated swiftly past the region he was purported to have observed with his own eyes.

Not surprisingly, La Condamine curtailed his discussion of native customs at precisely the narrative moment when he moved beyond the reach of Magnin's account of Maynas: "On the present occasion," wrote La Condamine, "I should not overextend myself [to reflect on] the habits and customs of those nations and a whole host of others that I have come across, unless they bear some relation to the physical sciences or natural history."[42] La Condamine chose to exclude any further discussion of Amerindian culture because he lacked any evidentiary material to back up his assertions. While the Jesuits served as the eyes and ears of many of La Condamine's downstream observations, their observations were either effaced (e.g., Magnin) or denigrated (e.g., Fritz) in the account that emerged of his journey. Magnin's presence was papered over not because his knowledge was dangerous or politically explosive, but rather because it detracted from the empirical image that La Condamine wished to provide to the Academy. When it came time to publish his account, it was not sufficient to suppress many of his sources; rather, as in the case of Samuel Fritz, La Condamine implied that many of the Jesuit's observations were inadequate as an evidentiary source because they failed to meet the standards of eighteenth-century empiricism, as we will see in the following section. Only after sweeping away his Jesuit predecessors would La Condamine claim his much-coveted status as triumphant prose conquistador. Only after effacing the sources from which he garnered his ethnographic conclusions would he be able to replace the Jesuits as the latest European source of empirical truth. He not only did this textually within the narrative structure of the *Relation abrégée*. He accomplished his Amazonian conquest with cartographic tools as well.

Graphic Rhetoric

Just as La Condamine used textual rhetoric explicitly to denigrate the contributions of previous explorers, he likewise applied graphic strategies to inscribe his own superior method onto the map he produced. To begin his narrative account, La Condamine once again followed Valleumbroso's instructions by inserting an abbreviated history of the European exploration of the Amazon prior to his own descent. This opening section contextualized his journey and provided a compendium of past explorers to which he referred throughout his own text. It also offered a pool of convenient references from which La Condamine later drew to compare his own mapping project with the circumstances and methods of previous expeditions; here, too, Valleumbroso's criticisms comprised an ample storehouse of preemptive munitions. For example, he pointed out that Guillaume Sanson, royal geographer to Louis XIV, based his map (fig. 9 above) accompanying the 1682 French translation of Father Acuña's *Nuevo descubrimiento del gran Rio de las Amazonas* on historical documents alone—echoing Valleumbroso's own observation—and that he did not make separate references to geographic surveys or eyewitness observations. Sanson's map, which La Condamine called "very defective," was subsequently "copied by all the geographers" and not superceded until Father Fritz's map appeared in France in 1718. Until Fritz, La Condamine argued, the cartographic record had been based on pure abstraction, bearing little or no relation to information gathered by firsthand observation.[43]

La Condamine's analysis of Samuel Fritz's map, which was chronologically the closest to his own project, exemplifies the strategy La Condamine used to discredit the works of previous cartographers and explorers. A native of Bohemia, Fritz (1650?–1725) had come to the western Amazon in 1684 and was active as a missionary and Jesuit superior for roughly four decades. The map he compiled of the Amazon basin, several versions of which were produced in print and manuscript form, became one of the most important graphic testimonies to the expanding missionary activities and proselytization efforts of the Spanish and Portuguese in the early eighteenth century.[44] La Condamine carried a large manuscript version of Fritz's map with him as he descended the Amazon, and he deposited this version in the king's library in France upon his return. But because Fritz's map incorrectly posited the Napo River as the true source of the Amazon (following Father Acuña's 1641 account) and portrayed the Marañon River as emanating from a lake near Lima, Peru, La Condamine questioned the veracity of all of the Jesuit's observations. In a particularly vehement attack against his predecessor, La Condamine asserted that

Father Fritz, without a pendulum and without a telescope, was unable to determine a single point of longitude. He had but a single wooden semicircle of three inches' radius for the latitudes; furthermore, he was ill when he descended the river to Para.[45]

Blind without his telescope, sick, and riddled with obstacles too numerous to recount, Father Fritz was depicted by La Condamine as a cartographic King Lear, banished from the scientific kingdom, whose account of the river's geography and the map drawn from it were fallible at best and useless at worst.

By criticizing the information gathered by earlier expeditions such as this one, La Condamine justified the need for a new Amazonian survey, discrediting previous cartographic knowledge in order to present his own project as more instrumentally accurate and his map as superior to those that preceded it. And to accomplish this purpose, he used the same arguments as Valleumbroso, transforming the marquis's manuscript account into a crucial source and then burying the evidence so that he could portray its conclusions as the result of his own close readings and independent bibliographical study.

To emphasize the distance between his own mapping process and those of his predecessors, La Condamine also presented a detailed picture of the instrumental method he used to measure and record the results from his journey:

In committing myself to draw up the map of the Amazon's course, I provided myself with a resource against boredom, which allowed me a tranquil navigation . . . I had to be constantly attentive to observe the compass and, watch in hand, any changes in the direction of the river's course . . . All of my moments were filled: often, I sounded and measured geometrically the breadth of the river and the size of the rivers that joined it; I took the sun's meridian height almost every day and often observed its amplitude upon rising and setting; in every place I visited, I also mounted the barometer.[46]

In what could serve as a mini-handbook for eighteenth-century cartographic methodology, La Condamine outlined the key constituents of his observational procedure. The cyclicity of his observations ("almost every day . . . in every place") and his "continual attention" to the compass, watch, and the river's direction emphasized the importance of a repetitive, verifiable process that was reliant on instruments for its accuracy and human effort for its thoroughness. La Condamine's account is replete with these observations, which punctuate the text and form a narrative bridge between one observational event and the next: "Upon passing Loxa, I repeated the latitude and barometric observations that I had already done there . . . and I found the same results." And a few pages later: "I measured geometrically the breadth of the river: I found it to be one hundred thirty-five toises, though diminished by between fifteen and twenty toises." Filling every free moment with a useful observation,

La Condamine's rhythmic repetition of geometric and astronomical measurements became his mantra in Amazonia: "I benefited from my forced sojourn at Sant-Iago to measure geometrically the breadth of the two rivers, and took the necessary angles to draw up a topographic map of the Pongo." Narrating the incessant process of measuring and mapping seems to have been La Condamine's primary objective in recounting his downstream journey; in terms of its overall importance to La Condamine's mission, it nearly surpassed the symbolic significance of returning home.[47]

La Condamine also privileged his own firsthand observations over those of his predecessors, calling into question the dubious reports of prior missionary, military, and cartographic expeditions and replacing them with his own testimony to justify the superiority of his map. Not only did this strategy serve to further discredit prior cartographic knowledge, but it appealed to an academic audience familiar with the language and method of firsthand observation:

The map of Father Fritz, who never entered the Rio Negro, and Delisle's last map of America, which was based on Father Fritz's map, show this river running from north to south . . . *I testify with my own eyes* that such [east-west] is its direction many leagues above its junction with the Amazon.[48]

In a broader context, La Condamine's appeal to firsthand testimony conformed to the development of a specific kind of experimental practice and a particular language of experience prevalent in seventeenth- and eighteenth-century academic culture. Within the discourse of the Academy, the personal testimony of a member held special weight, corresponding to a "predominance of the completely personalized scientific proof (*récit d'épreuve*)."[49] These norms had been carried over from earlier traditions in the latter third of the seventeenth century, and the privileging of an experiment that emphasized visual testimony as an acceptable, justifiable, and accurate mode of scientific observation was one valid current of empirical practice. Specifically, as the citation above demonstrates, La Condamine relied on his "own eyes" to question and dispute geographic information given by previous explorers and older cartographic representations of the Amazon. Firsthand, visual testimony, coupled with a narrative structure that emphasized repeated observation of experimental phenomena, were accepted modes of justifying empirical results in eighteenth-century academic culture.[50]

But if La Condamine saw fit to invalidate previous knowledge about the river by appealing to an academic audience epistemologically, he still needed to put something in its place. To be deemed a worthy successor to the Orellanas, Úrsuas, and Texeiras, he chose to make his mark cartographically, publishing alongside the *Relation abrégée* the *Carte du cours du Maragnon, ou de la grande rivière des Amazones* (plate 5). Creating a more accurate map of the Amazon in turn became the central focus and organizing principle behind his entire project. It was the way in which he justified his journey, organized his daily activities, and legitimated his choice of route. From the very first pages of the *Relation abrégée*, in which he described the genesis of his idea to travel independently down the Amazon, La Condamine presented his mapping of the river as the highest priority of the expedition:

> To increase the opportunities for observation, Mr. Godin, Mr. Bouguer, and I decided early on to return by different routes. I was determined to choose one that had been practically ignored and that I was sure no one would envy; it was the route of the Amazon River . . . *I intended to make the journey useful by drawing up a map of the river and by collecting observations of any kind that I would have the occasion to make in a country so little known.*[51]

La Condamine's alliance of the utility of mapping with the tasks of collecting and observing natural phenomena gave his cartographic activity a privileged place within the narrative structure of the *Relation abrégée*. He organized his itinerary in such a way as to maximize the opportunities for astronomical and latitudinal observations and, consequently, their cartographic representation on his map. Choosing routes explicitly to record the coordinates of a particular village or the intersection of two rivers, La Condamine often chose sites with political implications as well. At the junction of the Napo and Amazon rivers, for instance, La Condamine emphasized the significance of the site by referring explicitly to the "territorial pretensions of the Portuguese along the banks of the river."[52] These observations took on significance and carried implications for those who would use La Condamine's text to support a particular regime's subsequent territorial pretensions. But generally speaking, La Condamine justified his detours and diversions by referring to the comprehensive nature of his cartographic project rather than the political or diplomatic stakes of any specific location.

By presenting his excursion down the Amazon as yet another opportunity "to increase the opportunities for observation" and gather "observations of any kind . . . in a country so little known," La Condamine attempted to justify his entire mission under the rubric of discovery and utility.[53] In the process of criticizing other maps and the methods used by previous explorers, he outlined a scientific method that he then used as evidence of his own map's superiority. La Condamine not only cast himself as an empirical observer of the highest order, but he also portrayed himself as a skilled debunker of ancient myths and legends. The multiple forms of proof he used to accomplish this goal—including travel logs and missionary testimonies, Indian lore and manuscript maps—were shuffled and rearranged to provide watertight responses and unequivocal answers for persistent Amazonian myths. In this way, he was able to settle two separate and longstanding debates: the existence of the ancient city of El Dorado, and the disputed connection between the Amazon and Orinoco rivers.

The Enchanted Palace at El Dorado

The earliest European travel narratives to the Americas and the cartographic representations that arose in their wake conjured a host of provocative images to fire Europe's mythical imagination. In his description of Matinino Island, Columbus referred to a place "occupied entirely by women, without a single man," and this tale was soon echoed in the accounts of Amerigo Vespucci, Hernan Cortés, and other early explorers to the Caribbean.[54] Later in the sixteenth century, rumors of American Amazons continued to spread into texts such as Pietro Martire d'Anghiera's *De orbe novo* (1530), Giovanni Battista Ramusio's *Navigationi et viaggi* (1550–59), and Francisco López de Gómara's *Historia de la conquista de México* (1552). But it was the voyage of Count Francisco de Orellana from Quito to the "Land of Cinnamon" in 1541–42 that first propagated the myth of a community of unmarried women warriors living in isolation along the river that would eventually adopt their name. Late in the sixteenth century, the "River of the Amazons" shared center stage with its first explorer, Orellana, in the Spanish court where the count's discovery was considered an "important and marvelous" event. A branch of the river was even named in his honor on several maps from this period. But territorial sovereignty in the Amazon basin was still in dispute: Jacques de Vau de Claye's 1579 depiction of South America, for example, emerged in the context of a never-realized French plot to invade the region and rise up militarily against the Spanish and Portuguese (fig. 11). In Vau de Claye's map, a group of female "Almazones" were shown as potential French allies. They were represented as "industrious" arms-makers who lived in Arabian tents, constructing bows and arrows and conforming to a cartographer's Orientalist fantasy mixing legends

Figure 11. Orientalism in the land of the Amazons: these "Almazones" were thought to be possible allies for the French. Detail from Jacques de Vau de Claye, *Carte d'une partie de l'Océan Atlantique* (1579). BNF Res. Ge.D.13871. Reproduced with permission of the Bibliothèque Nationale de France, Paris.

from Asia Minor with the most recent discoveries in New World geography. Nevertheless, their cartographic presence reinforced the possibility of their physical existence and added to the conceptual vocabulary of a mythic civilization living hidden from view by the banks of the newly discovered river.[55]

Alongside the Amazons, the fabled existence of a "city of gold" known as Manoa El Dorado was also a well-established myth stemming from the earliest European exploration of South America. Orellana's expedition in 1541 had failed to come up with any evidence of a city of gold beside a golden lake, but his reports of American Amazons did push Europe's imagination eastward from the Andes into the continent's lowland region. It was here in the Amazonian lowlands, in fact, that both Pedro Texeira and Father Acuña testified to having seen native tribes engaged in trade using gold as the medium of commercial exchange. And it was here, two centuries later, that La Condamine attempted to confirm the veracity of El Dorado and inscribe its location cartographically on his map.

Although La Condamine expressed unrelenting criticism toward earlier explorers as he touted his own first-person observations, he occasionally

did accept third-party accounts when deciding on disputed geographical issues. Both written testimonies and oral accounts were important
sources he used to construct his map. La Condamine had earlier explained that several distinct sources were used for his *Carte du cours du
Maragnon*: the "Carte de la côte du Pérou" (plate 4); La Condamine's own
astronomical measurements and longitudinal determinations along the
river, detailed meticulously in the *Relation abrégée*; and "*mémoires*, journals, and notes that were communicated to us inside the country by diverse missionaries and intelligent travelers."[56] La Condamine made important editorial decisions about whose information he chose to accept
in compiling his data. These choices were shaped by the epistemological
principles of his academic culture as well as the perceived importance of
settling longstanding transatlantic debates. Such concerns also tended
to overshadow the credibility accorded to the specific groups with whom
he had intellectual commerce. The shifting standards he applied were
indicative of the social pressures he felt either to inscribe or refuse particular forms of geographic and nongeographic knowledge on his map.

But while he sometimes accepted accounts that corroborated one another, trouble ensued when he was unable to square third-party reports
with what he had seen with his own eyes:

So many conforming testimonies, and each of them respectable, removes all doubt
about the truth of these facts; nevertheless, the river, the lake, the gold mine, the
marker, and the village of gold itself attested to by the statements of so many witnesses,
everything has disappeared like an enchanted palace, and not even the memory of
these places remains.[57]

Despite the conformity of "so many testimonies," La Condamine was
still not able to confirm any of the "facts" to which his predecessors along
the river had testified—that is, until he left the site of his "enchanted
palace" and found corroborating evidence further downstream. While
the historical testimonies of Acuña and Texeira presented a strong case
for the existence of "a city whose roofs and walls were covered by sheets
of gold, [and] a lake whose sands were of the same metal," La Condamine
ultimately came to rely on the testimony of two Indians from the Manaos
tribe who had "penetrated all the way to the Orinoco."[58] They described
to La Condamine a lake, located five days from the Rio Negro via the
Yupura River, which bore a similarity to the accounts of Acuña and
Texeira and which La Condamine accepted as the most likely site of
the ancient city of gold: "[I]t seems likely that they forged the name
of the city of Manoa from the capital, Manaos."[59] Even La Condamine

admitted that these accounts did not in and of themselves adequately explain the geographic circumstances relevant to the El Dorado fable: only with the "help of exaggeration," he hastened to add, was one able to "make sense of the fable of the city of Manoa and the golden lake."[60]

La Condamine did, however, think it possible that American Amazons had inhabited the river's banks in the early days of European exploration. While many contemporaries scoffed at what they saw as La Condamine's credulity, eighteenth-century belief in the plausibility of American Amazons should come as no surprise. At the very moment that La Condamine was preparing his Amazonian voyage, the abbé Guyon published an account of the Amazons in France, the *Histoire des Amazones anciennes et modernes* (1740). Guyon had compiled several accounts penned by witnesses in America who had claimed to see female warriors who were similar in behavior to the famed Scythians of classical antiquity. He also put forth the hypothesis that Amazons from Africa could have made their way to the Americas at an earlier point in time. In the years leading up to La Condamine's journey, Guyon's account demonstrates that the idea of a group of female African warriors traversing the Atlantic and establishing themselves in the New World was not entirely implausible. As Guyon wrote, "I find nothing about this position that is untenable . . . on the contrary, [this hypothesis] is more probable than one might imagine."[61]

While he was often taken to task for believing what others saw as an outrageous fable, La Condamine felt that Amazons could have easily disappeared into the forest without leaving a trace. As he remarked early in his *Relation abrégée*, "the banks of the Marañon [Amazon River] were still populated a century ago by a large number of indigenous groups (*nations*), who disappeared into the backlands as soon as they saw [the arrival of] Europeans."[62] La Condamine was likely unaware of the scope of the catastrophic population collapse of indigenous communities along the Amazon following European colonization. But his argument in favor of the Amazon myth had to do with the retreat of indigenous peoples in the face of European encroachment, and he recognized at the very least the impact of a bellicose lifestyle on Amerindian domestic life.[63]

In the case of the El Dorado myth, however, La Condamine employed Indian testimony to bind together a heterogeneous mix of observed phenomena, reported travel narrative, conjecture, exaggeration, and fable. Rather than drawing on his own observations, he cited Amerindians as credible witnesses in order to corroborate the speculation of two previous explorers whose accounts he had criticized in other sections of the *Relation abrégée*. The differences in his approach to resolving the existence

Figure 12. La Condamine first heard of the connection between the Orinoco and Amazon river basins from an indigenous woman who had descended from her native village of Santa María de Bararuma on the banks of the Orinoco to the headwaters of the Amazon. The authority of her account gave way to Portuguese testimony in 1743, which he inscribed clearly on his map. The indigenous woman's testimony never became part of the cartographic record, however. Also visible in this image is Lake Marahi, the purported location of El Dorado. La Condamine never observed this lake firsthand, but trusted instead the eyewitness account of an indigenous informant. Detail from La Condamine, *Carte du cours du Maragnon* (1745; plate 5). Courtesy of the John Carter Brown Library at Brown University, Providence, R.I.

of Amazons and El Dorado are striking. In the case of the Amazons, it is clear from contemporary reactions that he might just as easily have sought to refute their existence as to confirm it. But with the myth of El Dorado, the agreement of so many witnesses and the fact that Manoa was possibly a *geographical* reality as well left him with fewer choices. He ended up inscribing onto his map the lake he *believed* to be Parime at the sight he *supposed* Manoa to be without ever having observed first-hand either the lake or any tangible evidence of El Dorado's existence (fig. 12). Rather than leaving a blank cartographic space in a region he had neither visited nor observed, he chose to incorporate several layers of reported testimony as part of the cartographic record, testimony that had at its base a fable from the earliest reports of New World exploration. What this act of cartographic "metamorphosis" shows is La Condamine's deployment of nonempirical data deriving from sources as diverse as Indian lore and European myth. It also points to La Condamine's deft use of third-party testimony to settle geographical issues along his journey. He used the reports of certain missionaries and "intelligent" travelers for observations that he could not personally observe and relied on two Manaos Indians to corroborate Acuña's and Texeira's conjectures about the location of El Dorado. Through narration, La Condamine rearranged native testimony when it was able to corroborate evidence that supported his already well-formed suspicions, and he disparaged these same testimonies when their aims did not coincide with his own. As Antonio de Alcedo y Herrera (1735–1812) emphasized in the entry for "Manoa" for his *Diccionario geográfico-histórico de las Indias occidentales ó América* (1786–89), the result was a shift from mythological fable to geographical fact: from the articulation of an "imaginary and fabulous city named El Dorado that did not exist except on maps made by certain geographers who lacked information and discernment" to the situating of El Dorado in a series of "mountains inhabited by a great number of Indian infidels." La Condamine's coarticulation of the myth of Manoa with a reported geographical reality served to perpetuate the centuries-old fable of El Dorado, whose origins lay in an intertwined historical helix of native mineralogical practice and unadulterated European fantasy.[64]

The Cartographic Fusion of Amazon and Orinoco

On the whole, La Condamine was far more comfortable using European testimony than Indian lore to corroborate his geographic theories. His attempt to resolve the long-disputed connection between the Amazon and Orinoco rivers offers another example of how he used third-party testimony to affirm Amazonian myths and assert geographic claims. In this instance, he set out to resolve an issue that had been "formerly established, later negated, and finally newly certified through decisive testimonies."[65] The ancient legend of a connection between these two massive South American river systems stemmed from the earliest accounts of New World exploration; the Orinoco and Amazon rivers were prominent features on early European maps and atlases and were often showed connected (plate 6).

La Condamine's "discovery" of the two rivers' connection highlights the ephemeral nature of cartographic knowledge as well as the manner in which information was intentionally suppressed by previous generations of mapmakers:

The connection of the Orinoco and Amazon, recently verified, can be seen as a discovery in geography all the more since, though the junction of these two rivers was marked without equivocation on older maps, all modern geographers had suppressed it in the new ones, as if concertedly; it was treated as imaginary by those who should have been best informed of its reality. It is probably not the first time that what is likely, and *purely plausible conjectures*, have gotten the upper hand over facts attested to by travel journals, and the *critical spirit pushed so far* as to negate decisively what should have only been doubted.[66]

Here, La Condamine implied something akin to a cartographic conspiracy to "suppress" the "proven" connection between these two rivers. At the same time, he demonstrated how geographic information was malleable and could be used to support a wide array of divergent cartographic purposes. Third-party observations, in this instance exemplified by travel narratives, were not only accepted by La Condamine but were privileged above what he called a "critical spirit pushed too far," which negated what he considered perfectly reasonable hypotheses. But where did the boundary stand between "plausible conjecture" and the use of third-party accounts? This was a dilemma that confronted many eighteenth-century mapmakers: how to reconcile conflicting textual and graphic information and how to reduce geographic features found in textual narratives to the material limitations imposed by the cartographic medium. In La Condamine's case, the epistemological standards by which he claimed superiority over previous explorers oscillated according to the particular object under observation.

While he admitted to using unverifiable information from travel narratives in order to settle particular geographic debates, he never explicitly communicated what his sources were: "I will not enumerate the different proofs of this connection, which I have carefully collected along my route." In the case of the connection between the Orinoco and the Amazon, however, he did make reference to the "unsuspected testimony of an Indian woman from the Spanish missions." This woman had traveled from the banks of the Orinoco all the way to Pará near the mouth of the

Amazon. La Condamine later discovered that the chief Jesuit missionary on the Orinoco had dispatched a letter with a group of Portuguese militia and that they had also returned to Pará via the Rio Negro and the Amazon "without disembarking" from their raft. Armed with testimony suggested by Indians and corroborated by Europeans, La Condamine asserted that "this fact can no longer be placed in doubt." Third-party testimony, once maligned by the intrepid cartographer, became the basis for a complete reevaluation of centuries of geographical knowledge, as well as the basis for a polemical debate following the publication of Gumilla's *El Orinoco ilustrado.*[67]

La Condamine admitted that although he had proven the connection between the Orinoco and Amazon "unequivocally," he still had "nothing but conjectures" about the precise manner in which these massive river systems actually came together. The copious amounts of material he consulted to determine the exact nature of their connection included "logs, memoirs, and maps, printed copies as well as manuscript versions...following accounts given by the most intelligent missionaries and navigators." But they only allowed him to construct an incomplete theory of how these two rivers came together. Out of three separate hypotheses that La Condamine laid out, "many reasons"—none of which were ever communicated to the reader—led him "to believe that the first system was the most likely." Indeed, this "first system" was the one recorded on La Condamine's 1745 map, inscribing without equivocation a geographical conjecture whose validity was only beginning to be affirmed at the time La Condamine returned to Europe armed with the "discovery in geography" upon which he rested his reputation as a geographer and explorer (fig. 12 above).[68]

La Condamine's lengthy discussion within the text of the importance of settling this longstanding dispute underlines its important rhetorical place in La Condamine's overall project. Connecting rivers, like finding the lost site of El Dorado, was proof of La Condamine's aptitude for condensing, distilling, and ultimately unifying information for an academic audience and a learned public alike. These activities required the skillful elision of radically diverse source materials into a single sheet that represented the Amazon River from its source in the Andes to the Atlantic Ocean. This unifying impulse was made possible by a narrative structure that gave La Condamine a storehouse of evidence with which he could criticize past images and, at the same time, lay the groundwork for a future performance of his own work at the Academy.

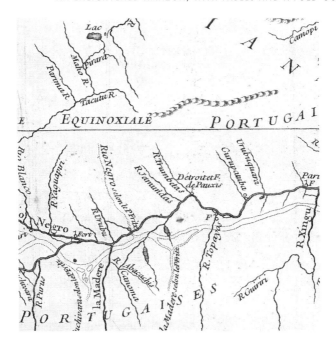

Figure 13. The dotted Fritz "underlay" is visible in this detail from La Condamine's *Carte du cours du Maragnon*, as is a lake near the Pirara River, reported to La Condamine by Nicolas Horstmann, from whom La Condamine received a map and other details of the region's hydrography. Detail from La Condamine, *Carte du cours du Maragnon* (1745; plate 5). Courtesy of the John Carter Brown Library at Brown University, Providence, R.I.

Conclusion

La Condamine published his 1745 map of the Amazon as part of his *Relation abrégée*. And alongside the unobserved lake marking the site of El Dorado and the forced connection of the Amazon and Orinoco rivers was an additional feature he included for his readers. Grafted onto the map along the entire width of the South American continent was a ghost image of La Condamine's Amazon, a duplicate version of the map produced by Father Samuel Fritz in 1707 (fig. 13 and plate 5). Wending its way inexorably toward an encounter with the Atlantic Ocean, the underlaid form of Father Fritz's map functioned as a graphic representation of La Condamine's rhetorical strategies within the *Relation abrégée*. Like the narrative techniques presented in this chapter, the dotted "underlay" of Fritz's map served La Condamine as a device to legitimate his carto-graphic methods vis-à-vis the Jesuit missionaries who preceded him. By using this kind of graphic device, La Condamine underlined in graphic form the extent of Fritz's errors while drawing attention away from the idiosyncratic elements in his own account. Just as he attacked Fritz's lack of proper scientific instrumentation in written form, so did he critique his predecessor cartographically.[69]

For the European scientific community, the period between 1730 and 1740 was marked by "an ethic of exactitude and a profound transfor-mation in the practices of measurement."[70] Responding to the immense variability of experimental data they were recording with instruments, eighteenth-century natural scientists sought to create empirical "order" and "uniformity" as part of a larger system of "geographic expansion and a material and intellectual appropriation of the world."[71] La Condamine and his fellow academicians played a central role in this experimental quest, a project that was echoed several years later in a different context. Denis Diderot, along with d'Alembert and their cohort of *Encyclopédistes*, were busy collecting and distilling data from a host of diverse sources to include in their ordered and unified dictionary of human knowledge. Diderot referred to these activities in his article "Encyclopédie," where he described the project as one that would "gather, display, and trans-mit (*rassembler, exposer et transmettre*)" new materials from throughout the globe to a European audience.[72] La Condamine's project to encapsu-late the Amazon in cartographic and narrative form corresponds in many ways to the project laid forth by Diderot and d'Alembert. La Condamine's gathering involved the collection and measurement of geometric data through an articulated and instrumentally based scientific method. The production of the map itself corresponded to the display stage, brought about through the representation of collected measurements in graphic form. Finally, the transmission occurred when the account was publish-ed and distributed alongside his written commentary. La Condamine's map of a region at the fringes of European geographic consciousness was a fusion of two contemporaneous methods: an attempt, on the one hand, to use scientific instruments to measure, compile, and situate a complex hydrological system through surveying and astronomical ob-servation; and, on the other, to create a holistic expression of a distant and little-known world by collecting data and resolving puzzles that had eluded previous explorers. Using methods both contained by and at the limit of eighteenth-century scientific procedures, the resultant carto-graphic depiction is a spectacular exemplar of the Enlightenment fusion of science and myth.

In one of the many suggestions he provided to La Condamine, the Marqués de Valleumbroso encouraged his friend to explore, examine, and map the Pongo de Manseriche, a narrow channel where the Amazon's current accelerates with great force. "The Pongo Strait in the course of the Marañon is very famous . . . It would be good if you could measure its longitude, its latitude, and if the currents allow . . . its depth as well."[73] Although it was impossible to take a measurement of the Pongo's depth due to the strength of the current, La Condamine did include a map of the Pongo alongside his map of the Amazon, which he published in the *Mémoires de l'Académie Royale des Sciences*. Within this map of the Pongo, La Condamine had himself engraved in situ, posing in his academic garb in what we might call the eighteenth-century equivalent of the "photo opportunity" (plate 8).[74] The engraving shows the academician, with his trademark tricorne hat upon his head, scribbling intensively and deliberately. On either side of him, the banks of the river come ominously close, threatening to disturb his academic work were it not for three native guides who, oar in hand, worked to ensure the small raft stayed properly aligned in the center of the channel. Despite these dangers, he appears neither dismayed nor preoccupied. Rather, calmly and serenely, he contemplates his text, with a chest at his feet and a scientific instrument (a compass?) upon the table in front of him. In this depiction, seemingly oblivious to what is taking place around him, La Condamine records for posterity the scientific observations he has made consistently along his journey, choosing carefully the words that will express to a distant audience his extraordinary adventure in America.

This portrait of La Condamine as an objective reporter of firsthand data, a specimen collector, and an unbiased scientific observer was the image he wished to promote to the Academy. It is as if La Condamine himself were proving, with all the tools at his disposal, the wisdom of their choice to send him as a representative of the new scientific order. In exchange for the trust they placed in him, he composed a treatise that went on to garner popular praise for both its utility to science and the literary talents of its author. As the reviewer for the *Mémoires pour l'histoire des sciences & des beaux arts* explained in his account of La Condamine's narrative, "The work he carried out on this first voyage, through nearly impassable routes, is described with such precision and exactitude that one has the impression of having joined the Author on his journey."[75] Balanced and composed, like the seated academic figure floating down the mighty river, the *Relation abrégée* carried Europe down the Amazon alongside its author. It introduced the scientific community and a broad audience to a new world, a tropical laboratory that over the two-and-a-half centuries that followed would become a powerful metaphor for the richness and diversity of the Earth's plant and animal life. The fables and unresolved geographical disputes drawn from the earliest European exploration of the Amazon were the historical scenery from which La Condamine could narrate a new, reconfigured, and enlightened Amazon with himself and his scientific instruments at center stage.

THREE

Armchair Explorers

Thirty years after La Condamine, a Portuguese astronomer who explored the Amazon and its northern tributaries, M. Ribeiro [de Sampaio] confirmed on the spot everything that the Frenchman had put forward. He found the same traditions among the Indians; [and] he collected them with even greater impartiality because he did not himself believe in the Amazons as having formed a separate people. HUMBOLDT, *RELATION HISTORIQUE DU VOYAGE AUX RÉGIONS ÉQUINOXIALES DU NOUVEAU CONTINENT*

Readers are travelers . . . MICHEL DE CERTEAU, *THE PRACTICE OF EVERYDAY LIFE*

In the decades that followed the publication of La Condamine's *Relation abrégée d'un voyage fait dans l'intérieur de l'Amérique méridionale* (1745), Spanish American–born Creole intellectuals such as Francisco Javier Clavijero and Juan de Velasco challenged La Condamine's observations on the native populations of the Amazon basin and sought to criticize the academician's representation of their ignorance and stupidity. Both referred to his limited knowledge of indigenous languages, and Velasco indicted his hearing as well, explaining that it was a "notorious lie" that only one bird in the Amazon had a pleasant song. According to Clavijero, "La Condamine knew as much about [Amerindian] languages as [the Dutch philosophe Cornelius de] Pauw, and he undoubtedly took his information from some ignorant person, as is often the case with travelers."[1] He also surmised that La Condamine's observations about Indians' "gluttonous" behavior may have resulted from only the briefest of interactions: "Perhaps La Condamine saw a few hungry Indians eating anxiously during his trip down the Marañon River," Clavijero conjectured, "and from this he was persuaded—as is often the case—that they were gluttons." Clavijero would have been surprised to learn that the "ignorant" person from whom La Condamine drew his material, as we saw in the last chapter, had been a Jesuit like himself, someone who had lived among the natives for decades and who knew Quechua (Magnin referred to it as "the ancient language of the Inca") as well as anyone. Clavijero had based his epistemological critique on La Condamine's printed text alone. Not having access to La Condamine's manuscript sources, Clavijero was limited in his knowledge, almost as limited as La Condamine was in his knowledge of the languages of those Amerindians that had rowed him securely downstream.

As Clavijero's and Velasco's critiques of La Condamine's *Relation abrégée* make clear, the journey of a philosophical traveler did not end with the publication of a printed text. Instead, after publishing a travel account, the author's attention necessarily turned to the ever-precarious itinerary of his literary reputation. Given the predilection of eighteenth-century authors to doubt and denigrate eyewitness accounts of overseas locations, a traveler-turned-author had to pay constant attention to readings of his work that would challenge his reputation, much like a vigilant sea captain recalculating his course through blustery winds and stormy seas. Indeed, seafaring was one of the metaphors sometimes used to describe the duties and obligations of an author returning from his travels. "The account of a great voyage," wrote one reviewer in the *Journal de Trévoux*, "is like the crew of a ship, where . . . there always remains something to do . . . [and] an intelligent and accurate traveler happily goes back to justify his details, even to avenge them, if someone happens to attack their foundation." In this portrayal, the narrator of a travel account had to defend his objectivity frequently and emphatically to avoid being marked as a spurious storyteller who beguiled his readers with tales from distant shores.[2]

The travel account under discussion in the *Journal de Trévoux* was none other than La Condamine's *Relation abrégée* (fig. 14). The text had been reviewed six months earlier in the same journal, and its author was now responding to this review with a robust defense of his work. Wary of criticism left unanswered, La Condamine took issue with the manner in which authors and journalists alike had interpreted what he had written. Two months after the *Journal de Trévoux* reviewed a French translation of Gumilla's *El Orinoco ilustrado* (a text that contradicted portions of the *Relation abrégée*), for instance, La Condamine responded with a fifteen-page missive to the editors of the *Journal*, insisting that his own observations

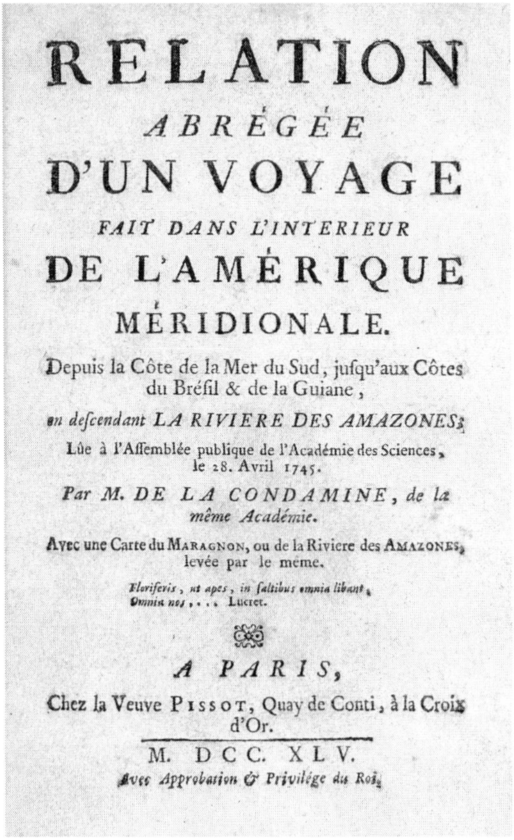

RELATION
ABRÉGÉE
D'UN VOYAGE
FAIT DANS L'INTERIEUR
DE L'AMÉRIQUE
MÉRIDIONALE.

Depuis la Côte de la Mer du Sud, jufqu'aux Côtes
du Bréfil & de la Guiane,

en defcendant LA RIVIERE DES AMAZONES:

Lûe à l'Affemblée publique de l'Académie des Sciences,
le 28. Avril 1745.

*Par M. DE LA CONDAMINE, de la
même Académie.*

Avec une Carte du Maragnon, ou de la Riviere des Amazones,
levée par le même.

*Floriferis, ut apes, in faltibus omnia libant,
Omnia nos, ... Lucret.*

✦

A PARIS,

Chez la Veuve Pissot, Quay de Conti, à la Croix
d'Or.

M. DCC. XLV.

Avec Approbation & Privilége du Roi.

Figure 14. La Condamine's *Relation abrégée d'un voyage fait dans l'intérieur de l'Amérique méridionale* (Paris, 1745) chronicled his journey from the Andes, along the Amazon River, up the South American coast, and back across the Atlantic. En route, he passed through Brazil, French Guyana, Suriname, and Holland before returning to France. Part travel narrative, part ethnographic description, and part geographical treatise, the compact text was conceived to appeal to a broad audience. Astronomical measurements dotted the narrative amid descriptions of the region's rich mythological history, while tales from the river's early European exploration were recounted alongside evocative, if often derogatory, accounts of the region's native inhabitants. Courtesy of the John Carter Brown Library, Brown University, Providence, R.I.

were, in fact, wholly legitimate: "[It] is plain from the facts that I have just established," La Condamine responded, "that the proofs provided by my account are valid in their entirety, and are not only exaggerated (*effleu-rées*) by the new edition of the *Orinoque illustré*, [but are] far from being answered, as the reader might be led to believe by reading your excerpt."[3] In order to remedy any potential blemishes on his reputation, La Condamine provided "new proofs" in support of his observations. He even pointed out various printing errors that the journal had committed. These defensive remarks were meant to buoy and stabilize the status of La Condamine's account before the ever-important tribunal of the reading public, and they seem to have had their desired effect. The *Journal de Trévoux* beat a hasty retreat by praising "the high degree of accuracy (*grande exactitude*) of this academician, his trustworthiness, [and] his concern for the public." In the words of the *Journal de Trévoux*, La Condamine's comments had "[made] the bulk of the controversy disappear" and had allowed the academician to preserve his reputation in the eyes of the salon-attending, journal-reading Parisian public.

In chapter 2, we saw how La Condamine emerged from the narration of his journey as a triumphant conquistador of geographical as well as intellectual territory. This chapter examines the reception of La Condamine's *Relation abrégée* beyond the Academy, not only within the broader sphere of the European periodical press but also among individual readers—from Amsterdam to Amazonia—who happened to read his account. If the public reaction in Europe was largely muted by La Condamine's proactive strategies in print, a rather different reception awaited his work in the realm of epistolary correspondence and transatlantic textual critique. Individual readers of remarkably varied backgrounds reacted passionately—and surprisingly univocally—to La Condamine's account. A Dutch savant named Isaac de Pinto, for instance, engaged La Condamine directly about what he considered some of the traveler's more dubious assertions, especially related to his portrayal of Amerindian cultures. Drawing on philosophical traditions as well as on the first-person testimony of a Spanish American Creole named Pedro Vicente Maldonado, de Pinto attempted to compare La Condamine's pejorative image of Amerindians along the Amazon with accounts and treatises he had read that challenged such a simplified view of native cultures. As to the legitimacy of La Condamine's on-the-spot observations, Maldonado had his own opinions, which he discussed with de Pinto who in turn communicated them back to La Condamine in epistolary form.

In the decades that followed, Portuguese explorers along the banks of the Amazon saw in La Condamine's account a foil for their own imperial aspirations. They were eager to praise his instrumental approach when it suited them, but they also took issue with other elements of his account when their individual and political interests so dictated. Two Portuguese travelers in Amazonia—the Brazilian-born vicar Jozé Monteiro de Noronha and a jurist named Francisco Xavier Ribeiro de Sampaio—frequently relied on La Condamine's measurements but offered critiques of his more "incompetent" observations. These varied responses to an individual text allow us to examine in a transimperial and transatlantic context what Lisa Jardine and Anthony Grafton have called, within the European sphere, a "transactional" mode of reading, in which "a single text may give rise to a plurality of possible responses [and] not a tidily univocal interpretation."[4] But despite the diverse array of religions, professions, and national affiliations represented by these readers of La Condamine's *Relation abrégée*, there were striking parallels in many of the ways they responded to his text. Yet even the most vehement and direct critiques of his work ultimately proved little contest for La Condamine's skillful and spirited ability to defend himself and his work. Returned to his native soil, the once-errant philosophe proved gifted at guiding his narrative cargo toward the safe haven of literary prestige.

All the Amazon That's Fit to Print: The Reception of the *Relation abrégée* in the European Periodical Press

During the months following its publication, the *Relation abrégée* was widely reviewed in the European periodical press. As the first account produced by a member of the Quito expedition, published as it was long before Jorge Juan and Antonio de Ulloa's *Relación histórica del viage a la América meridional* (1748), Pierre Bouguer's *Figure de la terre* (1749), or La Condamine's own *Journal du voyage fait par ordre du Roi* (1751), the *Relation abrégée* was eagerly anticipated by readers and reviewers alike. While an abbreviated Spanish version was initially published in Amsterdam in 1745, the public had to wait until early in 1746 to get its hands on the official French edition. Soon thereafter, the *Journal des Sçavans*, *Bibliothèque raisonnée des ouvrages des savans de l'Europe*, *Journal de Trévoux*, and *Mercure de France* all published reviews of La Condamine's account that in the words of one journalist were meant to "incite curiosity for reading the description [of his journey] in its entirety."[5] Often quoting directly from the text itself or paraphrasing discussions deemed worthy of public interest, the journals' editors brought the *Relation abrégée* before a much broader audience than the small elite group of naturalists, astronomers, and mathematicians that gathered in the Bibliothèque du Roi to listen to the semiweekly sessions of the Academy.[6]

One such journalist was Pierre Massuet, editor of the *Bibliothèque Raisonnée*, who was personally acquainted with La Condamine and favorably disposed toward his account. In a letter written to La Condamine some months later, Massuet recounted their first meeting: "I had never seen such sweetness, such affability, such an open heart, such sincerity, such integrity, so many talents, so much merit in a single individual. Shortly afterward, I found all of these characteristics clearly illustrated in the Account of your Voyage in America. I was enchanted by the text. I gave a summary of it in the *Bibliothèque Raisonnée*, and I made sure to provide a favorable account to the public."[7] Massuet's letter emphasizes the extent to which the Parisian social milieu directly influenced reception in the eighteenth-century periodical press. Massuet set out to defend the text against its potential critics, asserting that "equity requires that we defend the merits of this work from the attacks that some individuals have wished to perpetrate against it." While the identity of these individuals was never made explicit, Massuet argued that certain individuals had accused La Condamine of not "embracing a broad enough subject" and, in turn, of being tedious:

The assertion has been made that one does not find [within the text] what should be found there, that the Author... does not entertain his readers sufficiently, and that he only provides [descriptions of] things for which they have no interest whatsoever. They would have preferred other kinds of details, like those in Tournefort, Kaempfer, Chardin, or the Fathers Labat & Charlevoix.

Massuet countered these assertions by insisting that the merit of a work should be determined not by generic expectations but rather by the success of an author in executing his plan. And because of their previous encounter, Massuet made excuses for what La Condamine had and had not carried out.[8]

In his own favorable review, Massuet echoed some of La Condamine's more caustic assessments regarding Amerindians: that "insensitivity" formed the base of their character, apathy and stupidity were the only two categories for judging their behavior, and that they were cowardly, gluttonous, and lazy drunkards. By summarizing La Condamine's work in these terms, reviews such as Massuet's served both to amplify and legitimate La Condamine's observations, observations that we know from chapter 2 were based largely on unacknowledged sources. Massuet emphasized in no uncertain terms that La Condamine had superbly executed the narrative project he had set out to achieve: "Between these two poles [of instruction and entertainment]," he wrote, "[La Condamine] seems to have found an appropriate middle ground." Massuet presented La Condamine as an intrepid adventurer and a heroic philosophe, unyielding in his determination to provide useful information to his European audience. With reference to the "marvelous" exaggerations told to travelers along their journeys, Massuet concluded his article by stating that such "frightening stories did not even slightly diminish [La Condamine's] resolve."[9]

As Massuet's review makes clear, the *Relation abrégée* was acknowledged both for its value as a scientific tract and as a work of general public interest. Following its original publication it was seen in the French press as a welcome addition to the collective geographical understanding of South America, knowledge that now extended beyond the bounds of France's colonial settlements at Cayenne. One reviewer wrote that La Condamine had been "compensated for his efforts by the quantity and importance of the knowledge he brings back with him—literary treasures that are preferable to those other riches that our ill-conceived greed forces us to seek out in the New World, and whose acquisition required as much courage as Enlightenment."[10] Elsewhere, La Condamine came to be recognized as an authority on the history and geography of South America. Sections of the *Relation abrégée* were included in subsequent compilations of travel narratives about the New World. Even Rousseau, who in his *Second Discourse* famously criticized European explorers for their lack of objectivity, deemed the accounts of "les La Condamine et les Maupertuis" worthy of respect. At a time when writers were evoking the noble traits of the North American "savage" following the writings of Baron Lahontan and the Jesuit Lafitau, La Condamine's detailed descriptions of Amazonian customs positioned the *Relation abrégée* as a staple of the reading diet of a broad European audience. Through the periodical press as much as through the publication of the text itself, La Condamine's reputation as an accurate cartographer, talented author, and daring explorer was proclaimed with vigor both within France and beyond its borders.[11]

Amazonia-on-the-Amstel: A Dutch Patrician Responds

The widely acknowledged "utility" of La Condamine's voyage in print was not universally accepted beyond French gazettes and journals. Manuscript accounts and epistolary correspondents challenged the image of

La Condamine as an exacting and indefatigable eyewitness. A litany of readers, from Portuguese explorers to Dutch savants, found serious grounds upon which to object to La Condamine's geographic and ethnographic conclusions. In the case of Jozé Monteiro de Noronha, a priest from the Amazonian city of Pará, his own eyewitness testimony and a consideration of imperial geopolitics shaped many of his critiques. For Francisco Xavier Ribeiro de Sampaio, a Portuguese jurist and colonial administrator, his erudite culture kindled a vehement rebuttal of the mythical elements in the *Relation abrégée*. And closer to La Condamine's home, the Dutchman Isaac de Pinto used aesthetic and climatic theories, in combination with the observations of the Spanish American Pedro Vicente Maldonado, to contest the conclusions La Condamine had reached. Readers from a variety of backgrounds, in other words, offered wide-ranging critical assessments of La Condamine's prose rather than simple acceptance of the author's "discoveries."[12]

La Condamine made his triumphant return to Europe not in Paris but via Amsterdam. In 1744, while still in Cayenne, he had become convinced that traveling aboard a French merchant vessel could potentially put him at risk, especially given the resumption of hostilities between France and England during what would come to be known as the War of Austrian Succession (1740–48). While still in territory under French control, La Condamine wrote that "I had seen seven or eight merchant ships leave Cayenne for France, but I did not dare to go on board for fear of exposing the fruits of my work to the discretion of the first Corsair."[13] Rather than run the risk of losing his cargo to a pirate on the open seas, La Condamine chose instead to secure his safe passage by accepting an invitation offered by the governor of Suriname to travel aboard a Dutch ship. This decision afforded him the opportunity to share the first "fruits" of his decade-long labors with an eager circle of Sephardic savants who had made their home in the Dutch Republic.

Isaac de Pinto (1717–87) was undoubtedly the most enthusiastic of the group. A descendent of Jews who had emigrated from Portugal to Holland in the seventeenth century, de Pinto gained notoriety in later years for his views on free market economics, his extended debate with Voltaire over the status of the Jew in Enlightenment Europe, and his assuredness that the revolt in British North America during the 1770s had little chance of succeeding. Much earlier, however, de Pinto had welcomed La Condamine into a small "literary society" that met regularly to discuss the most important news from outside the Dutch capital, including significant debates within the European scientific community. How the two met remains unclear; indeed it is noteworthy that La Condamine made no mention of de Pinto and his literary society in any of his published works. But de Pinto chaperoned La Condamine through Amsterdam during the two months he spent there in 1744, and the two continued their discussions through epistolary correspondence following La Condamine's departure. De Pinto was captivated by the *Relation abrégée*, and the letters they wrote back and forth bear witness to de Pinto's especial fascination with Amerindian culture. For his part, La Condamine explained to de Pinto that he had greatly enjoyed his time in Amsterdam: "[T]he most pleasant memory for me," La Condamine wrote in one of the numerous letters they exchanged, "were the moments spent in your literary gatherings, where you did me great honor by extending an invitation."[14]

This brief but intense interlude in Amsterdam presaged an intense epistolary debate relating to the European representation of South America and the moral character of the New World's native, mestizo, and Creole inhabitants. At first, de Pinto had portrayed La Condamine as a prophetic and credible eyewitness, a bearer of revealing light that showered "a few rays of [his] luminescence" upon his intimate community at its regular Amsterdam meetings: "We have had the advantage," de Pinto wrote, "of hearing from your mouth details of what marvels nature contains in another hemisphere." Citing Voltaire's poetic representation of the academicians as "vanquishers of nature . . . [who] strip her veil," he proclaimed in elegant prose that "Newton has had apostles worthy of his achievements" and that Newton "could not have confided his interests to better hands." De Pinto presented La Condamine as enlightening the "vulgar" among the savants of Europe, and he used poetry and prose to emphasize the generous service La Condamine had carried out on behalf of an ungracious, if learned, public.[15]

But this initial series of accolades, which may have served a largely perfunctory rhetorical purpose, was followed by less flattering characterizations in later letters. Along with the host of ideas that "assailed [de Pinto's] imagination," a series of doubts began to plague the curious Dutchman after reading the *Relation abrégée*. These queries, which de Pinto delineated over the course of their epistolary exchange, challenged La Condamine to deepen his superficial assertions regarding Amerindian culture. De Pinto offered a series of broad reflections on the eventual impact of the European presence in the New World, observations that directly challenged the denigrating descriptions La Condamine had provided of native culture:

Your letter, both instructive and amusing, was the subject of one of our sessions. We could not understand, nonetheless, what you say about the Americans. You suggest in fact that they will never cause a revolution in the New World. *I would just as soon imagine,* you write, *that on a deserted Isle where a family of humans had established itself, that the monkeys . . . could rise up and change the face of the government.* But what then is the cause of this idleness, this indifference, this lazy and indolent temperament that dominates these people, or rather, these automatons?[16]

La Condamine's figurative transformation of Amerindians into monkeys catalyzed de Pinto's interest in their physical and moral predicament, which in turn hinged on the broader question of the adaptability of human nature. If what La Condamine said were correct, quipped de Pinto, the Amerindian would be "a kind of human that differs from our own species more than any other species of the most diversified animal differs from its own kind."[17] And if the climate were uniquely responsible for the indolence and lassitude of the Amerindian, he remarked, European Creoles born in the Americas would be subject to the same degradation. The Dutch savant was skeptical of some of La Condamine's more polemical assertions with regard to the inconstancy of the native character, and he pressed La Condamine in his letters to elaborate upon this topic more fully, especially what he had meant when he described Amerindians as a "savage people . . . that grows old without leaving childhood."[18]

De Pinto's interest in the relationship between climate and physical and moral characteristics was strongly influenced by the theories of abbé Jean-Baptiste du Bos, author of the *Réflexions critiques sur la poësie et sur la peinture* and an advocate of the importance of climate in human development. De Pinto was especially curious about the extent to which climate alone could account for the laziness and indifference attributed to Amerindians by La Condamine:

If this temperament is so universal, its cause must without question be physical, and inherent in the nature of the Climate; supposing this to be the case, such an inherent vice would then be transmitted to the European Children born in this country, whose temperament should consequently become more and more similar to those from the New World.[19]

Like Montesquieu and others before him, de Pinto used Du Bos's theory on the aesthetic development of society and the important role of air and climate for a people's artistic evolution to ground his theoretical positions.[20] When he wrote that the Spanish in the Americas were reluctant to use Creoles in positions of public authority, he echoed Du Bos's statement that "[Spain] has always followed the maxim of never confiding a single job of importance in America to Spanish Creoles, or those born in America."[21] De Pinto and his literary entourage seem to have taken a particular interest in the figure of the Creole, whom de Pinto defined as "those upon whose countenance nature has imprinted that they are foreigners in this climate."[22] Articulating the Dubosian notion that racial and temperamental characteristics mutated when groups were uprooted from their climate of origin, de Pinto wrote that "in Holland, . . . the children and the nephews of the Spanish, Portuguese, and French become Dutchified (s'hollandisent), so to speak; they take on the same coloring, they acquire the same temperament, [and] they adopt the same dispositions."[23]

De Pinto's beliefs about the deterministic force of climate ultimately led him to suggest the possibility of a political response on the part of a people so severely subjugated by Spanish colonial authority.[24] He contemplated a future in which the transformations taking place in the American colonies and among their inhabitants would create a new society and a renewed social fabric, one in which Spaniards would acquire the habits and features of the Americans and vice-versa. De Pinto foresaw an environment in which new generations within a mestizo society would bring a "Revolution" in the colonies, paraphrasing in the same breath Du Bos's earlier observation that "[t]he court in Madrid seems to sense this since they exclude Creoles from public service."[25] What de Pinto did not know was that the potential bearer of just such a revolutionary flame was soon to make a live appearance on the Breestraat: a Spanish American Creole who would confirm many of de Pinto's incubating opinions about the nature and culture of the New World. Pedro Vicente Maldonado, a Creole savant who found himself in the stately northern capital of canals and commerce, would join de Pinto in a concerted assault on La Condamine's authority. Together, they would combine philosophical rigor and first-person proof in an attempt to counter his contentious ethnographic claims.

An Eyewitness Authority in Amsterdam

For a Spanish American Creole from the Andean highlands, the city of Santa Maria de Belém do Grão Pará near the mouth of the Amazon River was both the literal and figurative gateway to European cosmopolitanism. A direct descendant of Spanish conquistadors who had spent his formative years in the provincial capital of Riobamba, Pedro Vicente

Maldonado (1704–48) arrived at this bustling port city in Portuguese America with La Condamine, the two of them having descended the Amazon together from the Spanish mission of La Laguna near the eastern slopes of the Andes. And while Maldonado had spent his youth traipsing and trotting across the province of Quito on foot and in the saddle, he was far less foot-loose about traversing the poorly defined border between Spanish and Portuguese America by raft, even with a telescope-bearing, compass-reading academician from France seated beside him. Upon arriving in Pará (present-day Belém), the city's narrow alleyways and broad public plazas became for Maldonado the stage for a daring act of drama and disguise. Fearing that he would be perceived as a secret agent of the Spanish king, Maldonado attempted to pass himself off as one of La Condamine's compatriots. He was aware that some fifty years earlier, Portuguese officials had detained the Jesuit Samuel Fritz for nearly a year out of concern that he was a spy in the service of the Spanish Crown. Maldonado, who had been named governor of the province of Las Esmeraldas near Quito, was not traveling to spy on Portuguese fortifications but rather had his sights on more "natural" objectives: he planned to purchase books, instruments, and other supplies in Europe with which he hoped to establish a scientific laboratory in the Andes. Whether or not Maldonado succeeded in convincing the authorities in Pará that he was a Frenchman is unclear. But despite his presumably thick accent, he did manage to secure from the governor-general of Grão-Pará e Maranhão a letter of recommendation for the Portuguese court (addressed to Anne-Théodore de Chavigny, French ambassador in Portugal) as well as safe passage on a schooner bound for Lisbon. As he set sail from Pará on December 3, 1743, his international European adventure was just beginning.[26]

Collecting instruments to observe celestial objects was not the only goal Maldonado would pursue while in Europe. Acting as a mediator between the province of Quito and the Spanish colonial administration in Madrid, he also petitioned the Crown to grant city status to his native town of Riobamba. Having been sent at the behest of Riobamba's *cabildo consejo*, or town council, he successfully argued his case and secured for San Pedro de Riobamba the privilege of being named a "city," which provided him and his family with a host of special honors.[27] During the time he spent in the Spanish capital, he also requested that the Crown officially recognize the overland route from Quito to Panama, a road that he argued was essential for the proper functioning of this far-flung imperial periphery.[28] And finally, after spending two years advocating on behalf of his South American *patria* in Madrid, Maldonado embarked on a grand tour of northern Europe, gaining acceptance into the highest echelons of European philosophical circles. He accompanied the Spanish ambassador to witness France's participation in the attack on Flanders during the War of Austrian Succession and was even put forward as a candidate for membership in London's Royal Society and the Paris Academy of Sciences. For the savants and academicians whom he met, Maldonado represented the best of what Spanish America could offer. Many commented on his warm character and his wide-ranging intellect. In Paris, he was invited to dinners and salons, where he was presented to the most fashionable socialites and intellectuals of his day. His work was likewise praised in periodicals ranging from the *Journal des Sçavans* to the *Encyclopédie*. In London, he was greeted at the Royal Society as "a person of learning and great curiosity, as a most diligent enquirer into philosophical and natural knowledge, and every way well qualified to be a useful member & a valuable correspondent."[29] And his opinion on the geography and culture of South America was regularly solicited as he traversed the continent from south to north.

But in Amsterdam more than anywhere else, his testimony served as the catalyst for a series of animated philosophical exchanges between him and his European interlocutors, debates that focused on questions of identity, authority, and indigenous culture. Maldonado had arrived in Amsterdam with special orders to purchase an atlas for the Spanish king, but he also carried with him a letter of introduction from La Condamine, embossed with the name and address of Isaac de Pinto:

I have given [Maldonado] this sole letter for [his trip to] Amsterdam and I am sure that he will need no other if, as I hope, you will be able to suggest someone who could show him the curiosities of your city and entertain him during the brief sojourn that he plans to make there.[30]

Despite the cultural gulf between them, de Pinto and Maldonado became fast friends. De Pinto was overjoyed at getting to know someone who could shed light on his unresolved questions regarding the character of the Amerindians. And when de Pinto wrote to La Condamine in response to his letter introducing Maldonado, his first response was one of deep gratitude for having facilitated this meeting:

You have shown me a man from the other world who is not only worthy of the inhabitants of this one, but could serve them as a model. If all of the Spaniards born in America are like him, those of Europe should transplant themselves to that soil . . . the European talents and education with which his spirit is so well-adorned have not at all altered the pure and sincere simplicity that is the hallmark characteristic of the inhabitants of the New World . . . Excess pride and cruelty are the faults to which the Spanish are susceptible, but the climate of America inspires a lazy insensitivity that stifles (*étouffe*) those sentiments that European vanity calls glory. These excesses corrected the one by the other are favorable to our species, judging by the pleasant example that we see in the figure of Don Pedro.[31]

De Pinto's paean to Maldonado referred to several themes that had been evoked in earlier letters exchanged with La Condamine: the natural character of the inhabitants of the New World; the role of the climate in modifying cultural behavior, in this case transforming the Spaniards' "defects" into "favorable" traits; and, for the first time, the singularity and exemplarity of Maldonado as a personification of the Spanish American Creole. Like Benjamin Franklin when he arrived in France in 1776, Maldonado was seen as both radically foreign and yet somehow strangely familiar to European eyes, a robust American varietal of a metropolitan Spanish vine.[32]

In praising Maldonado so unequivocally, de Pinto invested the Spanish American Creole with a profound degree of credibility. But it was Maldonado's presence and their conversations together that transformed de Pinto's earlier and mostly demure contestations of La Condamine's theories into a much bolder challenge to his overall scientific method, especially vis-à-vis the study of indigenous cultural norms. "Our discussion of the Americans is not yet finished," puffed de Pinto in defiant response to La Condamine's tone of certainty: "Monsieur de Maldonado has encouraged my reckless persistence (*témeraire obstination*)."[33] It is impossible to know for sure what Maldonado actually said; everything was filtered, mediated, and reported by de Pinto. Nevertheless, after being emboldened by the elixir of Maldonado's authority, de Pinto confronted La Condamine with his most explicit series of objections. Accusing him of distorting evidence and inaccurately interpreting signs, de Pinto undermined the very powerful images he had earlier used to describe La Condamine:

Monsieur de Maldonado has helped me to see the insurmountable obstacles that you and other Europeans confronted in understanding and deciphering the true character of these poor Indians. No, Monsieur, I cannot accept that they are the automatons that you have described.[34]

One of these obstacles was La Condamine's (and other European visitors') complete ignorance of native languages: "Since you had no knowledge of their language," de Pinto wrote, "you were not able to divine their thoughts." One can presume this to be Maldonado's criticism ventrilo-quized through de Pinto's prose. Together, de Pinto and Maldonado thus inverted the figurative poverty of the Amerindians ("ces pauvres Indiens") and suggested that the linguistic poverty of *Europeans* was at the root of misunderstanding native life in the New World. By contrast, Maldonado operated as a conduit through which de Pinto and his friends in Amsterdam came to understand the Indians of Quito: "M. de Maldonado... has kindly shared with me several interesting and ex-tended conversations he has had with the Indians, and also what they think about this topic."[35] By emphasizing the linguistic advantages of Creoles, Maldonado valorized his own culture and, at the same time, that of the Amerindians. With Maldonado serving as a translator for the Indians of Quito, and de Pinto as a cultural ambassador for Maldonado in Europe, the two together forged a collaborative chain of ideas about native cultures across the Atlantic, which they then used to confront the authority of the French academician's narrative account.

Three issues comprised de Pinto and Maldonado's collective challenge to La Condamine: the *conversabilité des mœurs*, or the ability of human beings to adapt to changing moral circumstances; the impact of climate on human behavior; and the possibility that revolution could be sparked by these cultural transformations on American soil. De Pinto explained to La Condamine that

Monsieur Maldonado... is of my belief on the subject of the malleability of morals (*conversabilité des mœurs*) in relation to both the natives (*Naturels du Pays*) and those who are transplanted; he is under the impression that as time goes by, through the im-pact of climatic conditions, the new Spaniards will find themselves *Americanized*, so to speak, [and] the example and habits of [the Spaniards] will Hispanify (*Espagnoliseront*) the Americans.[36]

Despite having grown up in radically different religious, political, and linguistic contexts, Maldonado and de Pinto found common ground in their shared status as "transplants" to a new culture and a new climate: the one in Quito, the other in the Low Countries. This mutual identifica-tion with the cultural mixing inherent in a new environment led them to make radical claims about the explosive nature of political culture deriv-ing from such crossings. Just as Holland had "Dutchified" (*hollandiserent*) the waves of immigrants who found a new home there in the sixteenth and seventeenth centuries (which included de Pinto's own ancestors), so the native peoples of the Americas had "Americanized" the Spaniards, who in due course sought to "Hispanify" the natives. In the cultural and climatic crucible of both New and Old worlds, residents on both sides of the Atlantic forged new bonds of social and political solidarity. Unlike Ira Berlin's model of the itinerant Atlantic Creole, who accreted layers of identity over time and expressed that identity solely in the context of the American continent, both de Pinto and Maldonado seem to show marked similarities in their intellectual dispositions even though they grew up on separate sides of an ocean. According to them, it was perfectly plausible that the Amerindians might rise up and assert their rights against their Spanish masters, a political revolt to liberate them from the yoke of their oppressors. Once united, Maldonado and de Pinto were able to challenge their Parisian interlocutor and distance themselves from an ethnographic tradition that had attempted to persuade itself of Europe's cultural superiority. They sought to suppress the universalizing tendency that divided the world between brutes and civilized society.[37]

Losing no time in countering these criticisms, La Condamine responded by challenging Maldonado's authority, portraying him as a naive partisan who was not qualified to make unbiased observations:

> While he is a great admirer of everything he has seen in our Hemisphere... I have always perceived in [Maldonado] a high degree of prejudice in favor of his own country—even for those things that he otherwise condemns. This kind of sentiment is so natural that perhaps nobody in the world could avoid it entirely. I have also found that he had an enthusiasm for odd opinions and extraordinary arguments, and he always took great care to distance himself from the more common ideas about the Indians.[38]

La Condamine did not attack the universal principles de Pinto and Maldonado had espoused but rather the nature of the evidence they attempted to deploy. His four-page disquisition referred explicitly to the state of political affairs in the Audiencia of Quito, the corrupt dealings of Spanish American Creoles, and Maldonado's excessive attachment to his native land. La Condamine also incorporated his specific criticisms of Maldonado into a broader critique of Spanish ethnography, alluding to the writings of two of the fiercest apologists for indigenous culture known to have written in the Spanish language, Bartolomé de las Casas and the Inca Garcilaso de la Vega. While he praised Las Casas for fulfilling "the duties of humanity and his ministry" by defending the natives against Spanish political tyranny, he claimed that Las Casas gave no more favorable a view of their mental faculties than he had. Conversely, he claimed that Garcilaso had painted too rosy a picture of Inca culture. And so, by criticizing the Spanish tradition from which Maldonado emerged as inherently contradictory, and by adding a layer of anti-Hispanic prejudice to what might have been a more universal discussion of the nature of ethnographic evidence, La Condamine undermined Maldonado's authority by implying that he was too close to the indigenous populations he was describing. He was not, however, willing to write such explicitly anti-Spanish criticisms in his printed texts. But by doing so in this letter to de Pinto, La Condamine managed to divert attention away from his own more questionable assertions. As he had done with the historical accounts examined in the previous chapter, he used bibliographic tools and references to past authors to criticize earlier observers. In the process, he denigrated his contemporary critics as well.

This manuscript episode of transatlantic intellectual exchange never emerged from the disregard to which La Condamine condemned it. Maldonado returned to Paris following his Dutch excursion and later traveled to London, where he died suddenly before being able to effect any significant changes in his native land. De Pinto became engrossed in other more weighty issues of commerce and finance, and ultimately served as a mediator between the French and the British at the end of the Seven Years' War.[39] La Condamine published subsequent editions of the *Relation abrégée* and *Journal du voyage* without modifying the conclusions he had reached about the nature of the Amerindian character, and his descriptions from these texts were eventually picked up by Buffon, de Pauw, and others to argue for the New World's natural and human degradation. Epistolary critiques by two readers from radically different cultural contexts, yet sharing surprisingly similar views, were not sufficient to overturn the image of South America that was sustained by La Condamine in manuscript and print.

The letters sent from de Pinto to La Condamine represent the vision of the intellectually vanquished: not oppressed subalterns but defeated savants. La Condamine effectively shut down de Pinto and Maldonado's collective assault on his own observational methods, just as he had done with Juan and Ulloa, Samuel Fritz, and numerous others before him. De Pinto's hostility toward La Condamine's conclusions, an antagonism that was bolstered by Maldonado's presence, did not in the end counteract the academician's fierce resistance to accepting criticism. As a young and ambitious savant, de Pinto was careful not to burn any bridges with someone who might one day facilitate his entry into a coveted milieu. In another of his letters to La Condamine, de Pinto wrote that "the greatest pleasure you could offer me is an introduction to those thinking Beings

(*Êtres pensans*) who are the delight of the philosophe. I know that your city is the seedbed."[40] De Pinto was himself both strategic and prescient. The dispute with La Condamine was not worth sacrificing the opportunity to meet Diderot, Voltaire, and the distinguished *salonnières* he encountered through La Condamine during his journey to Paris in 1761.

When he attacked the failings of South American Creoles in the New World, La Condamine also made a blanket statement about avarice and ambition that provides a fitting coda to his exchange with de Pinto and Maldonado. In discussing the greedy and corrupt servants of the Spanish Crown, La Condamine inadvertently shed light on the strategies he himself had used in pursuing his critics:

> Sooner or later, our ambition turns us into enemies of those who stand in our path. But the pretexts are lacking to declare an open war, and our own interests require that the war be waged in secret. Dissimulation thus becomes a necessary means to arrive at the proposed ends.[41]

La Condamine's advocacy of duplicity as a means of squelching dissent demonstrates how he managed to succeed rhetorically in a world where "enemies," or at least critics, abounded. He responded to criticisms in the periodical press with lightning speed and demonstrated an uncanny ability to profit from publicity when it suited his needs and just as conveniently to avoid such publicity when the potential damage was too severe. This oscillation between shining light on particular issues and strategically suppressing others echoes the model of scientific narration discussed in chapter 2, adding an additional layer to Grafton and Jardine's transactional model of reception. La Condamine's evasive rhetoric shows how an eighteenth-century savant adapted his defenses to guard against multiple critics showering barbs from many directions at once.

Contested Conjectures: Portuguese Travelers Consult La Condamine

The critiques of de Pinto and Maldonado contrasted significantly with the manner in which readers from Spain, Portugal, and Ibero-America interpreted and deployed the *Relation abrégée*. Cosmopolitan imperial capitals in northern Europe were not the only places in which La Condamine's text was debated. What is more, the purposes to which its geographic and territorial conclusions were put depended greatly on the geopolitical context in which the text was read. La Condamine distributed his account to friends and colleagues both in Europe and abroad, conscious of the importance of predisposing readers to his work by offering it as a gift.[42] But he also translated the text into Spanish in order to increase its circulation and impact. The *Extracto del diario de observaciones hechas en el viage de la provincia de Quito al Parà, por el rio de las Amazonas* was a significant abridgement of La Condamine's original text, produced during his stay in Amsterdam and meant especially for "those in America worthy of my esteem and affection, as a memento and small payment toward the debt of gratitude I owe [them] for the many favors [they] bestowed upon me"[43] (fig. 15). He had twenty-five copies of this Spanish edition delivered to South America by way of Pierre-Nicolas Partyet, the consul of France in Cadiz; one copy of the *Extracto* went to his friend the Marqués de Valleumbroso in Cuzco. In the Amazonian mission of La Laguna on the banks of the Rio Huallaga (in present-day Peru), a 1768 library inventory also lists the "Viage de Condamine."[44] And yet another copy of the text went East to India, where a Portuguese viceroy who had been a provincial governor in Brazil received the text from his son who was then studying in Paris.[45] The simultaneous presence of La Condamine's travel narrative on opposite sides of the globe is ample proof that the *Extracto*, a text that was half the length (121 pages) and two-thirds the size (only six inches) of the contemporaneous French edition, was made to travel. And travel it did, not only to monarchs' libraries and the sitting rooms of *les grands* but into the hands of Iberian explorers and statesmen as well.

Wherever these Spanish and Portuguese readers may have been, geopolitical considerations were usually foremost among their concerns. The military officers Antonio de Ulloa and Jorge Juan, who accompanied La Condamine on the expedition to Quito, offer instructive examples of how La Condamine's instrumental methods were subordinated to geopolitical conclusions in the eyes of his Iberian readers. Ulloa and Juan used La Condamine in order to push the territorial agenda of the Spanish Crown, although their strategy was a peculiar one. Published a year after their own official account of the expedition (the *Relación histórica del viage a la América meridional*, a text that will be discussed in chapter 5), the *Dissertación histórica y geographica sobre el meridiano de demarcación entre los dominios de España, y Portugal* (1749) emerged in the immediate run-up to diplomatic negotiations between the Spanish and Portuguese over the shape of their respective empires in the Americas, discussions

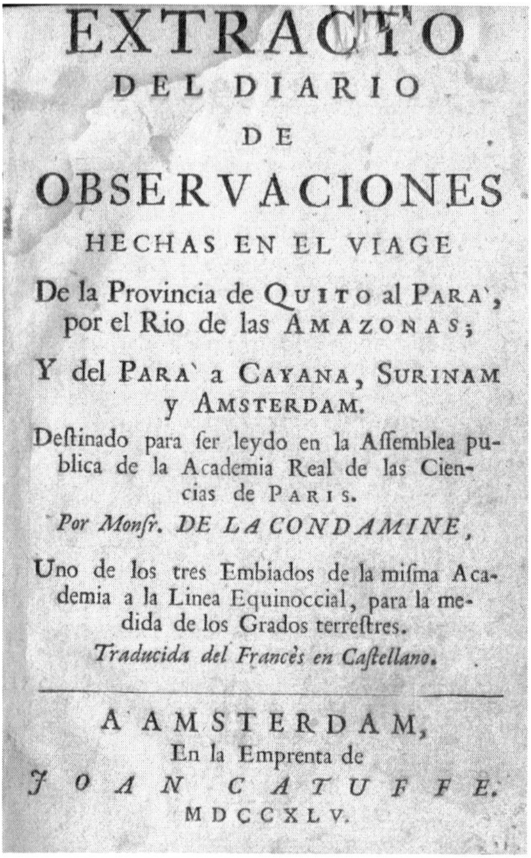

EXTRACTO
DEL DIARIO
DE
OBSERVACIONES
HECHAS EN EL VIAGE

De la Provincia de QUITO al PARA`,
por el Rio de las AMAZONAS;

Y del PARA` a CAYANA, SURINAM
y AMSTERDAM.

Deſtinado para ſer leydo en la Aſſemblea publica de la Academia Real de las Ciencias de PARIS.

Por Monſr. DE LA CONDAMINE,

Uno de los tres Embiados de la miſma Academia a la Linea Equinoccial, para la me-
dida de los Grados terreſtres.

Traducida del Francès en Caſtellano.

A AMSTERDAM,
En la Emprenta de
JOAN CATUFFE.
MDCCXLV.

Figure 15. The *Extracto del diario de observaciones* (Amsterdam, 1745) was published following La Condamine's two-month sojourn in the Dutch capital beginning in late November 1744. La Condamine claimed to have translated the *Extracto*, which amounted to a somewhat abridged version of his *Relation abrégée*, during his Atlantic crossing in 1744. In the preface, he explained that he hoped it would serve as a "small payment toward the debt of gratitude" he owed those South Americans who had assisted him along his route. Courtesy of the Trustees of the Boston Public Library.

that would culminate in the 1750 Treaty of Madrid. The text sought to show the geographical errors, historical manipulations, and ill-guided principles upon which the Portuguese had claimed dominion over a vast portion of the Amazon River's adjacent territory. Not since the 1494 Treaty of Tordesillas had the Spanish and the Portuguese discussed the borders between their dominions, and in the intervening centuries the Portuguese had moved to colonize and settle a far greater swath of territory than had been contemplated by the earlier treaty.[46] "This possession can in no way be valid," affirmed Ulloa and Juan in the *Dissertación histórica*, and they sought to take more "serious and efficacious precautions" against the creeping colonization of the Portuguese westward from the Rio Negro. To disprove the validity of the Portuguese pretensions, they relied on the bona fides of La Condamine, an individual whom they considered to be "an entirely impartial subject, who for his character and recommendations is worthy of the greatest faith, given that his institution and the purpose for which his Court sent him [to South America] was to ascertain the truth for the perfection of the Sciences (*aclarar la verdad para perfeccion de las Ciencias*)."[47] After a painstaking recapitulation of several astronomical calculations, including observations La Condamine had carried out at the mouth of the Rio Napo, in the city of Pará, and on the isle of Cayenne, Ulloa and Juan concluded that these measurements were worthy of the greatest confidence: "[T]o arrive at this degree of certainty, one could not wish for more exact [or] more commendable observations, nor better means to dispel mistrust than to have repeated [these measurements] in three distinct spots, so that the uniformity of their results vouch for their undeniable accuracy (*puntualidad*)."[48] For Juan and Ulloa, La Condamine's measurements were instrumentally and methodologically sound; by a happy coincidence, their results conformed to Spanish interests as well.

This was not the case with the Portuguese. During the reign of Dom José I (r. 1750–77), naturalists, military officers, and other agents operating under his auspices sought to transform the uncultivated territories of South America into a bounded colony. Using geographical maps, population charts, historical texts, and political treatises, they began to impose grids and graphs onto rivers, forests, and Amerindian settlements with the aim of controlling the untamed portions of Portuguese America.[49] Not only were these Portuguese travelers dependent upon empirical observation using telescopes and compasses; textual sources written by supposedly objective observers (i.e., of neither Spanish nor Portuguese extraction) came to play a fundamental role in this process as well. In the wake of the Treaty of Madrid, Portuguese administrative agents relied on La Condamine's account for this very reason. But even as they praised the geographical measurements he put forward in his account, they also attempted to dispute and dispel other elements in his text that they felt had been corrupted by faulty or biased information. Rather than accepting his observations as those of a credible eyewitness, the Portuguese applied local criteria as well as epistemological critiques to challenge many of La Condamine's conclusions as they made their way through the territory he had traversed several decades before.

When the Mercedarian priest and magistrate Jozé Monteiro de Noronha (1723–94) set out from his native city of Pará in 1768 to take up a new post as parish priest, his intention was to produce a panegyrical account of a region that might materially (as well as spiritually) reinvigorate the sagging Portuguese economy: the captaincies of Grão-Pará and Rio Negro. Noronha had passed his entire youth in the area around Pará, and on this westward voyage he waxed poetic about the glories of his native land:

Its climate is healthy and benign, and its lands are extremely fertile. It is abundant in springs, lakes, and forceful rivers; in open fields and thick forests; in trees that are ever replete with leaves . . . in cattle and wild animals; in birds of extraordinary size and shape, and . . . variety and vividness of color.[50]

To emphasize the potential lucre that could be gained by more active colonization, he enumerated the myriad commodities that could be cultivated there, explaining that the city's commerce consisted of "cacao, cinnamon, sarsaparilla, copaiba oil, coffee, sugar, tobacco, cotton, and leather, [all of which] are transported for trade in Portugal." But the glories of this land were not only industrial. In his guise as principal vicar of the Rio Negro captaincy, Noronha was also meant to assist Bishop Miguel de Bulhões in the spiritual governorship of the far-flung diocese, and it was his duty not only to remark on objects that would "interest the curiosity of travelers" but to provide a sacred guidebook as well. In so doing, he was confident that he could entice skeptical metropolitans with spiritual as well as economic pleasures and thereby encourage them to sample the delights of Portuguese America.[51]

Noronha's first mention of La Condamine was in the context of these florid descriptions. Hoping to evoke something of the sensorial experience associated with life in the "equatorial tropics," Noronha enumerated some of the natural products available near Pará. He wrote that "there are many oils in the state that are valuable and pleasing to the palate, such as the sesame, the chestnut, the bataua palm, and the ybacaba, which Mr. de Condamine calls in the language of the Maynas, Ungurave."[52] A footnote in Noronha's text directed the reader to "Condam. in the abridged version (*extracto*) of his diary, page 36," where La Condamine made reference to the "innumerable gums, balsams, and resins" that Amerindians cultivated by cutting into the trunks of palm trees. In this particular example, La Condamine referred not to the plants of Pará, however, but rather to those of the Maynas region of the western Amazon, more than a thousand miles away. Through Noronha's reading, the *Extracto* effected what might be called a biogeographical mediation. At a time of significant tension between the two Crowns, the narrative led Noronha to recognize the botanical links between Spanish and Portuguese America. The two regions of Maynas and Pará were of course geographically, culturally, and politically distinct, separated by miles and miles of forest and water. But Noronha's active appropriation of La Condamine's text brought them side-by-side in a textual "geography of plants" worthy of the nineteenth-century writings of Humboldt, Lamarck, and de Candolle. La Condamine's traversal of the political frontiers between Spanish and Portuguese territory thus served to reconfigure geographic and natural historical understanding two decades later and made possible a botanical linkage between two vastly different territories that were under distinctive political regimes.[53]

Much of La Condamine's text was useful in affirming Noronha's purposes, especially those examples that included instrumental measurements of rivers and channels. But Noronha did occasionally refer to La Condamine's "incompetent" observations as well, such as when he placed a stream in an inappropriate location and called it "Catabuhú" instead of Uacaburá.[54] And when it came to disputing Spanish claims to Portuguese territory, Noronha used sophisticated reading techniques to attack the foundations of La Condamine's credibility altogether. In the first instance, he took issue with La Condamine's reliance on Spanish Jesuits, sources whose knowledge according to Noronha would have seriously limited La Condamine's understanding of Portuguese activities in the Rio Napo region. Secondly, like de Pinto, Noronha pointed to La Condamine's poor understanding of indigenous languages. Without knowing the etymology of native terms, La Condamine would not have been able to understand the origin of particular names and would thus have erred in his geographical interpretations, as was the case with his faulty reconstruction of the fables of Lake Parima and El Dorado (discussed in chap. 2). La Condamine had written that "*Para-guari* means nothing more than River of the *Guaris*, or River that bathes the land of the *Guaris*." Attacking La Condamine's interpretation of Pedro Texeira's 1641 account, Noronha demonstrated that La Condamine's understanding of Indian languages was inadequate for such a task: "In the invariable phraseology of the general language (*idioma geral*) of the Indians of Brazil, when there are two nouns it is necessary to put the genitive before the nominative . . . for this reason, in order to mean River of the Guariz, it would be necessary to say Guari Paraná, and not Paraguari."[55] Noronha pointed out that Paraguari would mean "Guaris of the River" and not the "River of the Guaris," as La Condamine had interpreted it. According to Noronha, it was useless to base geographical understanding on the etymology of these names, since they depended "exclusively on a free and voluntary imposition of men." Noronha, in turn, insisted that the region's legitimate name was "Parauri" and not "Paraguari" at all; the Indians had named the area after a small parakeet that abounded in this zone, not after an Amerindian group's purported possession of the river.

Noronha's use of La Condamine's text in situ shows that travelers reading the *Extracto* as they floated along the Amazon River were able to craft sophisticated epistemological critiques similar to those who discussed and debated his observations in the salons of Amsterdam or Paris. La Condamine's account was considered required reading for Spanish and Portuguese travelers, but their interpretation was not necessarily preordained by their imperial affiliations. Like La Condamine's use of Valleumbroso's letters, the *Extracto* allowed Noronha to structure his own account as a response to another figure and to point out the ways in which his own knowledge was superior. But Noronha openly cited the *Extracto* in order to situate himself within a larger transnational debate, unlike La Condamine, who suppressed Valleumbroso's letter in his printed text. The *Extracto del diario* was clearly the most important text in Noronha's portable library, but it was deployed politically as well as philosophically. Within the protracted imperial rivalry between the Spanish and Portuguese in eighteenth-century Amazonia, La Condamine's text was read not only with an eye toward how it might contribute to the church's catechistic purposes but also how it could catalyze a program of economic and political conquest as well.

While Noronha portrayed the bounty of Amazonia as a way of furthering the Church's ecclesiastical mission, the Portuguese jurist Francisco Xavier Ribeiro de Sampaio (1741–1812?) had less pious goals in mind. Foremost among them was asserting the territorial rights of the Portuguese. He attempted to do this by contradicting the fallacious conclusions of previous explorers to the Amazon region, and he had La Condamine squarely in his sights. A native of the Portuguese village of Mirandela, Sampaio had crossed the Atlantic for his first overseas post in 1767, when he was named to a high-ranking juridical post—"Juiz de Fora"—in the city of Pará. His more permanent post was in the regional capital of Barcelos, from where he embarked two years later on a journey of territorial reconnaissance to the wild and unsettled lands along the Rio Negro, the very region that Monteiro de Noronha had visited six years earlier. Indeed, in preparation for his journey, he avidly read and annotated both Noronha's narrative and La Condamine's *Extracto*.

Sampaio's primary aim on the Rio Negro was to carry out a "correction" (*correição*), which was a detailed report on the present state of the indigenous populations who lived along the riverbanks. He also was charged with assessing the potential agricultural and industrial resources available for exploitation by the Portuguese Crown. His arguments figured implicitly as a counterbalance to continued Spanish territorial pretensions; from the earliest pages of his account, he attempted to refute the foreign travelers that had countered Portuguese claims by perpetuating historical and geographic myths. Sampaio bolstered his arguments with both on-the-spot observations and careful readings of earlier texts. And La Condamine's *Extracto* contained evidence he wielded as part of his arsenal.[56]

While Sampaio did use La Condamine's text as a guide for some of his own observations, he sought above all else to refute many of La Condamine's geographical conclusions. Not coincidentally, he focused on one of the same regions that had troubled Noronha in La Condamine's map: the complex fluvial network where the Japurá River drained into the Amazon (plate 9). Comparing La Condamine's observations with his own on-the-spot examinations, Sampaio pointed out the great extent to which the Frenchman had led so many geographers astray:

From the northern shore the following rivers drain into the Jupurá: the Maruá, a small river that Mr. de la Condamine mistakenly calls a lake and with equal error claims that it connects to the Urubaxí, which drains into the Rio Negro . . . I have personally seen and examined [these rivers] . . . and have dispelled the error that has been with us since the journey of Mr. de la Condamine, who with a decisive tone claimed that those [other rivers] existed.[57]

This skeptical attitude toward the "tone" of La Condamine's conclusions continued in a section entitled "Mr. de la Condamine's opinion on the boundaries of the Portuguese colonies on the Amazon River is refuted, and the inalienable right [of the Portuguese] is affirmed against the pretensions of Spain."[58] In this extended refutation, Sampaio again sought to dismantle the claims of the Spanish Crown through bibliographical consultation. By taking aim at La Condamine, a supposedly neutral third party, he undermined the standing of the French academician and diminished the source of much of La Condamine's own prestige:

The established reputation of Mr. de la Condamine could fool those who read his writings without access to better information . . . It is a shame that such a highly regarded man would want to discredit himself intellectually (*desilustrar-se*) in this way.[59]

Making reference to those who might read La Condamine's text without access to "better information" and personal experience, Sampaio illustrated the perils of reading a text in isolation. Several pages later, he like Noronha also stripped La Condamine of his linguistic credibility by undercutting the academician's geographical conclusions, which Sampaio claimed had been reached through faulty etymological reasoning. Based on his own knowledge of the *língua geral* and a thorough rereading of published and manuscript documents relating to the Iberian history of Amazonian exploration, Sampaio offered a vehement critique of La Condamine's "metaphysical arguments, [which are] useless to determine historical facts."[60]

But Sampaio further sought to debunk a series of ancient legends that had maintained their grip on the European imagination through the writings of missionaries, navigators, and conquistadors. In a section entitled "Brief Dissertation on the Name of the Amazon River and on the Existence of Amazon Women," Sampaio set out his own claims that contradicted two centuries of eyewitness testimony, including that of La Condamine:

Much has been said about the existence of American Amazons, about their Republic set apart from men excepting the specific moment of their encounters, and of their similarity with the Asiatic Republic. Everyone is aware of what Laet, Raleigh, Cunha, Feijo, Sarmiento, Coronelli, and Condamine have written.[61]

By citing a series of canonical authors who discussed the fabled existence of women warriors, Sampaio presented what he called "the facts upon which these discourses are based." He provided detailed citations from the testimonies of Orellana, da Cunha, and others in order to build a documentary basis for their conjectures about the Amazons. Once he had laid out their respective arguments, however, he concluded that it would be impossible for him to share this opinion:

If we examine this material using the laws of true logic and solid criticism, we must conclude that the existence of the American Amazons is one of those popular preoccupations that, finding a foundation in the marvelous, which people adore, take on a life of their own with extraordinary ease.[62]

Rather than evaluate the myth by consulting indigenous sources, Sampaio drew on the work of eighteenth-century philosophers to demonstrate the absurdity of the story:

What could be more difficult to imagine than a self-governing republic of women who live in the torrid zone without admitting men but for certain days of the year? What moral causes can we imagine that would be so effective as to conquer the almost irresistible force of the climate? Hot climates agitate the spirit (*animo*) with respect to the union of the two sexes: everything leads to this purpose, as a great juridical philosopher has said.[63]

This "great juridical philosopher" was Montesquieu, author of the *Spirit of the Laws*, and Sampaio responded to the Amazonian myth with a reading of his climatic determinism that negated the possibility of such an "unnatural" confluence of circumstances.

Sampaio's overall achievement was to combine eyewitness observation of geographical phenomena with on-the-spot consultation of philosophical texts. Like de Pinto, Maldonado, and Monteiro de Noronha before him, he constructed a coherent critique of La Condamine's method that was not solely based on eyewitness empiricism but involved his own internal standards of rational criticism as well. Reading for Sampaio also included checking for the implicit plausibility of a text. But this did not prevent the Portuguese jurist from attacking the foundations of an instrumentally based empiricist who had reached political conclusions that conflicted with the pragmatic charge he had been given by the Portuguese Crown.

Conclusion

The various cases examined in this chapter show that despite tremendous divergence in the political affiliations and cultural identities of La Condamine's readers, their reactions were often surprisingly similar in both tone and tenor. There were no distinctly "national" or even "imperial" reactions to the *Relation abrégée*, although certain readers were strongly influenced by ideological pressures due to their specific imperial affiliations. But even taking into account such predilections, each reader constructed and deployed his own epistemological matrix based on a series of factors, and it was rare that a reader's individual response could be subsumed within any recognizable category of collective identity, political affiliation, or communal belief. Readers of La Condamine's *Relation abrégée* did not passively absorb the texts they perused; rather, they used them to create new paradigms for understanding history, language, and cultures across oceans and between continents.

In an oft-cited passage, part of which serves as this chapter's epigraph, Michel de Certeau portrayed readers as literary wanderers and writers as sedentary beings, the latter digging wells and building houses from raw materials provided by others. But the dichotomy that de Certeau established between reading and writing—readers as nomadic parasites, writers as builders of civilization—belies the symbiotic interchange between these two modes of cultural production. Readers undoubtedly made signal contributions to writers' "soil of language" wherever their nomadic itineraries happened to take them. These "Ulysses of everyday life," as de Certeau called them, may have been limited to some degree by their social or class standing, but their participation in a larger system could not be reduced to the particular collectivity—whether social, cultural, or political—from which they emerged. Their reactions often diverged too far from an author's intended meaning—to the extent that such meaning could be discerned from an author's own work—and frequently coalesced around too narrow an interpretive critique to lend credibility to the idea that a reader's response was either predetermined by the text, on the one hand, or entirely free-floating, on the other.

What is clear is that eighteenth-century readers of New World travel literature responded in ways that generated important questions for, and may have even contributed implicitly to, transatlantic intellectual debates, both in and out of the public arena. Their reactions to what they read could influence the way texts were produced, edited, and even transformed in subsequent editions. La Condamine himself was an exemplar of how this might happen. His frequent insistence on the retraction of half-truths dealing with his own travels led him to criticize those who aimed to mock or critique him in a public forum. Even at the very end of his life, La Condamine continued to defend his work against readers' critical barbs. In a letter written to the editors of the *Journal Encyclopédique* in 1770, for example, La Condamine took issue with the sarcastic assessment by one Pierre-Jean Grosley of La Condamine's 1761 journey to London, during which La Condamine was followed around by a large crowd who made fun of the near-deaf academician's ear trumpet. La Condamine's reading of Grosley's text, and his subsequent protest in a prominent periodical, stimulated Grosley to change the parts that corresponded to La Condamine's excursions in a subsequent edition of *Londres* published four years later. Grosley also indicated in his preface to this new edition that La Condamine had sent him a copy of his *Voyage à l'équateur* as "proof" of his friendship.[64] Reading not only influenced the direction of travel by predisposing travelers to reach certain conclusions, as we saw in the previous chapter; it also influenced the public perception of travel through the texts and periodical reviews that were composed once the travelers returned home.[65]

On balance, then, how was La Condamine's text read and received? Or put another way: where are we to look for the reception history of a text that was read in such different contexts? The broad circulation of La Condamine's most celebrated narrative and the critical resistance it generated in individual readers demonstrate that reception almost never meant passive absorption. Library lists and reviews in contemporary periodicals alone cannot adequately reveal the manner in which a text was read and understood, either in Europe or beyond it. Even more importantly, the very scope of reception history needs to be expanded if we are to understand not only the transatlantic commerce of ideas but their transimperial circulation as well. The broad geographical range of readers covered in this chapter demonstrates that a French text (and its Spanish translation) elicited interest not only from those whose empires were involved territorially in the disputed regions it discussed. And the criticisms it received were shaped by, but did not always conform explicitly to, the imperial affiliations of its readers. The critique Sampaio offered of La Condamine's perpetuation of the Amazon myth, for instance, was a sanction of irresponsible observation as well as an affirmation of the way that "marvelous" fables could "take on a life of their own." The "dispute of the New World" may have pitted defenders of Creole culture in the Americas against numerous northern European authors, but resistance to these overly simplistic ideas about the New World was already brewing in a wide range of readers, including individuals with such utterly divergent backgrounds as Isaac de Pinto, Pedro Vicente Maldonado, and Francisco Xavier Ribeiro de Sampaio.

But it must also be recognized that the likelihood that a reader's comments would be heeded depended greatly on where that individual stood in the geopolitical hierarchy of the early modern world.[66] As La Condamine's ability to influence debates over his own work indicates, academic prestige, intellectual authority, and social standing within Old Regime culture could go a long way within both national and transnational contexts toward prejudicing these debates toward one side or another. Conversely, a figure such as Maldonado—of uncertain status as a Spanish American Creole within northern European sociocultural hierarchies, accepted as a member at both London's Royal Society and the Paris

Academy of Sciences yet whose testimony was curiously denigrated by his closest friend and most ardent supporter—went to great lengths to communicate his individual testimony to a wide audience but was ultimately heard most attentively by a Dutchman of Sephardic ancestry, a figure "between nations" himself who craved acceptance into Enlightenment culture but was consistently denied full participation.[67] Often the remarks of individuals who were poorly placed in the sociocultural hierarchy remained confined to these interstitial zones epistemologically as well. Frequently, their critical comments were left either literally or figuratively on the margins, whether of a debate or within a printed text. Reading may have been a generative activity, but readers' interpretations were ultimately dependent on their access to social and material conditions that provided a forum for their corrections and criticisms. It is these two activities, and the materiality of the transatlantic "spaces" in which they took place, that the following two chapters will address.

FOUR

Correcting Quito

It is regrettable that d'Anville, in making important corrections on the copperplates of his maps, did not mark the period in which these changes were made.
Geographers who do not know the circumstances could be inclined to err about
the dates of many discoveries.   HUMBOLDT, *RELATION HISTORIQUE DU VOYAGE*
AUX RÉGIONS ÉQUINOXIALES DU NOUVEAU CONTINENT

Is it not revealing . . . that a map can be disassociated and depersonalized from
its author and reappropriated by a new cartographer who will sign his name to
it after making a few small or large changes?

CHRISTIAN JACOB, *L'EMPIRE DES CARTES*

A fictive scene in a dramatic cartographic cartouche: on the
verdant banks of a river in the tropics, a village comes to life
through industry, indigenous labor, and colonial aspiration
(fig. 16). Along the shoreline, wooden poles extend in geometric precision, marking out a series of structures and showing the phased construction of thatched-roof huts. With a
saw, two simply clad natives cut logs to provide wood for
the dwellings. A church with its bell tower, the ubiquitous
symbol of ecclesiastical authority, is visible in the background, while further in the distance palm trees hint at the
demarcation between savage wilderness and civilized space.
Conspicuously at the center stands a youthful Spanish American in European dress, surrounded by several mestizos and
Amerindians who lean toward him in careful counsel; at
his feet a kneeling man sets a compass on the sandy shore
and aligns it carefully with the wooden poles. The scene
emphasizes not only the product of European colonialism

Figure 16. *Carta de la Provincia de Quito y de sus adjacentes* (1750), cartouche. Courtesy of the Geography and Map Division, Library of Congress, Washington, D.C.

in the form of a small settlement outpost but also the very tools and materials that were used to construct it piece by piece: the progressive stages of building an Iberian empire on American soil. Taken together, these elements present an idyllic image of industrious bliss at the heart of South America—but they only reveal half of the story.

Just as the village is coming to life, a portent of its demise appears imperceptibly from above. Swooping down unexpectedly, a wizened and winged figure does double duty as Chronos and the Angel of Death, a veritable Grim Reaper holding an hourglass in one arm and a scythe with the other. The fine writing carefully inscribed on his blade reads "Dum adhuc ordirer succidit me," a citation from chapter 38 of the Book of Isaiah making reference to the sudden vacuum created by a life ended at the height of its powers. The figure is shown pulling the corner of a drape across the cartouche's field-of view—a somber indication that this colonial theater merging science and industry would be closing its doors even before it properly began. And the protagonist of this dramatic scene, inscribed in a cartouche at the corner of the four-sheet map, was about to be permanently hidden behind this curtain. These elements portend the transformation of a scene of colonial prosperity into a theater of erasure and decay, and the image of the map's author within the cartouche provides an uncanny reminder of his own premature death and the subsequent evanescence of the glory he might have derived in life from a map that bore his name.

It was in April 1750 that Gravelot—Hubert-François Bourguignon d'Anville—put the finishing touches on this cartouche in his Parisian atelier. In previous years, Gravelot had masterfully designed several cartouches with anonymous allegorical figures representing continents, deities, and mythic ethnic groups. This time, however, the image was not meant as a flowery complement to a generic cartographic depiction or as an allegory of a region's natural products. Nor was it merely a flamboyant frame in which to place descriptive text. Instead, this cartouche was to function as an epitaph-in-print: a cartographic monument to Pedro Vicente Maldonado, whose life had been suddenly and mysteriously cut short. There were no trumpeting angels manipulating scientific instruments, no Amazon warriors subduing crocodiles beneath their feet; rather, the scene exuded harmony and industriousness, and it showcased the critical role Maldonado had played as a cultural intermediary and explorer. The road he had forged between Quito and the Esmeraldas region along the Pacific coast was visible just beside the cartouche, represented by a part of the map that owed the most to Maldonado's own geographic reconnaissance in situ. And it was Gravelot's job to connect these two sections symbolically by designing a graphic image in the short time available to him.[1]

Time was running out for the intaglio artisan as well. Although there was no scythe poised above the engraver Brunet's head, he was nonetheless being hastened to finish his etching work without delay. This pressure came directly from Gravelot's brother, Jean-Baptiste Bourguignon d'Anville, royal geographer to Louis XV and one of the most renowned cartographers of his day. D'Anville was responsible for bringing together the disparate elements of the map in tandem with the person to whom Maldonado had confided his papers before his death, La Condamine. Writing to La Condamine, d'Anville explained that he was unwilling to let the deadline pass unheeded, since it had been imposed by the Spanish ambassador who had financed the map's production. He thus urged his colleague to encourage Brunet to complete his work post haste: "I have made it my personal obligation [to the ambassador] to see that the plates are sent in twelve days. It is the duty of M. de la Condamine to assert great pressure for the etching (*gravure*) of the cartouche . . . I therefore request that M. de la Condamine expedite the etching . . . We must have the plates by Tuesday or Wednesday in order to have the proofs."[2]

This flurry of looming obligations and last-minute pressures may have been familiar to early modern authors racing to meet a publication deadline. But eighteenth-century documents from the cartographic atelier all too rarely reveal this behind-the-scenes backdrop in such detail. The process by which sketches, notes, and editorial revisions became fused into a two-dimensional cartographic representation was rarely observed by anyone outside the printer's workshop. But d'Anville's letter, like the cartouche described above, allows us to see in one frame the diverse cast brought together to create a single map: the artists, engravers, travelers, and compilers who were each individually responsible for a single element in a collective product. Rather than presenting the map as a fait accompli, d'Anville's description reveals an insider's perspective on the map as it came into existence: the contemporaneous elaboration of the plates and cartouche; the physical and material limitations imposed by the engravers; the printing process; and a hint of the trans-Pyrenean political dialogue with regard to a map of Spanish America produced in a Parisian atelier.

This chapter follows the *Carta de la Provincia de Quito* (1750) through its various stages of editorial elaboration, from the compilation of materials in a cartographic workshop to the distribution of the various states and editions of a map. It shows the importance of studying not only the content of a printed map's final version but also the distinct stages of its production. In the case of the *Carta de la Provincia de Quito*, the observational practices of a Spanish American Creole were facilitated by the power of print, but they were compromised in particular ways as well. The European fascination with Maldonado's printed map allowed his geographical observations to circulate far beyond the Andean highlands, but the editorial relationships inherent within French cartographic culture also placed unforeseeable and unintended restrictions on the direct transmission of Maldonado's cartographic agenda. The techniques of print production allowed for omissions and imposed editorial exigencies that did not necessarily correspond to the wishes of the map's author. The final version of the printed map of Quito does not show the territory as it truly is, or even as Maldonado wished for it to be portrayed, but rather reflects information that successfully traversed the Atlantic from Andean peaks to a Parisian printing house. Much was left behind.

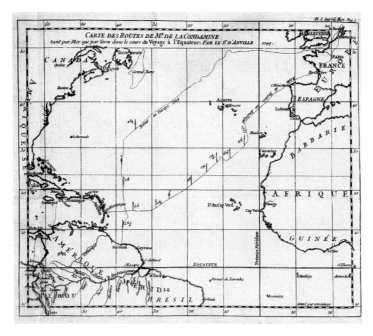

Plate 1. *Carte des Routes de M. de la Condamine* (1749). La Condamine's journey across the Atlantic began with his departure from La Rochelle in May, 1735. His travels took him through the Caribbean to the western coast of South America and then back to Europe via the Amazon River. Courtesy of the William L. Clements Library, University of Michigan.

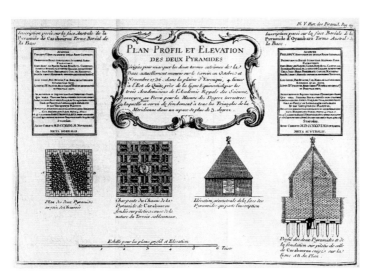

Plate 2. "Plan Profil et Elevation des Deux Pyramides," *Histoire des pyramides de Quito* (1751). The pyramids at Caraburu and Oyambaro were laden with structural and symbolic significance. Each was also adorned with a commemorative plate in Latin, which created a controversy over whose nation's monarch – France or Spain – deserved greater praise as sponsor of the expedition. Courtesy of the William L. Clements Library, University of Michigan.

Vue de la Base mesurée dans la plaine d'Yar[ouqui]

Sous un arc qui comprend

Dessinée du haut de la chute d'eau

NB. On a representé dans cette vue tous les objets
compris dans le demi tour de l'horison en supposant
que l'œil se tournoit successivement vers chacun
d'eux sans sortir du même point.

Renvois des Signaux qui ont servi pour la suite des Triangles de la Meridienne compris dans cette Vue.

A. Pyramide de Carabourou.
B. Pyramide de Oyambaro.
a. Signal de Cotopaxi.
b. Signal du Coraçon de Barnuevo.
c. Signal de Hilshalo.
d. Signal de Guapoulo.
e. Croix de Pithintcha.
g. autre Signal de Pithintcha pour les Triangles.
h. Signal de Casitagoa.
i. Signal de Tanlagoa.
k. Signal de Cotchesqui.
l. Signal du Sommet de Pithintcha.

On a suivi l'ortographe françoise dans les noms Indiens
et l'Espagnole dans les noms Castillans qu'on a soulignés.

Renvois des Objets remarquable de la Plaine comp[rise]

A. Pyramide de Carabourou, terme austral de la Base.
B. Pyramide de Oyambaro, terme boreal de la Base.
C. Manufacture de Draps d'Yarouqui.
D. Ravine dite de Carthagene.
E. Embouchure de la Ravine d. de St. Rose.
F. Grand chemin d'Yarouqui à Quito.
G. Chantac maison de Campagne.
H. Pifo Annexe.
I. Tchaupi-molino Ferme.
K. Tumbaco Paroisse.
L. Cumbaya Paroisse.
M. Notre Dame de Guapoulo, Chapelle celebre.
N. Nayon Annexe.
O. Sanbiça Paroisse.
P. Ruembo Paroisse.

Q. Quito.
R. Ferme d'Alban.
S. Ferme de Matis.
T. Ferme d'Aguayo.
V. Ferme de Siman.
X. Carabourou Ferme.
Y. Ferme de Monta.
Z. Cotchesqui lieu de
   nord de la Meridien.
J. Tabavela, Annexe d.
V. Pouh-hal Ferme.
&. Mançaouantas; F.
W. Chitche, Ferme.
AA. Tocatché Annexe.
Æ. Malchingui, Paro.

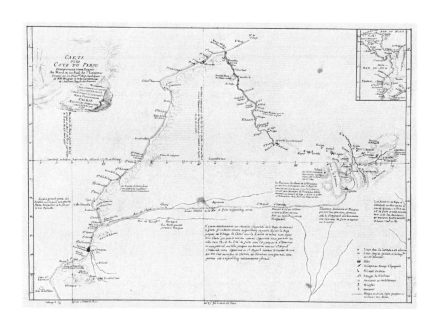

ouqui, *depuis* Carabourou *jusqu'à* Oyambaro,

) 180. degrez de l'horison ,

(du moulin à foulon d'Yarouqui .

ris dans cette Vue)   Renvois des Objets remarquables qui bornent l'horison .

1 . Sinchoulagoa , Montagne couverte de Neige haute de plus de 2560 Toises au dessus de la Mer
2 . Cotopaxi , Montagne couverte de Neige , Volcan qui s'est rouvert en 1742 . haut de 2950.
 Toises au dessus du niveau de la Mer .
3 . Roumignacui , Montagne .
4 . Passotchoa , Montagne .
5 . Iliniça , Montagne couverte de Neige , presumeé ancien Volcan haute de 2720 Toises
6 . Chongou ou Coraçon , Montagne couverte de Neige , haute de 2490 Toises .
7 . Hilahalo petite Montagne .
8 . Atacatso ou San Juan Ourcou Montagne .
9 . La Viuda Montagne .
10 . Pitchintcha Volcan de Quito Montagne couverte de Neige . Embrasee en 1577. et 1690. haute de 2430 Toises
11 . Casitagoa .
12 . Montagne à l'Ouest du Bourg de St. Antoine où passe la ligne Equinoctiale .
13 . Tanlagoa .
14 . Chaine de Montagnes qui court à l'Ouest
15 . Cotacatché Montagne couverte de Neige
16 . Moh-handa Montagne .
17 . Yana-ourcou de Moh-handa Montagne .

Plate 3 (*above*). "Vue de la Base mesurée dans la plaine d'Yarouqui, depuis Carabourou jusqu'à Oyambaro," *Journal du voyage fait par ordre du roy* (1751). A view of the plain of Yaruquí from Caraburu in the north (right) to Oyambaro in the south (left). The volcano Pichincha (letter "f"), which sits above Quito (letter "Q"), is visible in the center of the plate. The pyramids of Caraburu and Oyambaro are, respectively, letters "A" and "B" at either end of the plain. Courtesy of the John Carter Brown Library at Brown University, Providence, R.I.

Plate 4 (*left*). "Carte de la Côte du Pérou" (1736). La Condamine and Pierre Bouguer sent this manuscript map of the Pacific coast to the Academy in 1736. The cartouche shows an alternative version of the Palmar stone, upon which La Condamine chiseled an inscription after determining the site where the equatorial line crosses the Pacific coast (see also figure 6). Portions of this map were later incorporated into the *Carta de la Provincia de Quito* (1750) (see plate 10). Reproduced with permission of the Bibliothèque Nationale de France, Paris.

CARTE DU COURS DU MARAGNON OU DE LA GRAND[E]

Dans sa partie navigable depuis Jaen de Bracamoros jusqu'à son Embouchure et qui comprend la Province de QUI[TO]

Levée en 1743 et 1744 et assujettie aux Observations Astronomiques par M[...]

Augmentée du Cours de la Rivière Noire et d'autres détails tirés de divers Mémoires et Rout[es]

PARTIE DE L'AMÉRIQ[UE]
MÉRIDIONALE

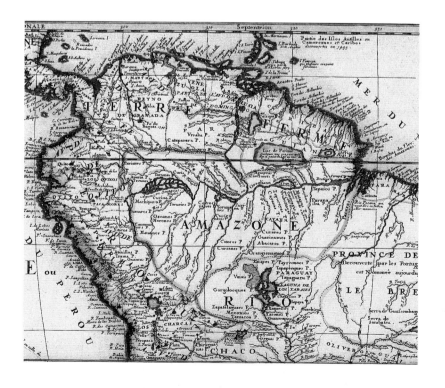

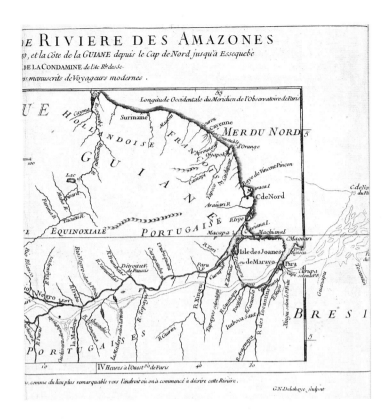

IE RIVIERE DES AMAZONES

_r__, et la Côte de la GUIANE depuis le Cap de Nord jusqu'à Essequebe_

DE LA CONDAMINE _de l'Ac Rl des Se-_

... manuscrits de Voyageurs modernes .

Longitude Occidentale du Meridien de l'Observatoire de Paris

U E HOLLANDOISE

Suriname

GUIANE

Cayenne

MER DU NORD

Cd'Orange

Oyapok

Baye de Vincent Pinçon

Lac

Maroni R.

CdeNord

CdeNor du Fh

Araüari R

EQUINOXIALE PORTUGAISE

R.Negro PORTUGAISE

Rhyo

Macapa

Guiana I.

Machiana I.

Isle des Joanes ou de Marayo

CMaguari

Para

IV Heures à l'Ouest 30 de Paris

BRESIL

... comme du lieu plus remarquable vers l'endroit où on a commencé à décrire cette Riviere .

G.N.Delahaye _sculpsit_

Plate 5 (*above*). *Carte du Cours du Maragnon* (1745). La Condamine's map of the Amazon River extended the range of his observations from Quito and the Andes all the way to Cayenne on the Atlantic coast. His project to encapsulate the Amazon in cartographic and narrative form represented the fusion of two contemporaneous methods: an attempt, on the one hand, to use scientific instruments to measure, quantify, and situate geodetically a complex hydrological system; and, on the other, to create a holistic expression of a distant and little-known world by resolving puzzles that had challenged previous explorers. The result is a spectacular exemplar of the eighteenth-century boundaries between the geographical sciences and South American mythology, in which claims put forth regarding the map's accuracy—including the underlying trace of Samuel Fritz's earlier map—lay side by side with lakes and mountains whose placement was inextricably linked to the Amazon River's legendary past. Courtesy of the John Carter Brown Library at Brown University, Providence, R.I.

Plate 6 (*left*). Vincenzo Coronelli, *L'America Meridionale* (1690). This map of South America contains several common features of seventeenth-century Amazonian geography, including the islands of the "Aguas or Omaguas" peoples (center left), the island of the Tupinambas (center right), the Parima Lake which was purportedly the site of El Dorado, and the direct connection between the Orinoco and Amazon rivers. Coronelli's map bears a strong resemblance to Sanson's 1680 map of the Amazon (fig. 9). Courtesy of the Fundação Biblioteca Nacional, Rio de Janeiro, Brazil.

Plate 7. "Carte du Cours du Maragnon" (manuscript, n/d). PRO MR 1/904. 62.5 cm × 141.5 cm. This colorful hand-drawn copy of La Condamine's *Carte du Cours du Maragnon* appears to be a presentation version of La Condamine's much smaller printed map. Aside from the size and color, its most striking feature is the incorporation of geographic, ethnographic, and mythological details about the region in textual form along both sides of the map. Legend number 22, for instance, at the intersection of the Amazon and Jamuendas rivers, is described as the "river where Father Acuña claims that Orellana was attacked by Amazons." Another text (number 13) describes the Yameos as "Savages who have a difficult pronunciation." The high concentration and meticulous presentation of these "remarks" add credence to the idea that it was used as both a visual and textual description of the region for those who did not necessarily have immediate access to the *Relation abrégée*. Reproduced with permission of the National Archives (UK).

DE LA GRANDE R^{RE} DES AMAZONES
de QUITO, et la côte de la GUIANE depuis le cap de Nord jusqu'à Essequebe
...DAMINE de l'academie Royalle des Science.
...s et Routiers Manuscrits de Voyageurs Modernes.

HOLLANDOISE

GUIANE FRANÇOISE

XIALE

PORTUGAISE

Cayenne

cap d'Orange

MER DU NORD

Baye de Vincent Pinçon

C. DE NORD

Macapa R.

Isles des Joanes ou de Marayo

PORTUGAISES

BRESIL

Rio Negro

IV. Heures à l'Oüest de Paris

Plate 8. This image is taken from the 1745 edition of the *Mémoires de l'Académie des Sciences* (Paris, 1749) where it was inserted next to the *Carte du Detroit appellé Pongo de Manseriché dans le Maragnon ou la Riviere des Amazones*, a map that shows one of the narrowest portions of the Amazon River as it descends from the Andes: the famed "Pongo de Manseriche." The Marqués de Valleumbroso suggested to La Condamine that he measure the river's longitude, latitude, and depth as he descended the Pongo, and La Condamine in turn portrayed himself at precisely this point. Courtesy of the Special Collections of the Sheridan Libraries, Johns Hopkins University, Baltimore, Maryland.

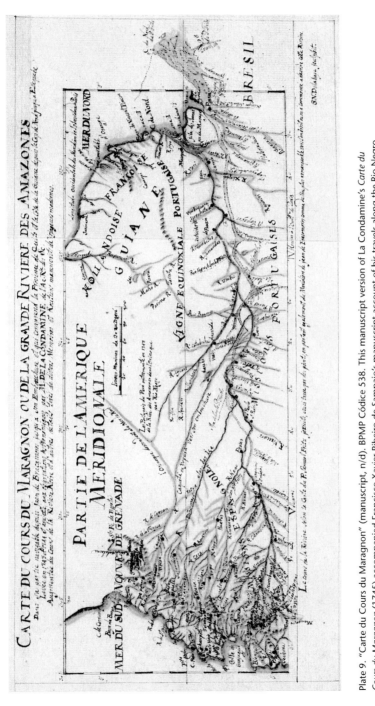

Plate 9. "Carte du Cours du Maragnon" (manuscript, n/d). BPMP Códice 538. This manuscript version of La Condamine's *Carte du Cours du Maragnon* (1745) accompanied Francisco Xavier Ribeiro de Sampaio's manuscript account of his travels along the Rio Negro in 1774–75. As La Condamine had done with Samuel Fritz's map before him, Sampaio used La Condamine's map to criticize the Frenchman's methods during his Amazonian journey. Courtesy of the Biblioteca Pública Municipal do Porto, Portugal.

Plate 10. *Carta de la Provincia de Quito y de sus adjacentes* (1750), sheet 1. Northwest quadrant of the province of Quito. Gravelot's elegant cartouche adorns the top left corner. The city of Quito, capital of the Audiencia, is located just beneath the equator on the righthand border of the map. Courtesy of the Geography and Map Division, Library of Congress, Washington, D.C.

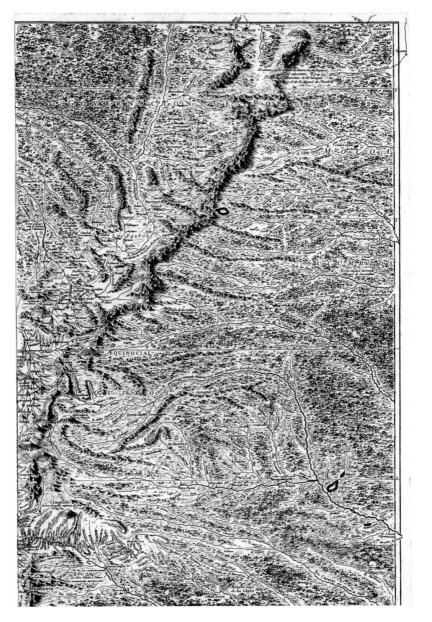

Plate 11. *Carta de la Provincia de Quito y de sus adjacentes* (1750), sheet 2. Northeast quadrant of the province of Quito. The spiny protrusion of the Andean cordillera cuts diagonally across this sheet toward the "Governacion de Popayan." At bottom, the Rio Pastaza and Rio Napo begin their descent from the Andes to the Amazon. Courtesy of the Geography and Map Division, Library of Congress, Washington, D.C.

Plate 12. *Carta de la Provincia de Quito y de sus adjacentes* (1750), sheet 3. Southwest quadrant of the province of Quito. The most prominent feature of this sheet is the large Gulf of Guayaquil in the map's top left corner. At right, a textual citation explains that the portion of the map from Cuenca to Jaén de Bracamoros follows La Condamine's account; it is this route he took when he descended the Andes toward the place where the Amazon River becomes navigable. Courtesy of the Geography and Map Division, Library of Congress, Washington, D.C.

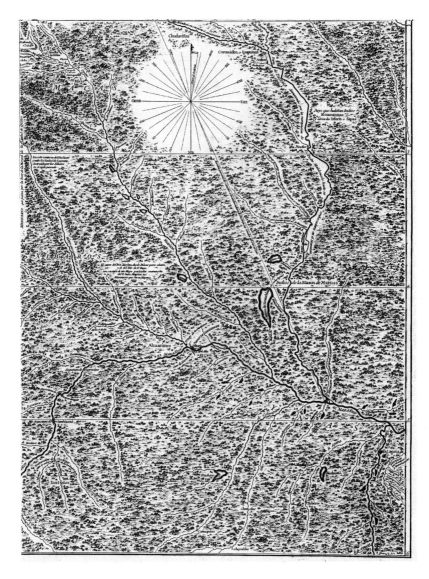

Plate 13. *Carta de la Provincia de Quito y de sus adjacentes* (1750), sheet 4. Southeast quadrant of the province of Quito. This sheet covers the area in the western Amazon basin known as Maynas, where Spanish missionaries carried out active proselytizing campaigns in the seventeenth and eighteenth centuries. When passing through Borja on his way toward La Laguna (bottom right corner of the map), La Condamine received a manuscript account of this region from the Jesuit Jean Magnin. A textual commentary in the center of the map explains that the course of the Rio Morona is based on information drawn from Magnin's papers. Courtesy of the Geography and Map Division, Library of Congress, Washington, D.C.

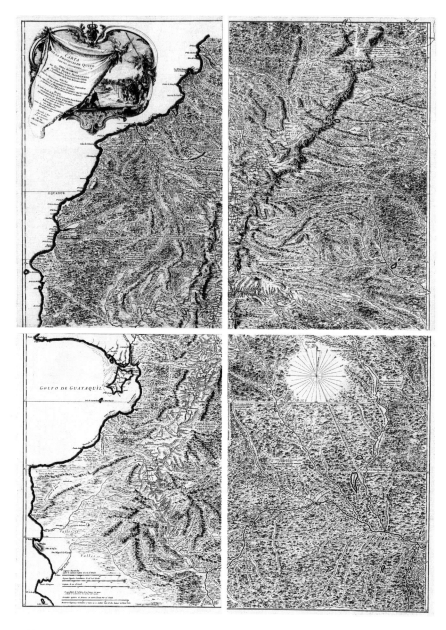

Plate 14. *Carta de la Provincia de Quito y de sus adjacentes* (1750), all sheets. The province of Quito. Courtesy of the Geography and Map Division, Library of Congress, Washington, D.C.

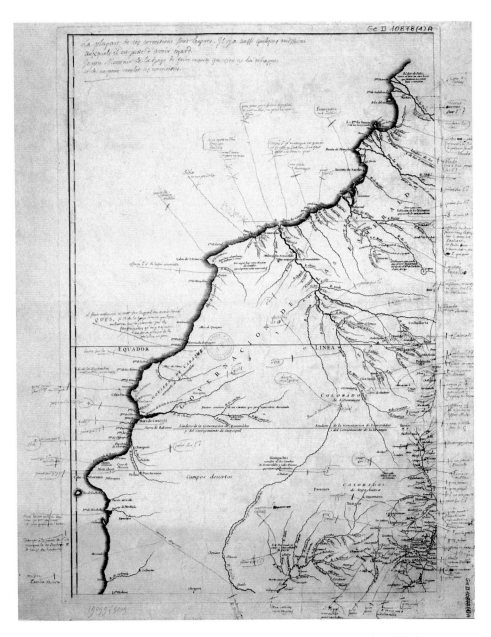

Plate 15. Proof of the *Carta de la Provincia de Quito* (sheet 1). BNF Ge.D. 10878(1)A. This sheet contains d'Anville's instructions to the map's engraver, Guillaume Delahaye. At this stage, the cartouche (see plate 10) has not yet been incised onto the copperplate. Reproduced with permission of the Bibliothèque Nationale de France, Paris.

Plate 16. Proof of the *Carta de la Provincia de Quito* (sheet 2). BNF Ge.D. 10878(2). Maldonado's penciled corrections are visible in the margins of this sheet. This was one of the earliest sets of corrections for sheet 2. Note that Maldonado suggested that the location of Lake Putomayo be changed; in plate 17, this change has been made. Other corrections, however, were not taken into account. Reproduced with permission of the Bibliothèque Nationale de France, Paris.

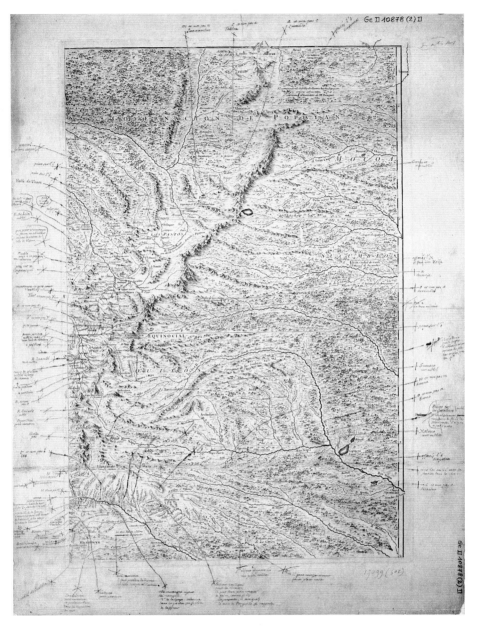

Plate 17. Proof of the *Carta de la Provincia de Quito* (sheet 2). BNF Ge.D. 10878(2)D. Compared to plate 16, this sheet contains added landscape features such as mountains and trees. Many of Maldonado's corrections, made prior to the printing of this proof-sheet, were not heeded. Note the presence of the "Sierra Nevada y minera de Oro" at lower left, a feature that Maldonado asked to be removed. The other corrections are in d'Anville's hand. Reproduced with permission of the Bibliothèque Nationale de France, Paris.

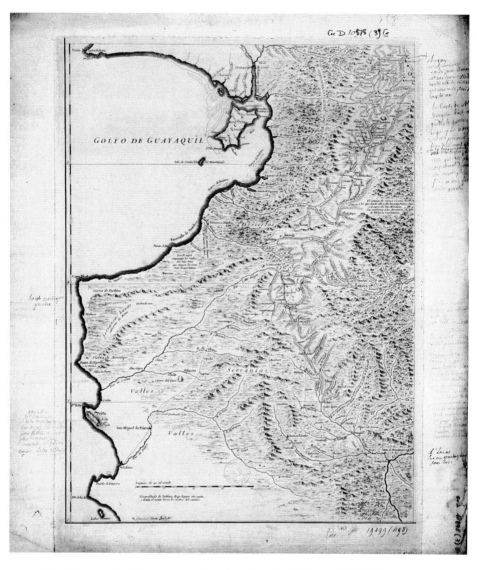

Plate 18. Proof of the *Carta de la Provincia de Quito* (sheet 3). BNF Ge.D. 10878(3)G. La Condamine's corrections are visible along the margins of this sheet, following the route he took from Cuenca to Loxa on the left side and from Cuenca to Jaén de Bracamoros on the right. His annotations range in topic from the shape of mountains to the placement of forests. Reproduced with permission of the Bibliothèque Nationale de France, Paris.

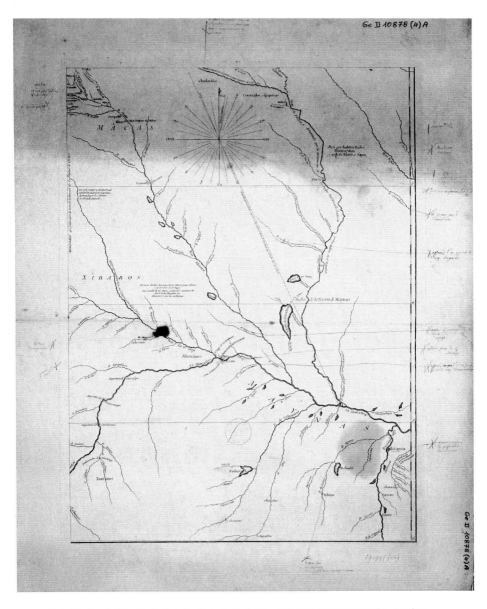

Plate 19. Proof of the *Carta de la Provincia de Quito* (sheet 4). BNF Ge.D. 10878(4)A. Another sheet prior to the insertion of landscape features. The few orthographic changes that are shown here were requested by d'Anville. Reproduced with permission of the Bibliothèque Nationale de France, Paris.

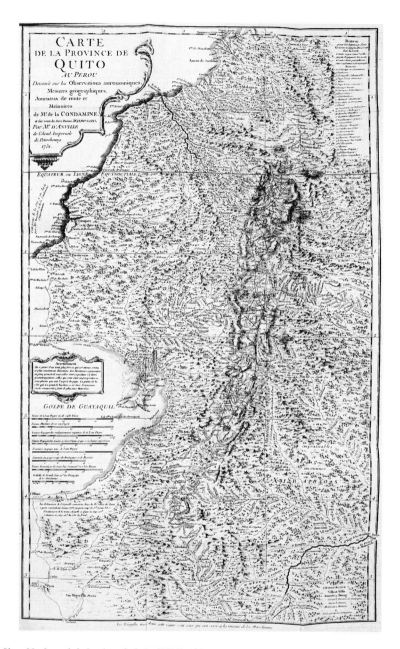

Plate 20. *Carte de la Province de Quito* (1751). This map accompanied La Condamine's *Journal du voyage fait par ordre du roy* (1751). La Condamine made this new map of Quito in an effort to regain cartographic control over the theater of his and his colleagues' geodetic work. Note the triangles that highlight their measurements from Quito to Cuenca. In addition, on the earlier Quito map of 1750, the route from Loxa to the port of Sechura (bottom left) was produced without a textual legend. In this version, however, the route was marked "Route taken by La Condamine on his way to Lima." Likewise, in the cartouche, La Condamine relegated Maldonado's contributions to a secondary plane by placing his name in a smaller font. Courtesy of the William L. Clements Library, University of Michigan.

Unraveling the layers of the map's provenance, production, and use allows for a new interpretation of the *Carta de la Provincia de Quito*, one that sees the map not as a fixed and burnished image of geographical space but rather as the product of social and material itineraries. Handwritten markings on the Quito map and their concomitant social referents demonstrate the importance of human relationships within a cartographic atelier often portrayed as rarefied and removed from the social realities of the day.[3] The sociocultural diversity of the map's many participants demonstrates that the creation of this eighteenth-century image of Quito functioned more like a transatlantic journey than either a European imposition or a particularly "Latin American" territorial vision. The success with which Spanish American Creoles projected a particular view of the land of their birth depended on their ability to affect or mediate the technical and editorial processes across an ocean and within European cartographic culture. Many scholars, however, still insist on reading Maldonado's map through the filter of narrow nationalist appropriations. Even the earliest descriptions of the *Carta de la Provincia de Quito* presented the map as if Maldonado's familiarity with the land of his birth had been mystically and seamlessly transformed into a printed image in four sheets. Historians of science and geography have followed suit by flattening out the tortuous channels through which Maldonado's map emerged from the workshops of Parisian printers and cartographers. Much recent scholarship has powerfully articulated the unique attributes of Creole, mestizo, and indigenous territorial conceptions that served as counterpoints to or foundations of models exported to the New World by Europeans.[4] But an attention to the practices of print elaboration can help clarify the processes of elaboration behind all of these models. By examining what D. F. McKenzie has called "the sociology of the printed text," we focus on social and material factors in order to understand the struggle for authorship and authority in maps of the Americas. We begin our journey through the worlds of the *Carta de la Provincia de Quito* with Maldonado's corrections to the map that would eventually bear his name.[5]

Maldonado in Pencil and Print

Maldonado was accustomed to expressing his geographical opinions. After traveling for two-and-a-half months on a ship bound for Lisbon, he arrived in Madrid early in 1744 and revealed his credentials both as a political emissary from the province of Quito and as a scientific ambassador with firsthand knowledge of Spanish American geography. His construction of a road from the Esmeraldas region along the Pacific Coast to Quito had garnered him notoriety and emboldened him to seek official recognition for his project at the Council of the Indies in Spain. He also sought acknowledgement for his fidelity to the monarch, depositing before the end of his first year in Madrid a "relación de méritos y servicios," or account of his service to the Crown, on his and his family's behalf. In addition, he composed and read to the Royal Council (*Real Consejo*) a petition to raise the status of his native town of Riobamba to "city." His description of Riobamba, presented as the "Memorial a nombre de San Pedro de Riobamba," exemplified the way in which a Creole could articulate his own territorial vision before the Spanish monarch: he of course mentioned Riobamba's utility to the Crown (remarking on its "profit to the Royal coffers"), but also emphasized the manner in which it had recuperated its history from before the Spanish Conquest (making reference to its "historic foundations," which were situated on the pre-Columbian ruins of an Incan royal palace) and the exuberance of its natural setting (a "delicious, fertile site abundant with all kinds of fruits"). Wherever he traveled, he was received with accolades and respect as both a Spaniard and a Creole, and he was ever eager to share his opinions about the geographical realities and cultural particularities of his native land.[6]

According to La Condamine, Maldonado arrived in Paris from Spain at the end of 1746, and for several months thereafter attended sessions of the Academy of Sciences alongside his French companion. Between journeys to Amsterdam, The Hague, Flanders, and London, he had two lengthy periods in Paris during which time he would have been able to collaborate with French cartographers on the production of the Quito map. The first was this initial period of roughly six months during which time he frequented the Academy with La Condamine; he was made a corresponding member on March 24, 1747, and departed for Amsterdam toward the end of June that same year. His second extended period in Paris was at the beginning of 1748, at which time Maldonado was forced to return from the Netherlands because of his inability to secure a passport to London. Indeed, he would have to wait for the cessation of hostilities following the Treaty of Aix-la-Chapelle (1748), which put an end to the War of Austrian Succession, until he was finally able to cross the Channel. He arrived in London in August 1748, three months before his untimely death.

What role did Maldonado have, then, in the production and correction of the Quito map during his time in Paris? To date, the only source for answering this question was La Condamine's *Journal du voyage*, which gave a year-by-year accounting of La Condamine's activities in South America. After enumerating the texts and maps he had used to construct his own map of Quito (discussed below), La Condamine explained in a footnote that Maldonado "had d'Anville compile under his [i.e., Maldonado's] supervision a Spanish map of the province of Quito in four sheets."[7] But d'Anville's mention of cartographic proofs in a letter to La Condamine makes Maldonado's role in this process come suddenly to life: "Maldonado made his corrections on proofs of the engraving," d'Anville wrote, "[and] I am convinced that most of these corrections match my own. Nonetheless, I will examine M. Maldonado's notes."[8] This allusion confirms that printed proofs were produced prior to Maldonado's departure for London in August 1748 and that Maldonado had the opportunity to examine these sheets and make corrections to them. In other words, he had the chance to intervene in the transition from manuscript to print and to mold the geographic image of Quito to his own specifications. D'Anville also examined the proofs and made corrections for the map's engraver, Guillaume Delahaye, as we will see below. But the truly extraordinary evidence these proofs reveal is the series of editorial remarks made by the author himself, in pencil, for corrections to the copperplates.[9]

An analysis of the proof marked up by Maldonado elucidates his role in the revision process; strikingly, however, most of the corrections Maldonado requested were left unchanged in the printed version of the map. This proof is clearly the one perused by Maldonado to which d'Anville referred in his letter: laden with penciled corrections and modifications, its script and grammar have more in common with the language of Lope de Vega than that of Molière (fig. 17; plate 16).[10] While it would be difficult to provide a precise figure for the dozens of editorial annotations that were entirely ignored, a few representative examples reveal the extent of these omissions: on the righthand margin of the proof, Maldonado instructed d'Anville and Delahaye to "delete the cross at la Concepcion" and to "delete the crosses of Loreto and S. Salvador." Despite these instructions, all of the crosses remained. Maldonado also made the suggestion to "place an anchor at S. Salvador," but the anchor was never added (figs. 17–18). "R. Aroana" remained "R. Araema"; "tabaquero" remained "R. Labaquero"; and so on and so forth. In the region of Popayán near the Rio Patia, penciled annotations show that the names of two rivers—the Rio Sn. Jorge and the Rio Huachiconò—were interpolated. Despite Maldonado's attempt to demonstrate this error by writing in the correct names, these rivers retained their cartographic misnomer.

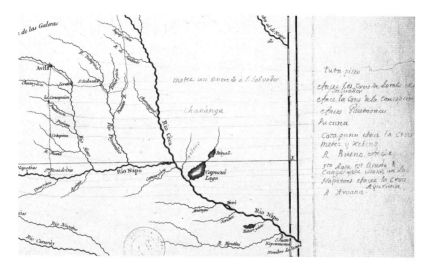

Figure 17. Maldonado's corrections (detail from plate 16). Note the three orthographical variations of the verb "effacer." BNF Ge. D. 10878(2). Reproduced with permission of the Bibliothèque Nationale de France, Paris.

Figure 18. San Salvador without its port. Detail from the *Carta de la Provincia de Quito* (plate 11). Courtesy of the Special Collections Library, University of Michigan.

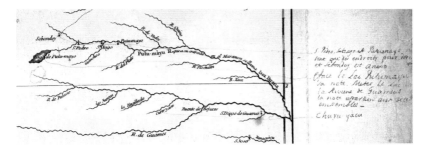

Figure 19. Maldonado's instruction to switch the lakes and maintain the same label for both (detail from plate 16). BNF Ge. D. 10878(2). Reproduced with permission of the Bibliothèque Nationale de France, Paris.

Figure 20. The new "Lake Putumayo" without a label (detail of plate 11). Courtesy of the Special Collections Library, University of Michigan.

These lapses seem to imply that *none* of the corrections requested were made, but other examples demonstrate that in fact some of the corrections were executed, if at times only partially. For example, Maldonado instructed the engraver in one instance to take out the "Lago de Putumayo" and instead place it as an extension of the northern branch of the Guamués River: "[D]elete the Lake Putumayo and its note. Place the Lake of the R. de Guames[;] the note belongs to both together" (fig. 19). Which note corresponds to which lake remains unclear—Maldonado's language is difficult to decipher—and this instruction may have been unclear to Delahaye as well. In the final version of the map, shown in figure 20, the label "Lago de Putu-mayo" was removed entirely, as was the lake, and another lake *without* a label was inserted at the end of the "R. de Guames" [*sic*]. Without a note that was supposed to be added, the lake remained without any textual referent, an annotation that was supposed to refer "to both [lakes] together" (fig. 20).[11]

As these examples indicate, many of Maldonado's changes were not carried out. Despite the fact that the Spanish Crown had commissioned the map and was paying for its production, the requests submitted for editorial changes by Maldonado were not always heeded. Nowhere is this more apparent than in Maldonado's request to remove mineralogical data from the map. His efforts to suppress particular labels on the map likely stemmed from the region's abiding economic resources. Mexico and Peru had always been important loci of mineralogical interest for the Crown, but more recently, in his *Relación histórica*, Antonio de Ulloa had provided a series of detailed descriptions of Quito's silver and gold mines as well.[12] And the Crown was also interested in one of Ulloa's own mineralogical discoveries: platinum. Ulloa made the first report of platinum to the Crown at around this time, commenting that it was "a stone of such resistance that it would not be easy to break"; Maldonado even carried several pieces of platinum with him to London when he visited the Royal Society.[13] The *Carta de la Provincia de Quito* should thus be seen in the context of this new mineralogical data pouring into Spain at a critical moment for the empire's finances; not surprisingly, Maldonado's corrections to the proofs illustrate his attempts to erase this potentially lucrative information from the map. As shown in figure 21, the mountain called "Llanganàte" (near present-day Baños) was represented with a brief explanatory text stating "Snow-capped mountains and gold mine." The proofs to this sheet show in no uncertain terms that Maldonado wished to remove this text and that he had asked d'Anville to erase the mountain and news of its supposed mine: "[E]rase the snow-capped mountains and gold mine. I had asked him to erase it" (fig. 21). The final printed version of the map, however, shows that the mountain and the text that accompanied it were left unchanged (fig. 22). Did Maldonado have concrete information relating to a mine in this region? No document provides a clear answer, but Maldonado certainly knew this region well given its proximity to his native Riobamba. And the Cerro Llanganate, from the time of the Spaniards' arrival in Peru, had gained notoriety as the supposed site of great Inca wealth. Another rumor had circulated that Llanganate was the tomb of Atahuallpa, son of the great Inca Huayna Capac. The provenance of Llanganate's mineral wealth on the *Carta de la Provincia de Quito* is not known, but Maldonado gave explicit instructions

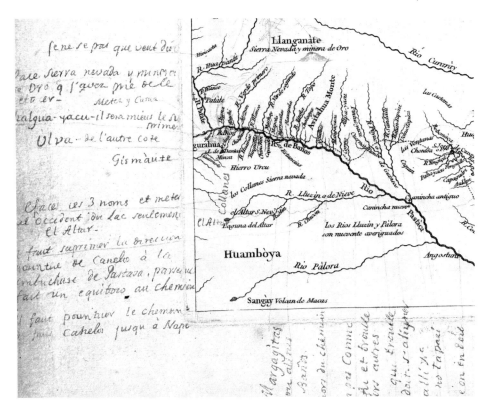

Figure 21. Maldonado's unheeded instructions to erase the gold mine at Llanganate: "Erase Sierra nevada and gold mine that I had asked him to erase" (detail from plate 16). BNF Ge. D. 10878(2). Reproduced with permission of the Bibliothèque Nationale de France, Paris.

to have the mine's label deleted. Without regard to who may have placed it there, the label was not removed.[14]

If these changes were so important for Maldonado (and, we can assume, for the Spanish Crown as well), why was the map not produced in Spain? Most likely, the technical skill to produce such a map was still lacking in the Spanish capital. A broad interest in renovating Spanish cartographical practice was only just beginning to emerge following the return of naval officers Antonio de Ulloa and Jorge Juan to Madrid in 1746. This period in which Ulloa and Juan arrived back in Spain was a period of political transition under Philip V's youngest and only surviving son, Fernando VI, who had just recently been installed on the Spanish throne. But it was also a time of ferment in the accumulation of technical knowledge, exemplified by Ulloa's travels throughout Europe from 1746 to

Figure 22. Llanganate in the printed version with its intact "Sierra Nevada y minera de Oro" (detail of plate 11). Courtesy of the Special Collections Library, University of Michigan.

1751 to collect information that would allow the Crown to establish a series of Spanish "scientific" institutions based on similar models found in other European capitals.[15] Ferdinand VI's plans included a new emphasis on "the elaboration of geographic maps, under the direction of Don Jorge Juan and Don Antonio Ulloa, with engravers from outside Spain or with Spaniards sent to Paris to learn the art."[16] Tomás López de Vargas Machuca and Juan de la Cruz Cano y Olmedilla were sent to Paris in 1752 for this very purpose, two members of a younger generation that was meant to make up for Spain's inadequacy in this regard.[17]

But even contemporaries were puzzled by Spain's unwillingness to carry out the production of this map on its own. La Condamine wrote:

I have no idea why the work of my deceased friend was not published in Madrid. I cannot see that politics, or reason of state, could have had a part to play in this mystery. The jungles of the Marañon do not contain mines or botanical specimens (*Generos*) that could motivate the greed of foreign nations. And the Portuguese, masters of the river and its shores from its mouth to more than five hundred leagues [downstream], have an overabundance of deserted lands that they could not populate or cultivate in many centuries' time.[18]

While it may have been true that the Amazon was devoid of mining areas, a mere glance at the *Carta de la Provincia de Quito* would have revealed the folly of not attributing any political value whatsoever to this land of such diverse and potentially lucrative lands. La Condamine knew the map's inherent value, which is why he and d'Anville conspired to keep its existence largely hidden as they raced to complete it; they wanted no competitors on either side of the Pyrenees to be aware of their surreptitious scramble to correct and print this cartographic treasure.[19] Following Maldonado's death, they no longer needed to heed a Spanish American Creole's corrections on a map for which they had become the responsible parties. In the end, the map was not edited in Madrid, where the Crown could attend to its mineralogical concerns. It was edited in Paris, where a host of engravers, intaglio artisans, geographers, and technicians had their own interventions to make on the surface of a copperplate. The final form of the map, evidenced by the insistent meddling of numerous correctors and cartographers, was a negotiated settlement between parties who had different if not conflicting agendas. And the map's Spanish American author was by that time already resting eternally in his London grave.

Quito's Parisian Editors, *post mortem*

Maldonado's unexplained death in November 1748 brought his scientific career to an abrupt end. A mere two weeks after his nomination to the Royal Society, Maldonado died of what President Martin Folkes called "a burning fever and chest congestion," shocking the membership and confounding Richard Mead and other London physicians called in to assist.[20] While his physical body was being laid to rest in Piccadilly, the corpus of his scientific work deposited on the other side of the Channel took on a life of its own. Eighteen months were to elapse between the date of Maldonado's death and the publication of the map in 1750, which allowed sufficient time for others to elaborate on and reconfigure his geographic work.

Maldonado had left his notes and sketches with La Condamine in Paris, and he and d'Anville were able to reconstitute Maldonado's cartographic vision by compiling these materials with information collected by Jesuits, members of the Quito expedition, and other sundry sources. Based on the editorial processes examined in the previous section, Maldonado's role in the correction of the *Carta de la Provincia de Quito* was limited; the map proclaimed by the elegant cartouche to be a "posthumous work of Don Pedro Maldonado" was engraved and printed without many of its author's corrections. But the activities of the editor and engraver, Jean-Baptiste Bourguignon d'Anville and Guillaume Delahaye, respectively, demonstrate that the process by which cartographic information was transferred from manuscript travel narrative to proof, from proof to copperplate, and from copperplate back to proof again for additional corrections, often yielded unexpected results. D'Anville accorded tremendous license to Delahaye, including permission to add or remove both graphic and textual elements from the map. He also gave him the latitude to decide the shape and content of numerous landscape features. Inadvertent omissions on the part of these two editors were thus not uncommon, showing how under the quotidian pressures of the printing house certain corrections were simply left unexecuted, their presence on the ephemeral proof left as the only indication of the shape that Quito may have taken but in the end did not. The cartographic proof also served as a tool for settling scores and proffering prestige through the textual citation of sources. Taken as a whole, these elements argue against seeing the *Carta de la Provincia de Quito* as a reification of Maldonado's geographic vision. All his untimely death did, ironically, was to shift these editorial processes into a higher gear.

In the editorial structure that undergirded the map's production, d'Anville was chief architect. Having accepted La Condamine's invitation to compose this map, he also accepted the ultimate responsibility for the map's production, culling both oral and written material and combining them to provide a unified cartographic perspective. During the previous five years, he had carried out two other major cartographic efforts related to South America: La Condamine's acclaimed *Carte du cours du Maragnon* (1745; plate 5) and the map entitled *L'Amérique méridionale* (1748).[21] As with these maps, d'Anville was involved in almost every detail of the Quito map's construction, attested by the letter requesting further information from La Condamine:

For the portion of Payta and Piurà, it is necessary to correct the latitude of Amotapè and some other local circumstances between Tumbez and Payta, according to the narrative of the voyage to Lima as provided by M. de Ulloa, without which [Ulloa] might seize upon this defect in order to slow down production of the work as a whole. M. de la Condamine would do me a favor by sending me a small extract of the voyage for this portion only; comparing it with the map, he will certainly see the need for making this correction.[22]

These corrections of the longitude of Amotape and the route between Tumbez and Payta offer vivid evidence of the direct relationship between textual description through narrative itineraries and cartographic inscription in the process of latitudinal corrections. Travel accounts, a linear and diachronic narrative form, provided the basis for much of the geographic information contained within the Quito map.[23]

But the use of specific narratives also occasionally raised issues as to how to cite or acknowledge their use, especially in a map woven together by overlapping and hence competing textual descriptions. The citation of certain narratives rather than others had social implications as well, since they could provide honor or prestige by acknowledging an individual's contributions in print. In this same letter to La Condamine, d'Anville explicitly mentioned the narrative of Miguel de Santisteban, a Spanish American naval officer turned colonial administrator whose many professional incarnations included a stint as *corregidor* of Cuzco, a peripatetic naturalist in the Andes, a judge (*juez de residencia*), and the superintendent of the royal mint in Santa Fé de Bogotá.[24] With reference to Santisteban's narrative of a journey from Lima to Caracas, d'Anville wrote that he had "no problem at all naming M. de St. Isteban [on the map]; his itinerary was such an important ally to me that it would be unfortunate not to seize upon this occasion to do him honor and to bring him in as an equal with another."[25] This "other" to whom d'Anville referred was Pierre Bouguer, whose "derrotero" or route description he had used to mark the path between Ibarra and Popayán (fig. 23; plate 17). It had been La Condamine who suggested that Santisteban's name accompany Bouguer's along the route from Pasto to Popayán, precisely at the moment in which the two academicians were locked in a fierce battle over their respective publications within the Academy of Sciences. Adjudicated primarily in the *Comité de librairie*, this feud had grown out of a dispute over La Condamine's and Bouguer's academic texts and occupied the Academy for several years, to the tedium of many.[26] Thus, La Condamine's suggestion that Santisteban's name be placed alongside Bouguer's was not merely innocuous recognition of his friend; it was vengeance against Bouguer. The addition of Santisteban's name turned the Quito map into a battleground for settling scores. By augmenting the prestige of Santisteban, he thereby diminished the prestige of Bouguer in a zero-sum game of apportioned recognition. D'Anville seems to have taken La Condamine's side: rather than naming Bouguer in his letter, he

Figure 23. Proof describing Bouguer's itinerary from Ibarra to Popayán, prior to La Condamine's suggestion to add Miguel de Santisteban. BNF Ge. D. 10878 (2)D. Reproduced with permission of the Bibliothèque Nationale de France, Paris.

Figure 24. Corrected proof with d'Anville's pencil instructions to add Santisteban: "y al de D. Miguel de St. Istevan." BNF Ge. D. 10878 (2)C. Reproduced with permission of the Bibliothèque Nationale de France, Paris.

called him somewhat disparagingly "un autre" and requested that Delahaye insert Santisteban's name alongside Bouguer on the northeastern corner of sheet 2, which Delahaye subsequently did (figs. 23, 24, 25).

The person who incised Santisteban's name on the plate, Guillaume Delahaye, was the map's engraver, responsible for physically correcting mistakes on the copperplates themselves.[27] D'Anville's generic instructions to Delahaye are emblazoned in large manuscript lettering at the

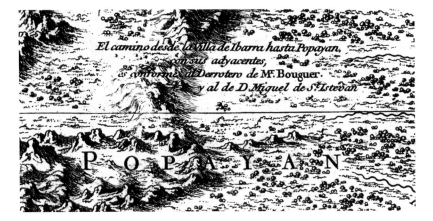

Figure 25. The printed version of the text describing the route to Popayán, which included references to both Santisteban's and Bouguer's texts. Note the italicization of "D. Miguel de St. Istevan," added later, as compared with Bouguer's name, which was incised at the outset in Roman script. Detail from plate 11. Courtesy of the Geography and Map Division, Library of Congress, Washington, D.C.

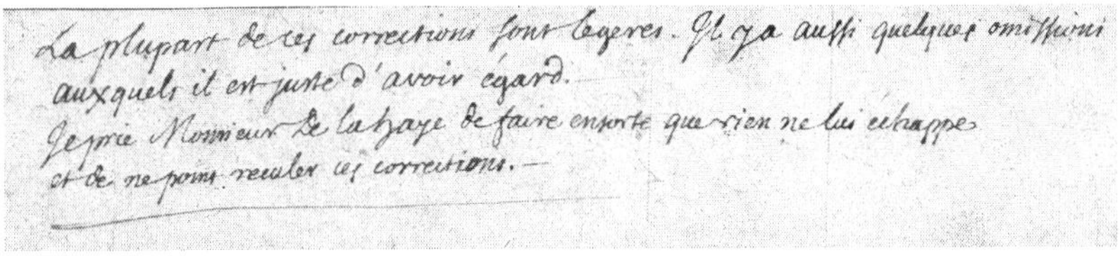
La plupart de ces corrections sont legeres. Il y a aussi quelques omissions auxquels il est juste d'avoir egard.
Je prie Monsieur De lahaye de faire ensorte que rien ne lui echappe et de ne point reculer ces corrections.—

Figure 26. D'Anville's handwritten instructions to Guillaume Delahaye, ensuring that Delahaye did not leave important changes incomplete. BNF Ge. D. 10878 (1)A. Detail from plate 15. Reproduced with permission of the Bibliothèque Nationale de France, Paris.

top of the proof of the map's first sheet (fig. 26; plate 15). Not to be easily missed, d'Anville's orders are authoritative and explicit but composed in a language of courtesy and respect:

> The majority of these corrections are minor (*légères*). There are also some omissions which it is important to take into account. I ask Monsieur De la Haye to make sure that nothing escapes his attention and to accomplish these corrections without delay.[28]

D'Anville gave his order to remedy errors and omissions with a firm but gentle hand, and he included a warning against further lapses or blunders. This handwritten note was a form of editorial shorthand, a behind-the-scenes message never meant for public consumption that stood in for

Figure 27. D'Anville's instructions to replace Cunzacoto with Canzacoto: "*Can* et non pas *Cun.*" BNF Ge. D. 10878 (1)A. Reproduced with permission of the Bibliothèque Nationale de France, Paris.

a more formal written correspondence between two collaborators, such as that between d'Anville and La Condamine. The exhortation not to let these corrections "escape" his attention was a tacit acknowledgement that in a world of minute details, instructions to make editorial corrections frequently *did* pass unnoticed and errors often were left unchanged. In his 1794 treatise on engraving, Delahaye admitted that because engravers were paid for the quantity and not the quality of their work, mistakes were inevitably made. D'Anville's note to "accomplish these corrections without delay" was a further indication of the out-of-the-ordinary circumstances of this particular cartographic project and of the pressures coming from a political sphere over which they had little direct control. The diplomatic world outside the printing house impinged, if sometimes subtly, on what would otherwise seem to be isolated technical processes of cartographic correction.[29]

In addition to this overarching set of instructions to Delahaye, there were hundreds of handwritten remarks huddled together along the borders and margins of the cartographic proof. Some instructions were objective: they included comments about orthography ("*Can* and not *Cun*" for *Cunzacoto*, fig. 27) and forgotten rivers ("Course of a forgotten river," fig. 28) as well as instructions for shading certain features ("this lake should be shaded," fig. 29) and adding in forgotten cities or churches. Hierarchies among civic and ecclesiastical towns were handled by adjusting the size of characters ("el Puyal . . . forgotten word . . . small characters," fig. 30), although no legend indicated what this hierarchy was or what symbols were meant to be used.

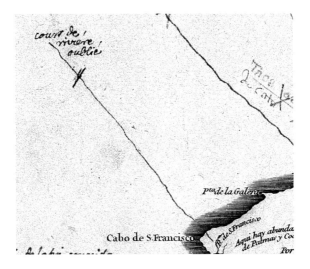

Figure 28. Instructions to include a forgotten river near the Cape of San Francisco: "cours de riviere oublié." BNF Ge. D. 10878 (1)A. Reproduced with permission of the Bibliothèque Nationale de France, Paris.

Figure 29. Fill in the blank lake: d'Anville directs Delahaye to shade a lake that had been left unshaded. BNF Ge. D. 10878 (1)A. Reproduced with permission of the Bibliothèque Nationale de France, Paris.

Despite the clarity of these instructions, d'Anville seems to have given the engraver a great deal of interpretive latitude, which speaks to the uncertainty and sometimes haphazard nature of the printing and engraving process: "[S]ee if the mountain has been engraved," d'Anville wrote, "and supply one if needed." Delahaye not only carried out many of the corrections as requested by d'Anville; he had a certain editorial license as well, furnishing details of a distant land in his own familiar graphic idiom. The shape of mountains and lakes, for instance, was left largely to the engraver's discretion, as was the insertion of two small hills near the Garrapatas River that d'Anville requested from Delahaye with a terse "two small

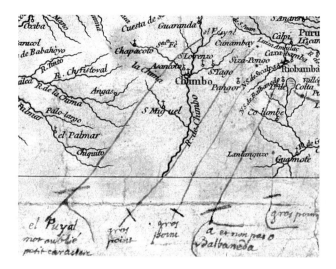

Figure 30. The addition of a forgotten place-name included d'Anville's instructions for the kind of characters to be used: "el Puyal . . . mot oublié . . . petit caractere." BNF Ge. D. 10878 (1)A. Reproduced with permission of the Bibliothèque Nationale de France, Paris.

mountains" and by connecting this text to their desired location with a short line. On more significant geographic features, such as the great Andean volcano Cotopaxi just south of Quito, d'Anville had recourse to an eyewitness—La Condamine—who reasserted control over the graphic modifications and offered specific instructions as to how to alter the drawing:

> The line of Cotopaxi C is too dark; it should be softened and lightened, the upper portion all white and the bottom in dental crevasses . . . X soften this middle mark X make a line to the left to mark the side of the large mountain beneath the writing quelandana make some marks (*tailles*) to link the large mountain to the smaller ones at right . . . a few lines in order to strengthen the chain of mountains in this place and to make a continuation.[30]

Softening here and strengthening there, the dramatic volcanic landscape of the central Andes was burned into the copperplate with careful attention to hues of light and dark, tones that could only be expressed through the skillful wielding of the engraving instrument. The proof corrections demonstrate a back-and-forth process involving multiple cartographic states to produce a final product that was a mixture of artistic license

Figure 31. Changes to the volcano Cotopaxi in La Condamine's hand. BNF
Ge. D. 10878 (2)A. Reproduced with permission of the Bibliothèque
Nationale de France, Paris.

and technical mastery, providing Quito with a mountainous backbone
worthy of its volcanic reputation (fig. 31).

Another aspect of correction involved the insertion of place-names
into regions already inhabited by other topographic, toponymic, or landscape features. This was a technique bemoaned by Delahaye, who felt
that it was the fault of bad engravers who "do not know how to arrange
the words, which are commonly very badly placed, running across the
woods, mountains, etc."[31] In order to place the "Peñas de los Serafines"

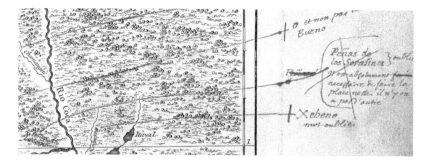

Figure 32. D'Anville's note to insert the "Peñas de los Serafines." BNF Ge. D. 10878 (2)D. Reproduced with permission of the Bibliothèque Nationale de France, Paris.

near the site of the intersection of the Rio Coca and the Rio Napo, however, d'Anville insisted that Delahaye clear away the forest and make room for this text: "[I]t is absolutely necessary to make the place clean. There is no other [space]" (fig. 32). Since there was no other available location in which the Peñas de los Serafines could be inserted, it had to be engraved in an empty space at this particular location, the expediencies of the engraving process overriding any concern for the "true" geographic location of the Peñas. Delahaye, in turn, was ultimately forced to scrape the copperplate clean in that place, removing the landscape features already present and transforming the territory from a forested riverbank to an open plain—all to accommodate the insertion of a place-name (fig. 33).

Delahaye completed most of d'Anville's corrections as requested. When made, they were most often checked off with an "x" or a slash mark to indicate that d'Anville's changes had in fact been heeded. However, certain corrections were *not* made and may have escaped Delahaye's attention. J. B. Harley has made the important argument that "[w]hat is absent from maps is as much a proper field for enquiry as that which is present," and this observation can be expanded to include corrections that were forgotten as well.[32] For example, d'Anville pointed out two omissions to Delahaye located just beneath the equatorial line: the "Passo de Ressavala" and "S. Miguel" (fig. 34). On the final map, S. Miguel appears clearly but there is no sign of the Passo de Ressavala. Indeed, the only trace of Ressavala's exploratory activities in this region is the "confused remnants of a road that Ressavala opened here." There is no slash mark over the editorial notation, Delahaye's standard shorthand for correction made. This lapse, intentional or careless, nonetheless affected the content of the printed map and may have inadvertently given

Figure 33. The new "home" of the "Peñas de los Serafines" (detail of plate 11). Courtesy of the Special Collections Library, University of Michigan.

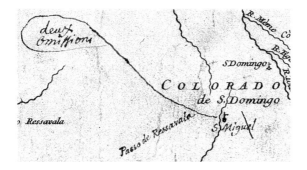

Figure 34. A note to include the "Passo de Ressavala," an omission of a feature that d'Anville corrected here but nevertheless remained absent from the printed version of the map. BNF Ge. D. 10878 (1)A. Reproduced with permission of the Bibliothèque Nationale de France, Paris.

added honor to Maldonado's explorations by silencing the activities of another explorer in the same region.[33]

While explorers' itineraries such as those of Ressavala and Maldonado were only occasionally inscribed within the limits of the *Carta de la Provincia de Quito*, graphic representations of Quito's history and its administrative districts were more frequent. Sheet 1 (plate 10) not only showed administrative divisions ("Lindero de la Governacion de Esmeraldas y del Corregimiento de Tacunga") and ecclesiastical markers through the use

Figure 35. A threatening notice about rebellious
Indians. Detail of plate 10. Courtesy of the Special
Collections Library, University of Michigan.

of text and symbols but also provided knowledge about particular regions
("From here on up, this river is poorly known"). Occasionally, the text
clued the unwary to potential dangers ("In this area there is a nation of
Gentiles that is barely known") or gave colonial administrators warnings
of possible unrest (fig. 35; "In this region lives the nation of the Malaguas
that rose up in rebellion some time ago").[34] Historical and ethnohistorical data were fused by marking early cities, such as the "Ancient site of
the Esmeraldeños," and by indicating the presence of native peoples and
other groups who resided within particular regions ("Mangachis name
of the Sambos of Esmeraldas and Cabo Passo that live [out] in the jungle
[retirados]").

But while it may appear that contemporary and historical annotations
gave equal weight to European and indigenous sources, the printed map
suggests that indigenous history was often ignored or suppressed by
compilers and correctors. In one such case, Incan history was exchanged
for administrative and political boundaries. The "derrotero" of Miguel
de Santisteban, one of the map's key sources, presented Incan history
as an essential companion to Quito's geography. Describing the region
near the city of Pasto, in what is today southern Colombia, Santisteban
wrote that

[t]he rio de Mayo is known in the history of Peru because it was the boundary of the
conquests made by Inca Huayna Capac, father of Huascar Inca, last possessor of this
vast empire and boundary at the same time of the Reino de Quito.[35]

For Santisteban, the early sixteenth-century conflict between Huascar Inca
and Atahuallpa, whom Pizarro vanquished at Cajamarca during the Spanish conquest of Peru, was inseparable from the geographical description

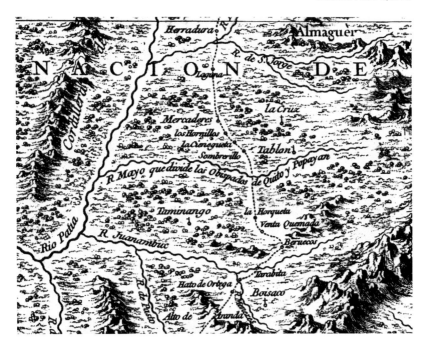

Figure 36. Ecclesiastical boundaries overwriting Incan history: "R. Mayo, which divides the bishoprics of Quito and Popayan." This illustration shows the route from Pasto to Popayán. D'Anville used Santisteban's travel narrative to inscribe place-names along this road. Courtesy of the Geography and Map Division, Library of Congress, Washington, D.C.

of the region he provided two centuries later. But despite the clarity of this historical information, which was included in a section of Santisteban's account that we know d'Anville read, the printed version of the map read simply: "R. Mayo, which divides the bishoprics of Quito and Popayan" (fig. 36). Here, Incan history was papered over by an ecclesiastical division.

The suppression of indigenous history within the map occurred alongside the glorification of one of Maldonado's own projects: the road he established between Quito and the Esmeraldas province. This achievement, whose glory has been echoed by South American historians and geographers since the eighteenth century, has come to be debunked by recent ethnohistorical research in the region near Quito. In Frank Salomon's words, Maldonado's project inscribed "history upon ahistorical space" by representing the indigenous settlements in the mountainous region west of Quito as populations that were poorly served by contemporary transportation methods and that could be linked through the construction of a carriage road. Opposition and indigenous resistance to this project stemmed primarily from the poor treatment of the Indians at the hands of Spanish officers under Maldonado's charge. The artistic conventions employed in the cartouche, and a cartographic methodology that erased vestiges of prior (Incan) civilizations and contemporary (Yumbo) transportation throughways replaced one version of history and geography with another. The printed version may have been more palatable for agents of the Spanish Crown interested in Maldonado's road-building projects.[36]

With all of these sources and changes, then, who was responsible for the map's final form? The exchanging, inserting, correcting, erasing, lightening, darkening, and modifying of cartographic features formed a process within which competing claims and conflicting authorial intentions jockeyed for preeminence in the scratches and grooves of a copperplate proof. But the precise motive for each correction is elusive. Every change, completed or not, was based upon variegated layers of reconnaissance on the spot as well as expedition travel logs of individual observers and annotations and corrections by d'Anville and Delahaye. These two editors made decisions regarding topographical, historical, and ethnographic features that significantly affected the way in which the landscape of Quito appeared to European eyes. In some respects, their editorial hegemony was supreme. One of the most fascinating textual commentaries on these proofs finds d'Anville explaining to Delahaye that "we must complete this word which we doubted by adding QUES. If M. de la Haye finds some obstacle (*quelque embarras*) on the plates because of the ornamental details (*fanfreluches*) we have added, he should do me the favor of getting rid of some of [these ornamental details]" (fig. 37). Such a comment supports the idea that d'Anville and his engraver—both at the pinnacle of their respective careers—served as mediators between a geographical space about which they had no direct experience and the copperplate, which was the medium of their métier and the yardstick of their fame and glory. "Doubt" and ornamental "obstacles" were some of the many motivating forces behind the cartographic correction of Quito. The precise form of these *fanfreluches* was never explicitly revealed and remains in a murky sociographic language. But this particular textual interaction between d'Anville and Delahaye highlights the ultimate "instability" of the cartographic image: a document that numerous actors with differing interpretations could attempt to bend to their own individual wills.[37]

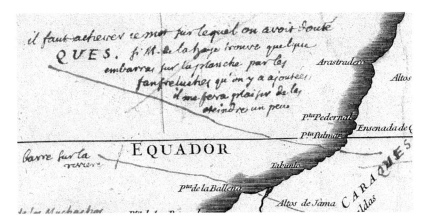

Figure 37. D'Anville's note to Delahaye warning him about the "ornamental details" (*fanfreluches*) they had added. BNF Ge. D. 10878 (1)A. Reproduced with permission of the Bibliothèque Nationale de France, Paris.

The Province after Printing

On April 27, 1750, La Condamine wrote to the Spanish secretary of foreign affairs to describe a special shipment that the minister would soon be receiving, and which he was certain would bring him much pleasure:

In carrying out the confidence of my deceased friend Don Pedro Maldonado . . . and following through on the intention which on many occasions he declared to me, I submit to Your Excellency the first copy that has been produced of the *Carta de la Provincia de Quito y de sus adjacentes*, whose execution [Maldonado] left in my care at the time of his departure for London, where he passed away in November of 1748.[38]

Following Maldonado's death, La Condamine had been instructed by the Comte d'Argenson, then secretary of foreign affairs in France, to return the completed map to Madrid along with the manuscript materials Maldonado had employed to construct it. La Condamine carried out this task according to the wishes of both Aragonese general Francisco Pignatelli (1687–1751), recently installed as Spanish ambassador to France, and the Spanish secretary of foreign affairs, José de Carvajal y Lancaster (1698–1754). With appropriate pomp and formality, La Condamine outlined in this letter to Carvajal y Lancaster the process by which he had published the map of his deceased friend. He also described the lengths to which he had gone to satisfy agents of the Spanish Crown, who saw themselves as the map's rightful owner and who not only expected to receive copies of the map but also the materials used to produce it. In this letter, La Condamine also included several details regarding the printing and distribution of eighteenth-century maps. He made reference to the weekly coach from Bayonne by which he shipped six additional copies of the map, including four loose copies—"not having had enough time for more"—and two special copies that were "bound and mounted," one destined for the King, emblazoned with his arms, and the other for Carvajal y Lancaster himself. La Condamine also discussed his lack of time for "perfecting the map," complaining that he had received the orders to submit the copperplates, documents, and Maldonado's personal notes to the Spanish ambassador on the same day that he had received "from the hands of the engraver" the sheet containing the frontispiece and the title of the map. But the most revealing portion of La Condamine's letter was his own detailed description of his role in the editing and correcting of the map:

I had them correct the burin errors that I could see in comparing the engraved map with the loose originals in my own hand, [maps] that I had loaned to the deceased to make his general map of the Province. This put [the map] into a much better state, given the short time frame I was allowed at the request of Don Ignacio de Luzán [secretary to the Spanish ambassador in France].[39]

The correction process of the Quito map, as we learned from d'Anville's urgent pleas to his collaborators, took place within a narrow temporal window, essentially two weeks' time. But what La Condamine emphasized in this letter was not only the expedited pace needed to complete the map but also Maldonado's reliance on La Condamine's "original" maps "in his own hand," materials he claims to have lent Maldonado in order for him to complete his larger work. Despite the insistence that his "greatest care was to ensure that the honor of the work be reserved for the memory of my friend, so that no one else could take advantage of his work," La Condamine acted through these subtle clues to imply his own stake in the honor accorded to Maldonado's map. Despite Maldonado's likely use of La Condamine's materials to settle issues of scale and proportion on the "general" map, La Condamine consistently repeated these claims to proprietorship in the years and decades that followed.

But the version of the map that left Bayonne for Spain was not the only edition of the map that was produced in Paris. There were actually two editions of the map that were made there, and these distinct versions

Figure 38. Version of cartouche prepared for the Spanish Court.
Note the slightly different weight of the characters in this image
and in figure 39. Detail from plate 10. Courtesy of the
Geography and Map Division, Library of Congress, Washington,
D.C.

demonstrate that imperial interests and politics were never entirely absent from the map's elaboration. While the topographical features of the editions remained identical, separate versions of the cartouche reveal each edition's distinct cultural and political stakes. The first was published with the text "Published by order and at the expense of His Majesty," without any indication of the place of publication or the cartographer responsible for its creation (fig. 38).[40] This was the version sent by La Condamine to Carvajal y Lancaster and the king of Spain. The second version contained an entirely different text: "By S. D'Anville Geographer of H. Christ. Maj. of the Imp. Acad. of Petersburg. Published by D.C.D.L.C. Paris MDCCL" (fig. 39). Both were adorned with Spanish text. But the first version was clearly designed to appeal to the Spanish Court, since the text covered over any and all elements relating to the conditions of its production in France. The second version not only erased any reference to the Spanish king but also recorded the names of the two key protagonists of the map's publication: the geographer d'Anville and the person responsible for coordinating the production effort, D.C.D.L.C.

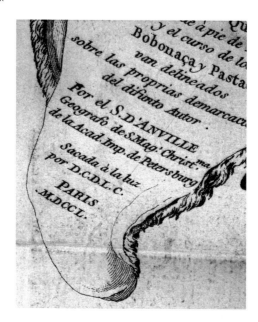

Figure 39. Version of the Quito map that La Condamine surreptitiously arranged to be printed and distributed without the knowledge of the Spaniards. La Condamine's name appears here in cipher. Courtesy of the Special Collections Library, University of Michigan.

(an acronym for Don Carlos de La Condamine). This latter edition also included a notation of its place of publication, Paris. These seemingly minor editorial changes were significant enough to warrant a change in the copperplates of the original map. They demonstrate the battle for authorial recognition within the cartographic text and indicate once again the malleability of the reproducible printed map.

As it turned out, La Condamine had dissimulated and delayed in order to produce this additional version of the map, copies of which he sent to Maldonado's relatives in Quito, the Royal Society in London, and the Academy of Sciences in Paris. The Spanish literary theorist Ignacio de Luzán (1702–54), then in Paris as secretary to the Spanish ambassador in France, had pressured La Condamine to turn in the materials as quickly as possible, and the French academician resisted by all means available to him. La Condamine had stalled in order to delay the eventual release of the map and the return of those materials used to create it, but in the end he managed to put aside several copies of the map for his own use. "I received another order to turn everything in to the Ambassador, which I did, and he reimbursed me for the cost of the plates," La Condamine later wrote. "[But] secretly I kept several copies, which I sent to [Maldonado's] family, [and] two others to the Royal Society in London and the Royal Academy of Sciences in Paris."[41] As we shall see, La Condamine used these copies of the plates to create a revised version of the Quito map, one that he eventually put forward to adorn his own *Journal du voyage fait par ordre du roi*.[42]

When at last it was printed and made its way into public circulation, the *Carta de la Provincia de Quito* was lauded as a watershed in the cartographic history of South America, and Maldonado was given special credit for its compilation and execution. In an article composed for Diderot and d'Alembert's *Encyclopédie*, Robert de Vaugondy gave special mention to Maldonado as an exception to the broader and (to Vaugondy) largely disappointing eighteenth-century practice of Spanish geography:

Another work to which we should pay close attention is the map of the province of Quito, surveyed (*levée*) by D. Pedre [*sic*] Maldonado, governor of the province of las Esmeraldas in America. This map in four sheets, whose plates are held by the king of Spain, was prepared (*dressée*) by M. d'Anville of the Académie Royale des Belles Lettres, also secretary to M. le duc d'Orléans. [The map] is the result of operations carried out in concert by the French and Spanish academicians to determine the true shape of the Earth.[43]

Vaugondy was doubtless well informed about the provenance and ultimate destination of the map, as his residence on the Quai d'Horloge was just across the river from d'Anville's atelier in the Galeries du Louvre and the two were in frequent contact. Bringing attention to two of the actors involved in the production of the Quito map, Vaugondy drew a common distinction between what it meant to "lever" and "dresser" a map, the former being the dominion of Maldonado (large-scale topographical maps requiring firsthand surveys) and the latter that of d'Anville (the preparation or compilation of small-scale geographical maps, which usually took place far from the location they were describing). The others mentioned in this brief account of the map's genesis were the participants in the expedition to Quito, since the map was a "result of operations carried out in concert by the Spanish and French academicians." La Condamine was included in this penumbra of honor as well, not only as one of the academicians who helped to determine the shape of the Earth but also as a catalyst for the execution of the map. Vaugondy completed his article by discussing the orders for the forthcoming "map of the [Spanish] kingdom," which was being planned and implemented by Jorge Juan and Antonio de Ulloa back in Madrid, a further indication of the close nexus between the enthusiastic reception of Maldonado's map and the efforts of the Spanish state to provide institutional support for their own geographical projects at home.[44]

The *Carta de la Provincia de Quito* not only trundled its way over the Pyrenees; it traversed the English Channel as well. At the Royal Society in London, the natural philosopher Martin Folkes (1690–1754) offered the most extended reflection on the importance of this map in revealing the contours of a land on the margins of European geographic consciousness. Folkes's text offered a rather bare and simple account of the ideas elicited by the *Carta de la Provincia de Quito*. His description essentially reiterated in narrative form many of the textual components of the Quito map:

It appears in general that the northern part of the kingdom of Peru is by far less cultivated, and more wild and uninhabited, than we for the most part imagine here in Europe, the whole country peopled by the Spaniards consists of little more than these very exalted vales, if one may call them so, which lie between the 2 chief ridges of the Cordillera Mountains, in which the Citys [*sic*] of Quito and Cuenca lie with several other lesser ones well peopled with Europeans and their descendants.[45]

This initial commentary by Folkes on the apparent absence of viable populations in the Andean hinterland innocently confirms the potency of graphic emptiness within the borders of a map. The description of the "wild" lands of the north and the "exalted vales" of European dominion imply a well-patrolled boundary between wilderness and civilization, a common trope in eighteenth-century texts: on one side, wild lands are portrayed as threatening and dangerous; on the other, majestic valleys, episcopal cities, and ecclesiastical boundaries represent the benefits of civilization.[46]

For Folkes, the Andean landscape as revealed through the *Carta de la Provincia de Quito* was a savage territory: daunting in its expanse and abundant in its gifts. His narrative evoked Maldonado's excursions into the interior as intrepid pursuits after new tribes and unknown lands, discoveries that were then communicated with transparency and ease to a European public overseas:

But all the rest of the Provinces . . . seem to be quite overgrown with immense forests, in which are however some scattered habitations of Indians hardly known to the Spanish inhabitants . . . at a considerable distance again to the Eastward of the same River, it is noted that here is an unknown habitation or *Poblacion*, which Governor Maldonado

Figure 40. Original label describing an unknown settlement that Maldonado had seen from Mount Tortolas. BNF Ge. D. 10878 (1)A. Reproduced with permission of the Bibliothèque Nationale de France, Paris.

discovered as he came down from the Mountains of Tortolas, and was searching the rivers thereabout in 1740 . . . many of the remarks he then made, I apprehend to be expressed in his map, particularly by the figures of the small anchors which I take to denote the places where he begun [sic] or ended his navigations, upon the several rivers near which they are placed.[47]

Folkes's description corresponded to a view that showed Maldonado as an active participant in the pacification of the Andean provinces. As we recall, the cartouche was conceived and executed to show Maldonado engaging directly with the land and its inhabitants: giving orders, directing traffic, provisioning tools and instruments, constructing shelters. Folkes also made the implicit observation that the small anchors placed throughout the map were direct allusions to the colony-building activities as represented in the cartouche, exploration that took place with Maldonado at the helm.

What Folkes did not know was that changes had been made to the proofs that brought Maldonado's active and purposeful role to the fore. The citation Folkes referred to highlighting Maldonado's adventurous search for the river had originally read, "Unknown settlement that Gov.r Maldonado saw from Mt. Tortolas in the year 1740" (fig. 40). Only afterward was the "exploring those Rivers" (*explorando estos Rios*) portion

Figure 41. Making Maldonado active: the label about the unknown settlement with "explorando estos Rios" added. Note the slightly different weight of the added characters. Courtesy of the Special Collections Library, University of Michigan.

of the text added. This additional participle, upon closer inspection, was printed in a lighter manner than its accompanying commentary (fig. 41). By adjoining a gerund and attaching these observations to the exploration of rivers, the editor emphasized the intrepid aspect of Maldonado's excursion to the top of Mount Tortolas. The identity of this editor is unknown, but the impact of these changes on the way the map was interpreted could not be more direct. Folkes's comments reflect the assumption that observations made in the field were transformed in a direct and unmitigated manner into features "expressed in his map." His observations also underline the range of heuristic possibilities represented by graphic icons—in this case, anchors—given that no legend was present to guide the reader of the map in interpreting them.

While Folkes's analysis before the Royal Society confirms the impact of some of the editorial operations used in the construction of the map, it was in a transatlantic context that the map had its widest significance. The broader legacy of the *Carta de la Provincia de Quito* enriched the annals of European academies far less than those who used Maldonado's map in situ, guiding themselves through the physical and historical landscapes that were inscribed on the map by a native of Riobamba. In the eighteenth century, then-governor of Maynas Francisco de Requena insisted on the map's historical importance for the eastern portion of the Province and lauded the "indefatigable zeal with which [Maldonado] acquired reliable and interesting information about these regions."[48] The exiled Jesuit Juan de Velasco, a Quito native and author of the renowned *Historia del Reino de Quito*, praised the western portion of the map, which included the province of Esmeraldas. He wrote explicitly about the importance of the road between Quito and the port on the Rio de las Esmeraldas, as well as Maldonado's larger project to connect the inland regions of Quito with the provinces up and down the Pacific coast. Despite its origins in a European printing house and the multiple authors and editors that contributed to its completion, the *Carta de la Provincia de Quito* came to be known as "Maldonado's map" and was explicitly associated with a territorial vision that configured this peripheral province as a strategic center, presented through the enlightened eyes of a native son.

Because it testified to the territorial unity of a geographic space as measured and interpreted by one of colonial Quito's cultural elites, Maldonado's map also came to be seen as an important symbol of Ecuadorian nationalism avant la lettre. The appropriation of Maldonado's work as a protonational symbol has its roots in the comments of such eighteenth-century figures as Don Antonio de Alcedo y Herrera, Francisco José de Caldas, and Eugenio de Santa Cruz y Espejo.[49] In a speech to members of the Escuela de la Concordia, Santa Cruz y Espejo (1747–95), a noted journalist and political firebrand, transformed Maldonado into the ultimate fusion of scientific genius and literary master: "Here is one of those rare and sublime souls who holds a compass in one hand and an artist's brush in the other; that is to say, a profoundly intelligent savant in geography and geometry and a gifted writer of History."[50] Notwithstanding the heroic tone of protonationalism, this image of Maldonado with a compass in one hand and a pen in the other evokes a fluid passage between geographical reconnaissance and the publication of its results that is belied by the map's history of multiple layers and interventions. The *Carta de la Provincia de Quito* became a battleground between diverse interests and came to represent not only the image of the territory surrounding Quito, but also a cultural and protonational patrimony that was used for intellectual as well as territorial appropriation. The contentious disputes over who had the right to profit from Maldonado's map were not only political disputes between nations and empires but contests between academicians and astronomers as well.

La Condamine's "New Map" of Quito

Through the extraordinary friendship between La Condamine and Maldonado, the posthumously published Quito map gained entry into the highest echelons of map culture in the eighteenth century and managed to preserve for itself a privileged place in the most widely read texts of the day. Through the contacts he provided and the positive publicity—to use a modern term for an ancient practice—with which he announced the arrival of his traveling companion, La Condamine gave special access to Maldonado's work and thus created a unique forum to present South American culture in the Old World. But his aggressive efforts to publish the *Carta de la Provincia de Quito* soon merged with his own project: to produce an entirely new map of Quito for which he could take authorial credit. By describing the conceptual labors as his own, he shielded himself from the potential criticism of profiting from his deceased friend's intellectual patrimony. He had employed the same process when he erected pyramids on the plain of Yaruquí—and then profited from their destruction by constructing monuments in print.

When the abbé Prévost described the Franco-Hispanic expedition to Quito in volume 13 of his bestselling *Histoire des voyages* (1746–59), he referred to the "attractive map of the province of Quito, based partially on [Maldonado's] memoirs." Drawing largely from La Condamine's *Journal du voyage*, the abbé explained that "we are obliged to M. de La Condamine for having collected the circumstances of [Maldonado's] return and those of his death."[51] La Condamine's description of his own efforts to fulfill the wishes of his deceased friend did not go unnoticed; instead, they seem to have earned him the praise and recognition of the leading literary figures of the day. In fact, the very person who claimed he would protect Maldonado's work may have been the first to use it to his own advantage. While La Condamine presented himself as duty-bound to see Maldonado's map through to completion, he lost no time in ensuring that the *Carta de la Provincia de Quito* would serve his own interests as well.

La Condamine accomplished this sleight of hand by implying joint custody of work that the two had accomplished in tandem. Rather than portraying Maldonado's surveying activities as comprehensive, La Condamine portrayed his and Maldonado's efforts as collaborative, emphasizing that they had worked together to produce new geographic knowledge and to extend the range of their experiments into Amazonian natural history and the history of the region's indigenous populations. In order to benefit from Maldonado's observations following his death, La Condamine attempted to blur the boundaries between his own work and that of his South American partner in order to garner the authority to reproduce Maldonado's map—with some key changes—in his own account of the Quito expedition. He also took advantage of having made copies of the *Carta de la Provincia de Quito*, which he had secreted away when it came time for him to publish his own map of the province. La Condamine's map of Quito, published in French and inserted into his *Journal du voyage*, came to be seen as his own work; he had successfully

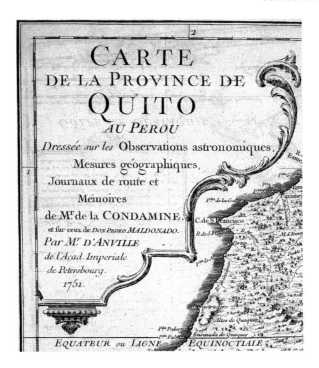

Figure 42. La Condamine's "new map": the cartouche of the *Carte de la Province de Quito* (1751). Detail of plate 20. Courtesy of the Rare Book and Manuscript Library, University of Pennsylvania.

appropriated the territorial mapping of Quito as an editorial achievement of his own making (fig. 42; plate 20).

La Condamine's description of the materials used to construct the northwestern portion of the Quito map provides a concise example of his subtle parsing of authorial credit. While he acknowledged Maldonado for the information gathered during the latter's reconnaissance work along the Esmeraldas River, La Condamine hedged on the credit Maldonado deserved in relation to their collective work:

At the same time, I worked with Don Pedro Maldonado on the map of the northern coastal section of the province of Quito, which he had just traveled; he communicated to me his routes, the estimated distances, and the wind currents he had observed with a compass made especially for this purpose, which I had shown him how to use. Based on his indications, and on the observations he had gathered in the country, we had enough [information] to trace the coast from the Rio Verde up to the mouths of the river Mira, and the course of the river Sant-Iago, by which Don Pedro had returned,

[and] which added a new fragment (*morceau neuf*) to the map that I had sent the Academy in 1736.[52]

By presenting the project as a joint venture, La Condamine emphasized several elements that would allow him ex post facto to claim authorial credit for a significant portion of Maldonado's cartographic work. He presented himself as the crucial intermediary between Maldonado's observations and the communication of these results to the Academy. It was La Condamine who had instructed Maldonado in the proper use of the special compass used to measure wind directions. Maldonado's contributions were presented as comprising only a fraction of their collective cartographic work, merely adding a "new fragment" to the earlier materials that La Condamine had provided his colleagues in Paris.

In affirmation of these rhetorical strategies, later reviews of La Condamine's *Journal du voyage* specifically referred to the map that adorned his account and ascribed sole credit to La Condamine for its creation. The January 1753 edition of the *Journal des Sçavans*, for example, discussed the map's central role as a guide to the narration of a scientific journey to Quito:

Even though the places he speaks of are clearly described, it would nonetheless have been difficult to have a proper idea of their position without the assistance of a map. He has added to his journal [the map] of the province of Quito, which extends roughly seven degrees in latitude, and approximately four in longitude . . . All of the materials that went into the construction [of the map] are indicated in a footnote on page 141.[53]

La Condamine's description of the "new map of the province of Quito" on page 141 of his *Journal du voyage* details the various elements of both his map, published in his *Journal*, and what he calls the "four-sheet Spanish map of the province of Quito," executed by Maldonado with the technical assistance of d'Anville. La Condamine's enumeration of the multiple sources that were merged together to create the final map was a common and necessary practice that was meant to narrate the successful fusion of cartographic diversity into a single sheet that he called his own new map:

My new map of the province of Quito extends roughly 7 degrees in latitude, and almost 4 in longitude. All the terrain occupied by the triangles of our meridian . . . was copied from the map elaborated by M. Verguin . . . The portion of the coast . . . was drawn up by M. Bouguer and me, jointly, when we disembarked at Manta in 1736, and I copied this portion from the map I sent to the Academy that same year. The rest of the new map was taken, 1. from my own observations during my various individual voyages in the provinces of Esmeraldas, Guayaquil, Loxa, Zaruma, Piura, Païta, Jaën, Borja, &c. 2. from what I have already said . . . that I borrowed from D. Pedro Maldonado . . . 3. From various memoirs and information that I collected from many places.[54]

In this description, La Condamine enumerated the various individuals from whom he had gathered his material, including the Jesuit Pablo Maroni, an anonymous French pilot born in Cadiz, Pierre Bouguer, Miguel de Santisteban, and other older narratives and manuscript maps. But even while giving credit to the Jesuit Jean Magnin for his map of Maynas, La Condamine clarified that he had "corrected all of the positions by a precise determination of the summits of the mountains of the eastern Cordillera, from which the rivers flow toward the Amazon." In the end, he explained that these were the same materials upon which the earlier "four-sheet Spanish map" was drawn up by d'Anville under Maldonado's supervision, "all of which I communicated to M. Maldonado."

La Condamine was singularly responsible for ensuring that Maldonado's map received the recognition it deserved, but it is clear that he profited directly from the observations of his friend and colleague as well. While he represented the map publicly as a posthumous publication of Maldonado, in private correspondence La Condamine emphasized that he considered himself to a certain degree the author of the work:

Most of [the map] was made based upon the very observations, notes, and demarcations that I gave to [Maldonado]; all of which, according to his intentions, were sent to his brothers (both of whom were still alive at the time), as I was named by the deceased as the executor of his will, his trustee, and even author of the work for the most part.[55]

In another letter, La Condamine discussed the fees he had paid for engraving the plates, later reimbursed by the Spanish ambassador, and explained that

I was already the trustee [of a particular plate] by virtue of the deceased's wishes . . . and, furthermore, author in large part since [the plates] had been elaborated based on my meridian observations in particular, and on my annotations, journal entries, and memoirs, all of which I had communicated to Don Pedro Maldonado.[56]

He went on to describe the "many rights to the map [I had] acquired," presenting his desire to publish it in the name of his friend as a benevolent act entirely at his disposition as the proprietor of the map.

On September 2, 1750, La Condamine at long last presented a copy of the *Carta de la Provincia de Quito* to the Paris Academy of Sciences, and he read aloud an academic *mémoire* written in praise of Maldonado.[57] Six weeks later, a regular public meeting of the Academy found La Condamine reading extracts from his new account, a "Journal historique du voyage à l'Equateur," which contained a smaller version of the *Carta de la Provincia de Quito* that had been modified to conform to La Condamine's new publication (plate 20). In the cartouche, shown in figure 42 above, La Condamine described the map as "elaborated based on the astronomical observations, geographical measurements, route descriptions, and memoirs of Mr. de la Condamine," relegating Maldonado's contributions to a secondary plane by placing his name in a smaller font on the line below. In addition to translating the titles and legends into French, other changes allowed La Condamine to regain authorial cartographic control over the theater of his and his colleagues' geodetic operations. In the southwest portion of the map, for example, the route from Loxa to the port of Sechùra was inscribed on the 1750 Maldonado map without any textual legend. In the version published with his *Journal du voyage*, however, the path included a textual commentary that linked the map explicitly to one of La Condamine's several Andean itineraries: "Route taken by La Condamine on his way to Lima."

The appropriation of Quito was complete. Split surreptitiously via editorial tools into three separate versions—one for the Spanish Crown, another for Europe's scientific societies (including the Royal Society in London and the Academy of Sciences in Paris), and a third for broad, public consumption to go along with La Condamine's *Journal du voyage*— the map of Quito was wielded by a host of actors for a wide variety of political, diplomatic, and intellectual purposes. Few recognized that two editions had emerged from a Parisian atelier, since most of these changes had been made behind the scenes and had left only the most subtle material traces. Only through close attention to the variegated layers of cartographic production is it possible to understand the minute editorial and print strategies by which Quito was molded and manipulated into its various forms. Only by recognizing the many voices represented by burin scratches and penciled scribbles on a cartographic proof of South America is it possible to comprehend an editorial epistemology that fused manuscript narratives and cartographic sketches into an eighteenth-century product of European print.

Conclusion

Historians of the book, literary critics, and scholars of print culture have emphasized the important role that manuscript documents and draft versions play in fixing as well as unsettling meanings in a printed text.[58] Their conclusions are equally valid, if not even more relevant, in the realm of eighteenth-century cartographic production. The geographical information and spatial understanding imparted by the eighteenth-century map were utterly dependent for both their visual and rhetorical presentation on the mechanical and material conditions under which a map was produced. This dependence becomes especially clear when the stages and structures of cartographic elaboration are revealed through an analysis of cartographic proofs.

The eighteenth-century map was often portrayed as a triumph of its own production, adorned with telescopes and compasses that papered over the document's condition as an object elaborated according to socially and materially defined circumstances. But the compilation and engraving of maps was a contentious, costly, and complex process, and the publication history of Maldonaldo's map demonstrates the sinuous and often unpredictable pathways that geographic information had to travel in the eighteenth century from the field to the print shop. In the map of Quito, geometric measurements carried out by European surveyors rested alongside markers celebrating early European expeditions, while destroyed villages and ethnonyms denoting Amerindian groups dotted the landscape amid the forests of the western Amazon basin. These myriad objects were sculpted and incised through material interventions, a stage in the process that has often been ignored. Scholars of the early modern European encounter with the New World have emphasized the social, legal, and discursive processes by which this knowledge was reported and exchanged, and these processes are of course absolutely central to understanding the epistemologies through which Europeans represented non-European peoples and places in an age of imperial expansion. But equally revealing are the internal workings of the *European* sites where that knowledge was produced and manipulated; these scholarly and editorial hierarchies that cull and compile information—from the erudite academy to the cartographic atelier—therefore deserve equally attentive scrutiny if the range and complexity of the Atlantic encounter is to be properly understood.[59]

In *The Practice of Everyday Life*, de Certeau put forward the idea that written narratives—be they travel literature or other discursive texts— were "journeys" that served to traverse and organize geographic space. In his words, "narratives select places and tie them together, making phrases and itineraries."[60] But the graphic language of maps also organizes space syntactically. In analyzing a map's unique fusion of text and image from a social and material perspective, we encounter skilled practitioners of graphic arts who adapted different forms of language and editorial techniques on the pages of cartographic proofs and who themselves made discursive "journeys" to portray South America within a single graphic frame. To comprehend the syntax that underlay their itineraries, it becomes necessary to see their narratives as divergent moments of geographical reconnaissance, fused together into a unified spatial-temporal image that often bore little resemblance to local knowledge of the territory.

The frustration with geographical errors on European maps of South America is evident in a letter sent by a resident of Cuzco named Joaquin de Lamo y Zúñiga to the French botanist Joseph de Jussieu in 1751. Lamo y Zúñiga, Count of Castañeda de los Lamos, had arrived in the viceroyalty of Peru along with José Antonio de Mendoza Caamaño y Sotomayor, the Marqués de Villagarcía, who was viceroy from 1736 to 1745.[61] He had composed a natural historical treatise on portions of Peru, and his letter to Jussieu expressed the preoccupations of someone who had little if any access to the European presses but who nevertheless wished to rescue older European maps of Peru from the "thousands of errors" they contained. In this letter, he criticized Guillaume Delisle's 1703 *Carte de la Terre Ferme*:

[W]ith this noble occupation I have come to understand how difficult Geography is: for the more care and attention one puts in the delineation of a kingdom or a province, it rarely comes out perfectly with the actual position of the places, [or] the origin and course of the rivers that run beautifully through it. Notwithstanding the differences [between the maps and their geographical referent] they manage to win the applause of the public.

Lamo y Zúñiga went on to lament the "work of fantasy and [the lack of] true speculation and care" that led cartographers like Delisle and other classical cosmographers to incorporate errors into their histories and maps. He ended his letter by comparing the representation of the region around Cuzco in Delisle's map with his own journeys of geographic reconnaissance and implored Jussieu to use "what experience had taught him" to correct the defective representations of South America then circulating in Europe.[62]

Like Lamo y Zúñiga, Maldonado also had recognized that distance could distort the European perception of South American geography. Emphasizing this relativistic perspective, he wrote to La Condamine that the separation between Europe and the New World "makes objects seem smaller than they actually are: America is far from Europe, and the mountains of Quito are very tall compared to those in Paris and Madrid."[63] Maldonado never had the opportunity to observe the finished version of his map of the province of Quito. But already before his death he seemed to understand that from the perspective of Paris, the Andes would appear flattened and their features diminished. The images of volcanoes, rivers, coastlines, and forests had to be compressed and manipulated in order to fit within a cartographic representation elaborated very far from their actual physical location. And while Maldonado's geographic legacy lived on through his map, the Quito map upon closer inspection reveals evidence of its own segmented fabrication, just as Gravelot's dramatic cartouche revealed the stages through which human labor constructed a colonial outpost using the raw materials of the forest. The interactions between the d'Anvilles and Delahayes, coupled with the physical and economic challenges imposed by the work of etching needles and burin engraving on a copperplate, remind us that cartographic knowledge-in-the-making was as dependent upon social and material processes in the printing house as it was on on-site reconnaissance in the field. In retrospect, Alexander von Humboldt's vision of a copperplate proof marked up with the specific dates in which individual changes were made—cited as this chapter's epigraph—appears utopian, if not technically impossible. The history of the annotations on the *Carta de la Provincia de Quito*—a map of Spanish America elaborated in France and tugged at by competing political and ideological interests—demonstrates that geographical knowledge, like other forms of scientific reconnaissance in the eighteenth century, operated simultaneously within many spheres, across various empires, and in several scriptural modalities at once.

A Nation Defamed
and Defended

[They are] populations possessed entirely by ignorance, full of rusticity, and only slightly removed from an uncultured Barbarity, like those who live scattered in the fields, almost like Irrational beasts, making their residence in the Forests and [other] most uncultured places ... I will now turn in my account to give notice of what can currently be observed of these Indians [of Quito], relating to their habits, customs, and character, according to what more than ten years' experience in commerce and communication with them has taught me.

ANTONIO DE ULLOA, *RELACIÓN HISTÓRICA*, BOOK 6, CHAPTER 6

It pains me greatly to see an entire nation, the Indians of the Kingdom of Quito, gravely defamed ... A passionate zeal for truth [thus] moves me to defend them ... It is my conscience that encourages me to return to them and to erase the black impressions that Don Antonio de Ulloa's history may have caused in the souls of its readers. ANONYMOUS CRITIC OF ULLOA'S *RELACIÓN HISTÓRICA*

Northern European presses dominated the production of eighteenth-century maps and texts representing South America for the European reading public, just as they did the *Carta de la Provincia de Quito*. But printed texts were not the only form in which knowledge about South America was expressed. Geographical descriptions often went hand in hand with other observations in a range of related fields and media, many of them avoiding the culture of print entirely. The objects that Maldonado collected during his travels with La Condamine down the Amazon exemplify the diversity of materials and observations that such explorations could yield. In addition to the cartographic proofs he left with La Condamine and d'Anville before traveling to London, Maldonado deposited another set of objects in Paris that history has largely forgotten. It was a set of natural historical specimens that would eventually join his map on the other side of the Pyrenees. Maldonado gave this trunk of South American "curiosities" to his compatriot, Pedro Franco Dávila, a wealthy native of Guayaquil then living in Paris who had been accumulating items of artistic, archeological, and literary merit since his arrival there in the early 1740s. The collection of objects Dávila had brought together, including bronzes, terra-cotta vases, and medallions, was significantly enhanced by Maldonado's contributions. In a letter written some years later, La Condamine explained that it was he who had "suggested [to Maldonado] to leave another case with his curiosities of natural history and papers... in the house of D.n Pedro Davilas [sic]."[1] According to La Condamine, Dávila became an enthusiast for the study of natural history in emulation of Maldonado after the latter's death. Dávila's enthusiasm, in turn, was fortuitous for the development of natural historical practice in Spain, for it was his collection that was purchased in 1771 by Carlos III, thus providing the basis for one of the greatest natural historical collections the Spanish monarchy had ever amassed: the Real Gabinete de Historia Natural in Madrid. Never cataloged separately, Maldonado's curiosities seamlessly and silently joined the ranks of one of Spain's most important institutions of eighteenth-century scientific knowledge.[2]

In addition to acquiring poison-dipped arrows, shards of platinum, and other natural curiosities that he used to woo curious observers at various scientific societies, Maldonado also collected numerous observations about the Amazon region's Amerindian communities, observations that he hoped would ultimately comprise a natural history of the Amazon River. The ethnographic description of native populations in the equatorial highlands and lowlands, including their customs, rites, and migration patterns, was a critical component of the Quito expedition's scientific bounty, even though the mission's official charge had little to do with observing and describing Amerindians. La Condamine, Bouguer, and Maldonado each attempted to catalog and assess the cultural characteristics of the various indigenous populations they encountered, and each sought to place these observations into a broader developmental framework as well. Portions of their work would come to form the raw materials, in fact, for the eighteenth-century debate regarding the nature of the American climate and its effect on the character of the hemisphere's inhabitants: the "dispute of the New World," which pitted American Creoles such as Juan de Velasco and Thomas Jefferson against their northern European detractors.[3]

But long before this dispute erupted, critical assessments of how visitors to the New World described Amerindian cultures were already brewing in the writings of Iberian and Ibero-American authors. Spain's official account of the Quito expedition, the *Relación histórica del viage a la América meridional* (1748) by Jorge Juan and Antonio de Ulloa, may have been the most costly, ambitious, and self-conscious publication effort the Spanish Crown undertook in the eighteenth century. From the time of its publication, the *Relación histórica* was portrayed as a triumph of Spanish empiricism, setting the standard for those texts and narratives that followed it and garnering near-universal praise within the European periodical press. Its appearance in print had been carefully managed and executed by royal supporters who saw a special opportunity to vaunt Spanish participation in a widely admired scientific project upon the European stage. But it still became the target of attacks, and one in particular that created an ethnographic firestorm within the effervescent manuscript culture of eighteenth-century Spain.

The polemical debate over the appropriate manner to observe and portray characteristics of Amerindian culture erupted between an anonymous critic who took as his target a section of Juan and Ulloa's account and another anonymous author who sought to defend Juan and Ulloa from the slings and barbs of this manuscript critique. The "Juicio Imparcial sobre los Indios de Quito" blasted Antonio de Ulloa's ethnographic portrait of Quito's native populations as revealed in a single chapter of the *Relación histórica*. For the anonymous author, who fashioned himself as a defender of the Indians of Quito, Ulloa's analysis was "facile," "malicious," and "calumnious." He set aside Ulloa's descriptions through ironic jabs, semantic maneuvers, and semiotic analysis, and sought to offer his own observations in their place. In response to this manuscript, an anonymous apologist for the *Relación histórica* launched a counter-critique, the "Defensa à favor de los Indios de Quito," a text that explicitly censured the anonymous author of the "Juicio Imparcial" for his "extravagance," his "temerarious contradictions," and other flaws of character.[4] The "Defensa" attacked the critic's credibility and accused him of "wanting to disturb the truth [about the present state of the Indians] with stratagems of falsehood." What is more, the anonymous counter-critic made reference to the first critic's "impertinent fire of voices," using a synesthetic image of cacophony to question the merits of the "Juicio Imparcial" and to claim in turn that the "impartiality [of the

Relación histórica] leaves no doubt as to the accuracy of the information [it contains]."[5] In the end, the fierce manuscript volleys that emerged between the "Juicio Imparcial" and the "Defensa à favor de los Indios de Quito" show that efforts by the Spanish Crown to control the results produced through Spain's overseas scientific expeditions were both contested and supported across a number of eighteenth-century modalities, including in this case a dueling dyad of anonymous manuscript treatises.

Whether cleric, Creole, or colonial administrator, the critic devised sophisticated arguments to deconstruct the legitimacy of what Ulloa had written in the *Relación histórica*. These included both a critique of ethnographic distance, that is, the physical and figurative separation between observers and the cultures they described, and an assault on the use of scientific instruments as a basis for making determinations about cultural traits. In addition, the anonymous critic argued that the potentially nefarious impact of Ulloa's observations on his metropolitan readers was reason enough to object to its publication, making an inherently moralistic argument that echoed early defenders of Amerindian culture such as Bartolomé de las Casas and António Vieira. But the production of the "Juicio Imparcial" also set in motion a vigorous project of rhetorical counterpoint. And while we know little about either author's background or beliefs, save in the case of the first critic his familiarity with administrative archives and ecclesiastical practices in the Quito region, their manuscript exchange opens a window onto the arguments that were possible as part of an eighteenth-century ethnographic debate. This veil of anonymity in both cases means that authorial identity cannot be used as the sole or even the most significant prism through which to understand the authors' analytical categories. Instead, the frustrating lack of knowledge about their respective identities forces us to focus on the epistemological, as opposed to identity-driven, foundations of each author's conceptual framework. All the while, in the midst of this polemical exchange stands the figure of the Amerindian, serving as a catalyst for diatribe but remaining a mute observer to a conflict in which the credibility of eyewitness observation frequently trumped the contested conditions of colonization debated within the pages of these two anonymous critiques.

What is even more striking about this exchange, however, was its almost absolute suppression from the public stage. Nowhere was the anonymous critique published; nowhere were its criticisms subjected to public scrutiny; and not until now has the ingenious strategy of attacking Ulloa's authorial integrity in the guise of protecting the Indians of Quito been brought to scholarly attention. Indeed, little can be surmised from available sources as to the precise circumstances under which these two anonymous ethnographic critiques were penned and to whom they might have circulated within eighteenth-century networks in Iberia or beyond.[6] But the polemic between Ulloa, the anonymous critic, and an equally vehement counter-critic nevertheless has broad implications for understanding how Amerindians in and around Quito came to be construed and contested as an object of anthropological assessment. In addition to showing how struggles over ethnographic description reveal conflicts over power and authority in a colonial context, this manuscript exchange captures divergent perspectives over the appropriate ways to observe, interpret, and write about non-European cultures using intellectual criteria developed against the backdrop of an Iberian tradition. In a larger sense, the anonymous critic's perspective as someone who at least claimed to have spent many years in situ also brings into sharp relief the ideological presuppositions of travel and observation by criticizing the transient nature of the anthropologist-as-*passeur*. While anthropological critiques frequently emphasize the extent to which relations of power, the cultural status of the observer, access to technical instruments, and the tools of representation infuse a particular interpretive and ideological framework in the field, this much earlier debate over the uses and abuses of ethnographic observation demonstrates that epistemology could not always be clearly linked to particular ideological positions within a colonial setting. Nor did these epistemological critiques always emerge in "patriotic" contexts that could be seen as moving Latin American societies inexorably toward liberation from colonial Spanish rule.[7]

What is clear is that Quito in the 1740s was an unacknowledged epicenter of enquiry into the nature and status of the Amerindian, just as Mexico and Lima had been and would continue to be for decades to come. And while it may seem counter-intuitive to posit the existence of a coherent critique from an apparent absence of clear political or ideological identities—due to the anonymity of the individuals involved—the discovery of these anonymous sources raises important methodological questions as well. By formulating an analysis of the anonymous critique based on the range of its arguments rather than on the matrix of ideological interests that coalesce around an author's individual or collective identity, the scope of *epistemological* possibilities widens, especially during a period where the rules and accepted practices of ethnographic description were still inchoate and thus largely open for discussion, interpretation, and debate.[8]

Binding "Reparos": The Material and Editorial Construction of the *Relación histórica del viage a la América meridional* (1748)

Published in Madrid following almost a decade during which its authors traversed the region stretching from Cartagena de Indias (present-day Colombia) to Lima, the *Relación histórica del viage a la América meridional* (1748) was written to reclaim for Spain what was considered by Iberian observers to be their prerogative: the geographic, ethnographic, and natural historical description of South America. Consciously conceived as a textual monument to Iberian science that would be observed and admired throughout Europe, the *Relación histórica* became a capacious vehicle for the recuperation of Spanish prestige and honor, especially after damaging episodes for the nation's military and cultural pride in the late-sixteenth and seventeenth centuries. The five-volume text was seen explicitly as an Iberian riposte to French pretensions in a new imperial realm that was not territorial but intellectual: scientific observation. After all, the French academicians in Quito were allowed to make their observations on the continent only at the privilege of the Spanish monarch, and early eighteenth-century visitors to Spanish America such as Amédée-François Frézier and Louis Feuillée had been prohibited from venturing far from a few prescribed venues on shore. With the threat of publication looming from Ulloa and Juan's French counterparts, the two naval captains and their ministerial sponsors were eager to see a textual account emblazoned with a Spanish seal. Their Herculean efforts paid off: they beat the Paris Academy of Sciences to the punch and were able to savor the thrill of being the first to publish the results of their observations, much to the elation of those who were intimately involved with the conception and execution of the *Relación histórica* in Madrid (fig. 43).

The myriad actors who participated in the well-orchestrated and costly publication of the *Relación histórica*—from political advisees to printers, from editors to censors—were acutely aware of the importance of the project, as was the Crown, which supported the undertaking financially and considered it a matter of high state interest. In the elaboration and editing of the volumes, close attention was paid to the papers, inks, typefaces, and copperplates that were imported to Spain from throughout Europe to ensure that such an important publication venture, carried out by the renowned printer Antonio Marin, would benefit from the highest-quality materials available. These material concerns were not only germane to this particular project, however; they were seen as pivotal to

Figure 43. The *Relación histórica del viage a la América meridional* (1748) was Spain's studied response to accounts from northern Europe's eighteenth-century scientific expeditions to the New World. Courtesy of the John Carter Brown Library at Brown University, Providence, R.I.

the trajectory of Spanish scientific publication on a much broader scale. The *Relación histórica* was merely the central monument representing this new drive to raise the status of Spanish publications: its attractive material appearance would function as window dressing to showcase Spain's enlightened management of its overseas territories. Both materially and metaphorically, the vellum bindings and elegant fonts would demonstrate the Crown's increased investment in an independent program of scientific investigation.[9]

This effort to develop an independent tradition of Spanish science did not emerge from nowhere. From early in the eighteenth century, philosophers such as the Benedictine friar Benito Jerónimo Feijóo (1676–1764) drew on the works of Descartes, Gassendi, and Bacon to advocate for new philosophical approaches to Spanish scientific pursuits. Their goal was to expunge the tarnished reputation Spain had acquired over the previous two centuries, largely on account of Protestant propaganda that represented atrocities Spain had committed against native populations in the Americas. One goal of Feijóo's program was to provide new models for the observation and treatment of non-European populations in Spanish American territory, and in many ways the *Relación histórica* was conceived as a material manifestation of Feijóo's philosophical project. But it was also more than that: the *Relación histórica* was an apologetic text that emphasized the more harmonious elements of Spanish culture in the New World. Judging by the nearly unanimous chorus of positive reviews, at the academies and within the European periodical press, the *Relación histórica* achieved the Spanish monarchy's ambitious goals with resounding success.[10]

The political and military upheavals following the accession of a Bourbon king to the Spanish throne in the early eighteenth century had brought a new and invigorated Spanish foreign policy under the leadership of José Patiño (1670–1736), who sought to strengthen Spain's presence on the international stage through a broad series of institutional reforms. Patiño was a conservative administrator of Milanese extraction with close ties to Elizabeth Farnese, queen of Spain and wife of Philip V, and he brought together many of Spain's administrative institutions in a movement known as the Bourbon Reform, which was meant to rebuild Spanish strength through naval, industrial, and fiscal means. As part of this broader mobilization, Patiño saw merit in sponsoring two young naval captains—Jorge Juan and Antonio de Ulloa—to participate in one of the most significant scientific enterprises of the early eighteenth century. Their primary obligation may have been to keep an eye on their foreign guests, since the French academicians were some of the first foreigners in centuries to be allowed more or less free access to Spanish America. But the instructions Patiño sent to Ulloa and Juan also included the charge to construct maps of cities and ports alongside their French colleagues, and to undertake a close study of "the motivation, industry, and abilities of [the region's] native inhabitants (*naturales*), and the bravery or friendliness of the unsettled Indians (*indios irreductos*)."[11] Cartography, natural history, and ethnography would form three interlocking segments of the reports these young officers were to pen. Their observations, so it was hoped, would come to play a central role in Europe's understanding of the natural history of South America, its mineralogical resources, and the indigenous populations of Spain's American territories as well.

But projects conceived under the reign of a given ruler did not always confine themselves to the lifespan of the monarch. The church bells mourning King Philip V of Spain (d. 1746) were still echoing in the streets of Madrid when Antonio de Ulloa and Jorge Juan returned to the Spanish capital after an eleven-year absence serving in his name. Since Patiño had died only a year after Ulloa and Juan set out from Spain in 1735, the two young officers found an entirely changed cultural and political landscape upon their return. The decade or so after 1746 corresponded to a transitional period in the history of the Spanish monarchy, a pivotal moment when momentum began to gather for significant changes that would emerge most clearly in the institutional and cultural policies of Charles III (r. 1759–88). The thirteen years during which Ferdinand VI (r. 1746–59) spent on the Spanish throne, by contrast, would reflect a more diffuse and inchoate project to bring Spain out from the shadows. The publication of the *Relación histórica* was, by all accounts, a central spoke in this strategy, and Ulloa immediately set to work on convincing the Spanish authorities, and above all the minister plenipotentiary, the Marqués de la Ensenada, of the importance of his project to produce a complete narrative of the voyage to South America.

One reader whose prose succinctly captured the project's aspirations was Antonio Alvarez de Abreu (1688–1756), the Marqués de la Regalía. After perusing an early draft version, he explained that "[the *Relación histórica*] will serve as public testimony that the Spanish have lost neither the appetite nor the capacity to take on great projects."[12] Regalía, who had spent time in South America as governor of Caracas and had also been a member of the Council of the Indies, had been asked to serve as one of the censors of the *Relación histórica* to determine whether anything contained within the text would be "harmful to [the Spanish] nation or monarchy." He took his charge seriously. But Regalía also acknowledged the absurdity of Spain's previously rigid policy of secrecy, likening it to the example of the Romans in their first century of power. According to Regalía, this situation had long since changed for the Spanish empire: "Time has made [this policy] vain and ridiculous in the Indies, especially in this century, since we ourselves have revealed our ports by way of the asiento [of African slaves] and the factories established in the principal French and English ports."[13] Transatlantic commerce, and particularly the lucrative slave trade, had changed the need to protect natural knowledge, and the newfound openness with which Ulloa's text was produced had been made possible by the changing geopolitical realities on the Spanish American mainland. In his suggestions for improving the text, Regalía took the liberty of "altering some terms and voices" so that the style as well as the content of the text would be perfected. Ultimately, however, his corrections were largely cosmetic, and Regalía strongly endorsed the production of the text. He even went so far as to point out the importance of highlighting certain aspects of natural knowledge through print by suggesting that "the author put in italics the names of birds, fishes, animals, rivers, lakes, and other notable items of interest."[14]

In addition to these encouraging remarks from Regalía, Ulloa also received enthusiastic support from Andrés Marcos Burriel (1719–62), key architect of the editorial revision and publication of the *Relación histórica*. Burriel was a polymath scholar trained in the grammatical and rhetorical traditions of the Jesuit novitiate, and he believed that Spain should come into line with the scientific traditions and contemporary perspectives of other European nations. According to Burriel, Spain not only needed to transcribe books that taught "the true method of study" into the Spanish language but also had to take a more aggressive role in producing "great works of profound wisdom."[15] His own *Apuntamientos de algunas ideas para fomentar las letras* (1749) outlined a program for artistic, cultural, and scientific reform in Spain. Other writings he produced, many epistolary and a majority of which were never published in his lifetime, also envisioned an ambitious role for Spanish men of letters. Not surprisingly, he believed the Society of Jesus should play a central role in this intellectual recuperation: "Without the Society [of Jesus] in Spain," Burriel wrote in the *Apuntamientos*, "no good can emerge."[16] His readings were eclectic and went far beyond the typical range of eighteenth-century Jesuit erudition. In addition to being a regular reader of the *Mémoires de Trévoux*, the Jesuit periodical published in France by Berthier, he was familiar with Descartes, Gassendi, Grotius, and Leibniz. During the gestation period of Ulloa's *Relación histórica*, Burriel was director of the Seminario de Nobles in Madrid; as such, he was well positioned within the administrative bureaucracy to pass judgment on high matters of state. The publication of the *Relación histórica* was understood as being of the utmost importance. Indeed, it was Burriel who recognized the potential impact of Ulloa's text for Ferdinand VI's program of institutional and cultural reform, a program that Burriel vigorously supported.[17]

Like Regalía, Burriel reflected optimistically on the role the *Relación histórica* could play in recuperating the pride and glory of Spain's imperial past: "This work is one of the best and most useful that has ever been published in our language: and I do not doubt . . . that it will satisfy the expectations of Europe with incredible glory to the Nation, His Majesty, Minister [Patiño], and its authors." Burriel called its method "the most appropriate to historical travel narratives," its language "proper and pure," and its style "sweet and florid." In summary, the text was designed in such a way that would elicit a positive reception from throughout the Republic of Letters, from Rome to London:

We could not invent something as appropriate to redound to the credit of our literature as this work in the present circumstances. No protection by His Majesty will be put to better use than that afforded to the authors [of the *Relación histórica*] for their inimitable work, which has no equal in our nation. Everyone will yearn to cultivate the sciences, [so that we may] acquit ourselves of the opprobrium laid upon us by foreigners, serve the nation (*Patria*) and the public good, and increase the glory of this happy reign.[18]

Despite such encomiastic rhetoric, Burriel wrote nineteen pages of "reparos," or corrections, which ranged from the reordering of chapters to suggestions as to how the authors might justify the recent Anglo-Spanish conflict, known as the War of Jenkins' Ear (1739–42). Such constructive criticism was meant to secure the standing of the *Relación histórica* as a showcase for the glory of the monarch, the longevity of the nation, and what Burriel called the "marvelous fruits" of Spanish science. This was a work that in Burriel's eyes would be both "perpetual" and "universal."[19]

The speed with which Juan and Ulloa had managed to produce their *Relación histórica* added to Burriel's overall enthusiasm as he discussed their work. Portugal, England, and France were all hoping to receive accounts resulting from the expedition as soon as possible following the mission's return and had been equally frustrated by the delays. Burriel insisted that the printing of the *Relación histórica* occur at the earliest possible juncture, but especially "before the French move ahead" with their own publication efforts. Rather than seeing the publication of the *Relación histórica* merely as an opportunity to record the results of the survey measurements produced in Quito, Burriel argued that the text's diverse subject matter would cause it to "make its way swiftly to all the countries of Europe . . . [since it] combines important information on navigation, routes, and poorly known itineraries [with] variations of the compass, up-to-date descriptions of coastlines and ports, and precise latitudes of many places."[20]

The "useful notices" contained within the *Relación histórica* were drawn from the Spaniards' voyages from Cartagena to Portobelo, from Portobelo to Panama, and from Callao back to Europe. Each of the nine books of the *Relación histórica* was divided into smaller subchapters treating discrete segments of their overland journeys, including panoramic

Figure 44. An ethnographic portrait with a strongly encyclopedic flavor. Several representative figures—from Spaniard to "Rustic Indian"—are displayed together against the dramatic backdrop of a roaring river. From the *Relación histórica del viage a la América meridional* (1748). Courtesy of the John Carter Brown Library at Brown University, Providence, R.I.

descriptions of colonial cities, treatises on the climate and inhabitants of towns and villages, typical foods eaten by the local population, generic details about their commerce and trade, and geographical descriptions of the administrative regions through which the two Spaniards had passed. Alongside many of the descriptions were sumptuous engravings of South American topography: nautical charts portraying entrances to bays and inlets, city plans showing the urban layout and buildings of interest, as well as typical scenes and clothing of the local inhabitants (fig. 44). The *Relación histórica* also included an extensive description of the city of Quito, including its climate, customs, commerce, and its connections to outlying regions of the Audiencia. One book widened the scope to include the larger Province of Quito, its administrative divisions, and the intelligence, customs, and character of the region's indigenous inhabitants, to which we will return shortly.

In addition to the meticulously produced plates, the use of four different covers to enclose the printed sheets—from paper and parchment to vellum and lustrous leather—was designed to clothe the text in accordance with the status and prestige of those who would receive it. The king of Spain, the Dukes of Parma, and "Royal Persons" in Naples and Paris received the finest "juegos," or sets, bound in fine leather, as did the two most eminent royal ministers, the Marqués de la Ensenada and José de Carvajal y Lancaster. Those bound with a kind of treated lambskin, "en pasta," went to prominent diplomats and members of other European academies. William Stanhope (1683?–1756), who had earlier served as a British ambassador in Spain, received his texts with this binding, as did Bouguer, La Condamine, and Bologna Academy of Sciences secretary Francesco Maria Zanotti (1692–1777), who praised Juan and Ulloa's "caution, prudence, nimbleness of mind, skilled background, [and] knowledge and understanding of the most recent discoveries." Likewise, after reading one of twelve copies sent to London's Royal Society, William Watson expressed his "extraordinary thanks" in the name of the society for "the gift of a book filled with such fascinating, well-researched, and interesting materials."[21] In total, according to Ulloa's meticulous accounting, 7,200 volumes were distributed throughout Europe for sale or as gifts.[22]

Ulloa's list also included a lambskin-bound set that went to the "Diaristas de las memorias de Trevoux," who would review the text in glowing terms in March of the following year. The six-part review praised the thoroughness of the text, commenting that "nothing is absent from this Account that might excite our curiosity." Plants and trees found in the Andes were omnipresent throughout the review, in addition to discussion of the peculiar characteristics of bats, monkeys, and tigers from Cartagena and Portobelo to Guayaquil and Quito and beyond. The third installment referred specifically to the description of the residents of Quito found in book 6, chapter 6; in it, the reviewer announced that "[t]he Spanish account deals extensively with the intelligence and customs of the Indians of the province of Quito," and affirmed that "[t]his description will not disappoint."[23]

But occasionally the *Relación histórica* did disappoint. Don Josef Borrull, a correspondent of the great Valencian philologist Gregorio Mayans y Siscar, complained to his friend that he had wasted eleven pesos on the purchase of the first three volumes (the last two had not yet been published):

After His Majesty spent so much on their journey, [Juan and Ulloa] wrote three tomes . . . whose contents can be summarized as follows: they were not able to afford anything and [they] were laughed at by the French. I am fed up and disgusted at having purchased [these books].[24]

The humiliation suffered at the hands of the French may have caused certain Spanish readers to bristle. But other critics took the authors of the *Relación histórica* to task for far more serious issues. While the reviewer from the *Mémoires de Trévoux* echoed Ulloa in wondering how "a nation that in centuries past [had] constructed such marvelous structures ... could degenerate to such an extent that it seems to be caught somewhere between man and beast," Ulloa's description of the cultural degeneration of Quito's native population infuriated another reader with a very different set of experiences and credentials. Unlike the demure reviewer from the *Mémoires de Trévoux*, the author of the "Juicio Imparcial" seized upon these denigrating remarks to launch a tempestuous indictment of Ulloa's work: not in the public pages of a periodical journal but rather under the protective veil of manuscript anonymity.[25]

The "Reparos" of the "Juicio Imparcial": From Textual Improvement to Moral Critique

If criticism was considered a "dominant fashion" of the eighteenth century and useful in "preventing an individual error from slowly infecting the public [arena]," as the French Minim friar Louis Feuillée commented in the preface to his South American travel account, then textual criticism might often appear camouflaged as a benign, even welcome addition to a given text.[26] The "Juicio Imparcial" began its own critical exegesis of Ulloa's ethnography in precisely such a vein, reflecting a tone that in many ways echoed Burriel's favorable assessment of what Ulloa had composed in the narrative portion of the *Relación histórica*. Writing "with sincerity and on friendly grounds," the anonymous critic at first gently suggested that he would have liked to see certain "improvements" made to Ulloa's text, and he used a familiar term implying both criticism and improvement: the "reparo." The use of the term "reparo," the same expression used by Burriel in his own critique, suggested the molding and redefinition of a text based on critical, yet amicable grounds. To whomever the critic may have been addressing his remarks, he used this ameliorative term to define his critique: "I will make [my] points clear with the improvements (*reparos*) presented in this paper" (fig. 45).[27]

But the neutral tone implied by the term "reparo" quickly gave way to a more indignant voice as the critic acknowledged his "great pain" at witnessing the "grave infamy" committed against "an entire nation." Unlike Burriel, the critic applied a moralistic valence when discussing these "improvements" that merged textual correction with moral censure: "The circumstances oblige me to expose the *reparos* that I would have made, and to work with the promise and sincere desire to correct all of the errors that Ulloa's wicked reports (*siniestras ynformes*), lack of experience, or

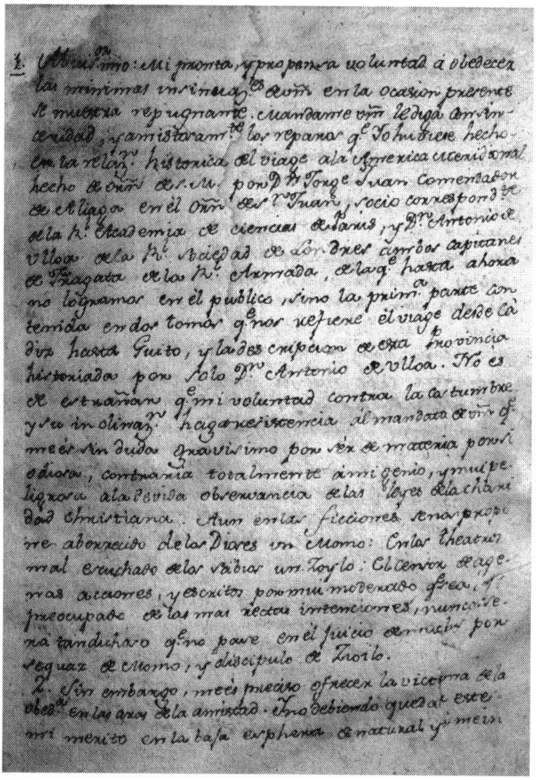

Figure 45. A page from the anonymous treatise that attacked the *Relación histórica*. In it, the critic discusses the "changes he would have made" to Ulloa's account. Courtesy of the Special Collections Library, University of Michigan.

excessive zeal might have caused."[28] This moral and emotional outrage, a leitmotif throughout his manuscript, was at the core of the critic's robust defense of indigenous culture. According to the critic, the "unfortunate abandoned Indians" (*tristes desvalidos Yndios*) were incapable of defending themselves against rhetorical or ethnographic attacks, unlike Creoles and other castes within Spanish America. The critic thus saw it as his duty to protect them through these editorial improvements: "The *reparos* that I have made up to this point," he explained, "rise to the level of charity and holy zeal [and they have] moved me in righteous defense of the truth and honor of a helpless nation, who lacking support also lacks the spirit and the tongue with which to defend itself."[29] It was as an advocate for the disenfranchised natives that the critic justified his intervention, since according to the critic they were incapable of expressing their own indignation. And it was the "conscience" of the critic that lay at the root of his support for their cause.

The author of the "Juicio Imparcial" was also concerned by the "black impressions" that Ulloa's text might cause in "the souls of its Readers."[30] His concern for their souls merged religious symbolism with an uncannily modern notion of reception theory. At the center of the anonymous author's moral critique was the potential impact Ulloa's observations might have on those who read his text, and he presumably had an audience clearly in mind. The critic's consideration of the subsequent impact of the text's ethnographic content on its *object* of study, the Amerindian, suggests that the critic understood the potential ramifications of this text within an imperial as well as a local context. That the critic recognized both the defenseless state of the Indians and the moral state of Ulloa's eventual readers argues strongly for his familiarity with what the metropolitan reaction to such a text might be. His rhetoric suggests that he saw himself as a go-between, or cultural mediator, between these two different populations, Spanish and Amerindian. And the fact that he couched his critical endeavor in moral terms gives some credence to the hypothesis that in addition to his experiences in Quito, he may have had ecclesiastical training either in Spain or Spanish America.

Who might the author of the "Juicio Imparcial" have been, then? First, the critic made several distinctions between his own identity and that of the Creoles and mestizos of Quito, arguing against identifying him with either of these two groups. And while he legitimated his perspective by acknowledging having spent many years in the province of Quito, having traveled regularly throughout the region, and having had ample opportunities to know and observe the customs of its indigenous residents, the critique lacked many of the features that would later be seen as comprising a Creole view of Spanish America: in particular, there was little if any exaltation of South American nature and few discussions of the region's pre-Columbian heritage. He did not indict Ulloa's perspective because the latter was born in Spain (a common Creole critique of the metropolitan perspective) but rather for not having spent enough time in the province. The critic derided Ulloa's lack of *experience*, not any collective political or cultural identity Ulloa represented. The author of the "Juicio Imparcial" did not appear to compose his account intending to provide Creole patriots with a legitimating narrative based on a response to northern European detractors, nor were clerical observers necessarily portrayed as trustworthy witnesses for ethnographic information. Rather, the critique provided by this anonymous author cut against the grain of later "clerical-Creole" epistemologies, and bore only a slight resemblance to the analyses offered by the likes of Juan de Velasco and Francisco Clavijero several decades later.[31]

The "Juicio Imparcial" appears to have been written contemporaneously with the *Relación histórica* itself, and possibly in Madrid. The first section of the anonymous critique provides a crucial clue in dating the manuscript. According to the text, only the first portion of the *Relación histórica* was circulating at the time of the critique's composition: "until now," the critic wrote, "only the first part in two volumes, which refers to the journey between Cadiz and Quito and the historical description of this province, has been released to the public."[32] In another letter, Burriel also indicated that there was a lag between the release of the two parts; he read the first part to compose his "reparos" in June 1747 and his assessment of the second part was only released in November 1747. By May 7, 1749, both parts of the text—including Juan's astronomical observations, for a total of five volumes—had been made available to the public. The anonymous critique was therefore composed sometime between June 1747 and May 1749, making it nearly contemporaneous with the production of the *Relación histórica* itself. What is more, the critic made reference to the "defenders of the Indians that have now dispersed and can be found anywhere, *including this court in Madrid*," suggesting that he may have been present in the Spanish capital at that time as well.[33]

Whether or not the text was actually composed in Madrid, the critic at the very least presented himself as someone writing back *to* the metropole. The audience that the critic addressed in his manuscript comprised individuals who were just as likely to be colonial administrators as they were curious readers: "Suppose that . . . one of the many [individuals] . . . who have read this history [i.e., the *Relación histórica*] were to travel to Quito, as part of a government, *corregimiento* or any other service [to the Crown]."[34] The critic seemed also to be faithful to the idea of a greater Spain, calling the regions of Quito "our provinces" and vindicating the absolute authority of the Spanish monarch, whose "sovereign name" he never lost an opportunity to praise.[35] He demonstrated as well an intimate knowledge of administrative records and archives, even in the most outlying areas of the province. In one instance he referred to Caloto, a small town in the district of Popayán whose bell towers had previously been destroyed along with the rest of the city. On this occasion he wrote that "[Ulloa] gave us an extremely defective account when, as a precise, prolix, and critical historian, he could have provided us with a far more complete story by examining what . . . is archived in the town council (*cabildo*) of Popayan."[36]

This intimate knowledge of the regional archives of an important administrative center such as Popayán suggests another possibility: that the critic was an administrator or Crown agent within the Audiencia of Quito. In sections of the "Juicio Imparcial," he attempted to dismantle Ulloa's observations on the political functioning of the Audiencia with reference to the towns and parishes of San Miguel de Ybarra, Latacunga, Riobamba, San Sebastian, Cajabamba, and Itambato. Demonstrating a thorough understanding of local bureaucratic processes, he criticized Ulloa's status as a passing outsider with reference to the political, as opposed to cultural or ecclesiastical, operations of Spanish colonial administration: "In any case, we understand that what Don Antonio Ulloa meant to say is that the town of Riobamba (by which he means the body of *regidores*) does not need recourse to the president of Quito in order to confirm its elected mayors (*alcaldes*)."[37] Through these references to specialized, administrative knowledge of the Audiencia, the critic diverted authority and legitimacy away from Ulloa and gave himself credibility as an "ocular witness" to speak to the political attributes of the Audiencia.

There is no question, however, that this individual was also profoundly familiar with the ecclesiastical culture of the Audiencia. The critic was keenly aware of the number of parishes in a given ecclesiastical jurisdiction and, subsequently, the number of priests present in each parish. Discussing the intersecting worlds of indigenous and religious life, he wrote that "the Indians are very inclined to all functions of the Church, [including] masses, parties, processions, etc.," and he regularly peppered his accounts with episodes from Quito's ecclesiastical history, such as the extirpation campaigns of a former bishop of Quito, Don Luis Francisco Romero. He was also familiar with the architectural structures of the various orders, including a Bethlehemite convent about which he claimed that Ulloa had misspoken in his account.[38] And finally, when criticizing Ulloa's insistence that Spanish be made the common language of discourse among the Indians, the critic pointed out that this rule had already been applied to the priests of the province, who were forbidden from taking confessions, preaching, or teaching Christian doctrine in a language other than Castilian. Because the instruction had had little effect, the critic explained through a rhetorical remark why this measure had been so impracticable: "Could a priest in full conscience follow this command, given that he knows quite well that his faithful lack the basic instruction in Castilian that would allow them to understand?"[39] These citations suggest both affinities toward metropolitan concerns, on the one hand, and a deep understanding of ecclesiastical practice in the province of Quito, on the other.[40]

All possibilities considered, the Iberian tradition of clerical critique going back to Bartolomé de las Casas argues strongly in favor of a priest having penned the "Juicio Imparcial." While he did lambast clerics as often
as he praised them, he also made frequent references to observing both
"natural and divine laws." It cannot, of course, be presumed that a cleric
would necessarily defend other clerics in all circumstances, nor that a
colonial administrator would perforce believe that the Jesuits wielded too
much power in the provincial environment, although this was certainly
a widespread view. But the arguments produced by the anonymous critic
lend credence to the idea that anonymity itself was a catalyst for intellectual flexibility: coupled with in situ experience, and brought together
through the formulation of an intellectual critique, these characteristics mixed with moral outrage (at least rhetorically) to fuel the critic's
explosive attack on Ulloa's prose.

Not knowing the identity of the critic, his intellectual origins must
of course remain obscure as well. If he was educated within a traditional
Jesuit curriculum, he would have followed Claudio Acquaviva's *Ratio
Studiorum* of 1599, a plan of study that comprised topics in the classical
languages, the humanities—including history, literature, and drama—
as well as philosophy and theology, not to mention scientific subjects
based on a Scholastic or Aristotelian model. Had he had the privileged
access of someone like Burriel to a broad range of European texts, for
instance, he might have been exposed to the historical and pedagogical
treatises of French clerics such as Philippe Labbé (1600–67) and Bernard
de Montfaucon (1655–1741), as well as more progressive European scholars including Luigi Antonio Muratori (1672–1750) and Johann Albert
Fabricius (1688–1736).[41] But while his early reading necessarily eludes
us, the one author he did mention explicitly in his treatise was Benito
Jerónimo Feijóo, whose *Teatro crítico universal* (1726–40) was a required
text for anyone interested in gauging Spanish reactions to intellectual
debates taking place throughout Europe. And the text had a profound effect not only on Spain's keeping abreast of what was happening over the
Pyrenees but also, more specifically, on the anonymous critic's criteria
for ethnographic descriptions far beyond Europe's shores.[42]

Feijóo's "Intellectual Map": A Critique of Geographic Distance

For nearly a decade before Patiño instructed Jorge Juan and Antonio de
Ulloa to "discover . . . the motivation, industry, and abilities of [Quito's]
native inhabitants," the Benedictine Benito Jerónimo Feijóo had already

been shaping Spanish opinion as to how populations far from the Iberian Peninsula should be discussed and debated. A native of Galicia, Feijóo broke with the obscurantist and dogmatic mode of Iberian scholasticism and cast a long shadow over the eighteenth century with his *Teatro crítico universal*, an extended series of essays with which any literate Spaniard who had intellectual aspirations would have been familiar. Both Ulloa and the author of the "Juicio Imparcial" were readers of Feijóo's work. In "Intellectual map and comparison (*cotejo*) of nations" (vol. 2, chap. 15 of the *Teatro crítico universal*), Feijóo wrote that

[o]ur intellectual perspective suffers the same defect as that of our bodies, in that it represents distant objects as smaller than they actually are. There is no man, as gigantic as he may be, that from a distance does not look like a pygmy. That which is true for the size of bodies is [also] true for the stature of souls. In those nations that are distant from our own, men are represented in such a diminished manner that they are not even considered to be rational beings. If we looked more closely, we would reach a different conclusion.[43]

What Feijóo offered in this passage was a critique of ethnographic distance, that is, the cultural distance that separated European observers from the objects they described. On a primary level, he pointed out how observation itself could distort the object being observed. Likewise, "interpretation"—what Feijóo referred to as analysis according to an "intellectual perspective" and the corollary of observation—fundamentally changed the proportions of a figure that had been perceived visually by the human eye. By moving from the physical level to a spiritual or metaphysical plane, Feijóo transformed this indictment against the distorting power of analysis into an epistemological critique: human beings of "distant nations" had been stripped of their innate humanness by the disfiguring nature of human perception itself. From Feijóo's perspective, distance was anathema to verisimilitude. The more distance between an observer and his object of study, the less truthful his representation of those observations would become. The pygmy and the giant, then, were part of the same human continuum, but the distance from which one observed the two made all the difference between seeing a beast and seeing a man.

By insisting that observers look more closely at their object of interest, Feijóo was implicitly impugning the superficial observations provided by early modern travelers both in Europe and overseas. His "intellectual map" was especially critical of those who denigrated non-European peoples and insulted their intellectual qualities. "Those nations that are commonly taken for rude or barbarous," Feijóo explained, "yield nothing and in some cases even exceed in intelligence those who consider themselves more cultured."[44] Feijóo aligned intellectual and cultural history with protoanthropological observations on a global scale, even while he criticized "those travelers who . . . either because they had not spent sufficient time in situ or because they did not understand the language [of their interlocutors] were not able to penetrate their minds."[45] The term "intellectual map" itself implied a form of ethnography conceived in a spatial mode: a cultural geography of difference from a bird's-eye perspective, by which customs could be compared with one another and judged in a coordinated and interconnected manner. His employment of the term "mapa," or map, presumed that a cartographic system could usefully represent behaviors, customs, and beliefs and that a conceptual ethnographic map could function in similarly mimetic ways to a graphic document used to demarcate cities, countries, and continents. The notion drew an implicit correlation between the inchoate practice of textually describing cultures at a distance—or at least for a distant audience—and the mapping of geographical spaces with the use of instruments and measures.[46]

Above all, the idea of the "intellectual map" permitted Feijóo to censure Europeans for characterizing the indigenous inhabitants of America as irrational and uncultured, a discourse that had emerged from the earliest New World accounts and continued among commoners and other ignorant folk—those whom Feijóo called "the plebeians"—up to his own day. According to Feijóo, these plebeians believed that "[Amerindians] govern themselves by instinct rather than reason," a notion that Feijóo considered patently absurd, "as if some Circe had passed through those vast lands [and] transformed all the men into beasts."[47] Feijóo appropriated Greek mythology for philosophical purposes, using the shape-changing, cannibalistic figure of Circe as a metaphor for ethnographic beguilement. He implied that early European travelers performed a Circean role, wittingly or not, transforming human populations into monsters and turning European readers of travel literature into witnesses to a prodigious spectacle. Feijóo concluded his "intellectual map" with the admonition that a new observational method—in the footsteps of Bacon, Boyle, and Newton, whom he cited at the end of his text—would lead to a new science "denuded of all artifice," bringing Spain into line with other nations of Europe and thereby refusing to allow heretical thinking to "[bury] such beautiful lights in such sad shadows."[48]

But the sad shadows to which Feijóo referred were to live again in the prose of Ulloa, at least according to the "Juicio Imparcial," which sought to exploit what it saw as significant differences between the respective projects of Feijóo and Ulloa. While Feijóo's promise of a scientific method without artifice found fertile ground in Ulloa's use of instruments to carry out many of his observations, Feijóo's critiques became fodder for invective in the hands of the critic, who turned Feijóo's analysis on its head. The author of the "Juicio Imparcial" contended that not only had Ulloa perverted Feijóo's project through a Circean manipulation of his own; he also challenged Ulloa's use of geographical metaphors to describe indigenous customs in Quito. In the *Relación histórica*, Ulloa wrote that "if one contemplates their barbarity, their rusticity, the extravagance of their opinions, and their way of life, it would be easy to place [the Indians of Quito] not far from the parallel of the brutes."[49] The "Juicio Imparcial" rejected Ulloa's use of a universalizing, geographically informed metaphor to describe the characteristics of these people, as if its author were attempting to draw a stark contrast between observation at a distance—subject to the kind of distortion discussed by Feijóo—and a more omniscient form of analysis in the form of an intellectual map. By moving into a geographical mode, Ulloa had transposed the very meaning of Feijóo's project and betrayed those who viewed the "intellectual map" as an idealized form of ethnological organization, one that differed radically from the kind that transformed humans into pygmies and Indians into irrational beasts. The anonymous critic cited Feijóo as the explicit authority on the irrationality of non-European peoples, and used Feijóo's "intellectual map" and its discussion of Circe wandering through the vast lands of America as an illustration of the distorting effects of the brutish conquistadors. For the critic, it was the conquistadors, and not the "irrational" Indians, who should have been sanctioned for their lack of "rationality"; texts written by other Spaniards in the colonial Americas, including viceroys, bishops, ministers, and priests, gave testimony against "false imposters" like Ulloa and provided ample evidence of laws and ordinances that dignified indigenous populations, unlike what emerged in Ulloa's account.

Ulloa's reference to the "parallel of the brutes" points to the complex relationship between cartographic and ethnographic discourses in the mid-eighteenth century. For Feijóo, conceiving of ethnology as an "intellectual map" provided a conceptual model through which all humans—brutes, pygmies, and others—could be compared with one another. For Ulloa, the geographic "parallels" and climatic lines were ethnographic frontiers that pitted the brutes in one region against the civilized cultures of another. Following Montesquieu's climatic theories and prefiguring Adam Smith, Ulloa placed "brutishness" within a particular geographical zone and attached irrational and rustic characteristics to inhabitants of these regions. The anonymous critic picked up on these deprecatory comments and criticized the acrimonious conclusions that resulted from their use. These geographically conceived zones of contact between Europeans and the uncivilized brutes who inhabited them became figurative battlefields where the debate over how best to observe and characterize human cultural diversity was to be waged. Whether Ulloa was distorting the Indians through his observations at a distance or providing a fair and balanced assessment of a society in a retarded state was the question the "Juicio Imparcial" would address in its extensive and polemical critique.

Be he a priest, an administrative official, or a Creole patriot, the anonymous critic grounded his critique of Ulloa's *Relación histórica* with impressive methodological rigor. After unleashing an initial volley of moral invective in the earliest pages of his text, the critic proceeded to bemoan several elements in Ulloa's account: the lack of a fair and balanced perspective; the text's problematic and unreliable sources, including Ulloa's own observations; and its use of universal descriptions to depict phenomena that should have been attributed to particular groups within the restricted ambit of Andean society. Taken together, these criticisms formed the initial methodological assessment offered by the anonymous author of the "Juicio Imparcial," reflecting a decidedly moralistic intervention into the evidentiary underpinnings of Ulloa's text.

In the first instance, the critic indicted Ulloa for having offered a panorama of indigenous life in the Andes that left out many of the Indians' most favorable cultural attributes. For the critic, a balanced assessment would have included both "agreeable" and "detestable" characteristics, leaving the learned reader to interpret among multiple possibilities. With disdain, the critic accused Ulloa of employing a strategy of conscious omission, although he later suggested (in a somewhat contradictory fashion) that Ulloa suppress certain characteristics attributed to the Indians where there was not adequate evidence to justify their inclusion. If Ulloa did not wish to falsify the Indians' characteristics, the critic argued, he should have either omitted this material or carefully studied the foundations of its truth in one of three manners: "[T]hrough the most deliberate and mature investigation of events; by studying [other] authors and their motives in discussing the Indians; or by choosing one of the numerous defenders of the Indians that have now dispersed and can be found anywhere, including this court in Madrid."[50] Instead, according to the critic, Ulloa's arbitrary choices as to what he did and did not include served to distort rather than reveal the significant features of Amerindian culture and beliefs.

In addition to the issue of balanced reporting, the "Juicio Imparcial" also impugned the credibility of the texts that Ulloa had used to assess Indian behavior and intelligence. According to the critic, these sources had been written by arrogant tyrants that were "blinded by greed" and who wished to hide their own misdeeds behind accusations of Amerindian irrationality. It was in this context that the critic invoked Feijóo as an authority on the purported irrationality of non-European peoples, since Feijóo also criticized the authors of earlier accounts for their exaggerated observations. The critic assailed Ulloa's choices of source materials, calling his *Relación histórica* a "mere compilation of the old imposters of more than two centuries past."[51]

But Ulloa's choice of sources was only one error he had made in construing his ethnographic observations. The critic also took issue with Ulloa's reliance on eyewitness testimony by those whom Ulloa had called "reliable persons" (*personas fidedignas*). According to the critic, these sources provided a weak foundation for establishing a competent description of native life:

In order to offer condemnations of this nature, [an author's] experience should consist of far greater duration and a much more mature age. Observation and reflection should arise from a neutral state of mind (*animo libre*), [one that is] not preoccupied by contrary objects nor by other serious concerns.[52]

Maturity and an unspoken number of years spent in situ comprised adequate experience in the eyes of the critic. But far more significant than the duration of eyewitness experience was the individual's character. Those eyewitnesses whom Ulloa called "educated and intelligent" (*doctos e inteligentes*) observers were, according to the critic, from "unhappy parishes" (*infelices curatos*) and had ulterior motives in providing evidence to Ulloa:

[They] are called mountain priests (*curas de Montaña*). In order to receive sacred ordinations, or to exchange those they already have, they dedicate themselves to hard work and exile in the rugged mountains, heartened by the hope and practice of being promoted . . . Of these, there are many who disparage their parish in order to exalt their own service, and exaggerate the barbarity of the Indians in order to praise their own work.[53]

Ulloa, who according to the critic would have had frequent contact with these mountain priests, likely misunderstood that the opinions he received from these supposedly objective observers had been skewed by their own rhetorical strategies: "What were actually exaggerations made by aspiring priests (*pretendientes*)," wrote the critic, "[Ulloa] took as truths expressed by intelligent and educated individuals." The critic thus indicted Ulloa's familiarity with local actors as well as his own experience. The "Juicio Imparcial" undermined the originality of any contributions Ulloa made by questioning the motivation of his informants and by portraying his account as a two-bit compilation: "Nothing one reads in this chapter is new," concluded the critic, "nor is it particular to the Indians of the province of Quito."[54]

For the critic, the "particularity" of Amerindian culture in Quito was the ultimate proof that Ulloa had misinterpreted its customs and rituals. Ulloa's totalizing rhetoric had attributed specific characteristics to all Indians, claiming in one instance that the native populations of the entire region were superstitious soothsayers. The critic responded with outrage and reprehension. This flattening out of a people's character led him to suggest a less egregious form of description. According to the critic, if Ulloa would "replace [the term] 'everyone' in the first clause with [the term] 'many' from [the second clause]," then the Indians would look increasingly similar to other nations: "[N]ot all of them are good Christians, not even the majority, but many certainly are." In the eyes of the critic, then, Ulloa's overgeneralization of the Indians' behavior and his need to make absolute as opposed to relative pronouncements on the moral stature of a people ultimately undermined the efficacy of his observations.

As proof that his own method was superior, the author of the "Juicio Imparcial" offered an example by which Ulloa attempted to link Indian superstition to the regularity of their attendance at mass. Ulloa had written that "[i]n the practice of religion, [the Indians] are consistent, if superficial, and [they] attend mass only out of obligation." The critic responded with a more nuanced conception of cultural geography, a counterpoint to what he called the "universal sign" of "everyone" by which Ulloa had taken an entire nation and diminished its diversity. The critic's attempt to distinguish between the complete description of a particular region and the overall portrait of a larger territorial unit— what the second-century Greek astronomer Ptolemy might have called the difference between chorography and geography—found analogous terrain in the realm of ethnography, by which the critic defended the Indians of Quito by pointing out that few cases of idolatry had actually been reported by ecclesiastical authorities in the specific regions Ulloa had discussed:

If one were to make this comment about the recently converted [Indians] in Sucumbios, Mosca, Maina, Napo, etc., it would be one thing. But how is it possible to say such a thing of the Indians of Quito, Cuenca, Riobamba, &c. where by divine grace there is such a flowering of Christianity?[55]

The religious reference to divine grace and the flowering of Christianity alludes to the moral and spiritual dimensions of the "Juicio Imparcial" and to the uses and abuses to which Ulloa's account could potentially be put. The anonymous critic was not only interested in the textual implications of the *Relación histórica*—the unbalanced reporting, the problematic sources, the attribution of universal characteristics to particular groups—but was also keenly sensitive to how individual readers might actually use the hefty tomes to justify cruel acts against the hapless victims of Ulloa's account. He felt that Ulloa should have considered the "consequences and grave injuries" that might follow in the wake of his narrative, drawing a direct correlation between the deleterious effects of faulty ethnographic description and the exercise of colonial power. The critic did acknowledge the "novelty, style, beauty, and variety of plates that adorn the text, including the attractive printing and binding" as features that would contribute to the "credibility" (*asenso*) of the text.[56] But the potential uses to which the *Relación histórica* might be put and the series of questions to which the critic felt he had received no adequate response amplified the pernicious aspects of these material considerations: "What kind of treatment can these poor Indians expect," he asked "from someone who is persuaded by Don Antonio de Ulloa's description?"[57] For the critic, the practices of the print shop were equally implicated within Ulloa's broader project. To him, luxurious covers and costly inks served to denigrate the image of the Amerindian in the Andes as well, and he forcefully admonished a system that had led directly to such disgraceful ethnographic description.

Further "Reparos" on the Temperature and Physiognomy of Quito's Indians

As was common in many eighteenth-century travel accounts, Juan and Ulloa separated their numerical data from the descriptive account of their journey. The lists of longitudes and latitudes, compass variations, trigonometric measurements, and distances were included in the fifth volume of the *Relación histórica* written by Jorge Juan, who was responsible for chronicling the mission's astronomical observations. But one tool so caught the fancy of Ulloa that he included its measurements in the body of his own narrative text. It was Réaumur's thermometer. In November 1730, René-Antoine Ferchault de Réaumur (1683–1757) had described to the Academy of Sciences the most recent developments of an alcohol-based device that used an eighty-degree scale to determine the rise and fall of ambient temperature in air and water. In his published *mémoire*, Réaumur evoked with poetic zeal the myriad observations that the thermometer had already facilitated:

> Without the thermometer, we would not know that liquids that were mixed together increase in heat . . . We would never have discovered that certain salts cool the water in which they melt . . . We would not know that certain kinds of ice are colder than other types of ice . . . In short, the physicists are aware that an infinite number of experiments remains to be accomplished with a thermometer in hand.[58]

But Réaumur likely never imagined the ire that his thermometer would elicit from the pen of the anonymous critic.

During his travels, Ulloa had largely employed the thermometer as its creator might have intended, comparing temperatures of South American cities with readings that Réaumur himself had calculated in Paris. In his description of the route between Guayaquil and Quito, for example, he wrote that "the Thermometer showed 1016 in Caluma" while "on the 17th day at six in the morning the Thermometer showed 1014 $^1/_2$ in Tarigagua, which we thought seemed rather cool, accustomed as we were to warm climates." At Tarigagua, a strategic climatic zone between the dry air from the Andes and the coastal humidity of Guayaquil, Ulloa recognized that "one can feel opposite temperaments at the same time," depending on whether one was ascending to the highlands or descending to Guayaquil. "These [observations]," Ulloa concluded, "prove the celebrated opinion that the sensations are subject to as many apparent alterations as there are senses with which to speculate."[59]

Recognizing the myriad effects of temperature and climate on passing observers and long-term residents, Ulloa made an explicit connection between a town's temperature, its climatic features, and the temperament of its local populations. Taking the temperature of a given region and building a repertoire of comparative climatic measurements was linked to a program of comparative description that sought to debunk long-standing myths associated with the Aristotelian notion of a torrid zone, a region sandwiched between temperate and glacial zones extending northward and southward toward the respective poles. In the period following Columbus's and Vespucci's earliest journeys to the New World, the torrid zone was transformed from a zone of sea monsters and burning lands unfit for habitation to its being known as the "tropics," a region from which Europeans could extract knowledge and material specimens. In his description of Quito, Ulloa explicitly evoked the idea of the torrid zone:

Who could judge the climate (*Temperamento*) which Quito enjoys by using imagination rather than experience . . . Who would dare to persuade himself, when the light of [experience] or history leaves him, that in the center of the torrid zone, at the equator itself . . . not only is the heat tolerable, but there are places where the cold is irritating . . . The mildness of the climate . . . and the constancy of its days and nights renders a country that many have conjectured through discourse [to be] inhabitable rather pleasant.[60]

This transformation from an uninhabitable land of extreme climatic zones to one of temperate bucolic wonder, which echoed similar observations made by José de Acosta in the sixteenth century, was accompanied by the instrumental verification of the thermometer:

Quito's situation is so moderate, that neither does the heat bother [its residents] nor are its frosts unpleasant . . . [T]he following experiments of the thermometer will give sufficient proof of this.[61]

By recording the fluctuating temperatures at the equator, Ulloa attempted to update a geographical tradition that since the Renaissance had posited the torrid zone as inhospitable for human habitation. His reliance upon instrumental measurements allowed him to communicate his findings to an audience that would have been familiar with Réaumur's thermometer. But Ulloa did not attempt to modify the accepted wisdom that the torrid zone's inhabitants were lazy, a carryover from some of the earliest European accounts of Spanish America's tropical climes.

In turn, the critic responded to Ulloa's overarching characterizations of the Indians as deceitful and indifferent by attacking the instrumental basis of Ulloa's ethnographic observations:

[Ulloa] is not satisfied with saying that this is a general characteristic, but he has to add the customary word *all*, in order to inform us that his general rules are infallible. It pleases me that this extreme degree of deceitfulness is, undoubtedly, measured with the infallible Thermometer of Mr. Reaumour [*sic*] that he carried with him on his observations.[62]

The critic ascribed to the thermometer a deterministic role in Ulloa's judgment of the Indians' character. But by sardonically referring to the thermometer and its fallibility, the critic also implied that Ulloa's determination of the Indians' laziness was fallible as well. According to the critic, Ulloa consulted his thermometer in order to determine why the Indians did not comply with the demands of their masters: "[H]e returns his hands to Reaumur's thermometer to measure [their] laziness, and since [the thermometer] has no line with which to indicate [laziness], he qualifies it as indeterminate."[63] Nonetheless, Ulloa's judgments regarding the character of the Indians in some ways gained legitimacy through the appearance of instrumental measures.[64] By punctuating his account with the temperatures and climatic observations in and around Quito, he built his reputation as an infallible observer, but the conclusions he drew with these observations had little to do with the intended purpose of the instruments he was using. As the critic explained, Ulloa's conclusions could not have been farther from the truth: the Indians Ulloa would have observed during his extended stay in Quito were in constant movement throughout the province, hiding only from the malevolent gaze of the Spaniards so as to avoid the castigation and harassment of colonial rule.

Temperature was not the only factor Ulloa used to measure the culture of Quito. He also made observations on the physiognomy of half-castes and mestizos, individuals of mixed birth whose heterogeneous appearance was one of the hallmarks of colonial Spanish American society and the subject of the famous *casta* paintings outlining the classificatory order of a controlling and hierarchically minded regime.[65] Ulloa had written that "the nose [of a Mestizo] is small, thin, and protrudes slightly at the bone, from which, although slightly pointed, it bends somewhat as it turns down toward the upper lip."[66] These physical characteristics were an attempt on the part of Ulloa to provide a categorical assessment of the social classes and ethnic groupings in Quito, moving as he did from one "class" to the next: "Spaniards or Whites, Mestizos, Indians or natives, and Negroes, with their progeny." To the anonymous critic, Ulloa's physiognomic distinctions were patently absurd:

The signs that Don Antonio provides so abundantly to distinguish Mestizos [from other groups] are so arbitrary that they lead one to laughing rather than helping one to recognize a Mestizo. The variety of noses in the Mestizos is like that in most people: there are flat ones and long ones and puffy ones and crooked ones.[67]

The anonymous critic placed what Ulloa considered to be distinctive features of the inhabitants of Quito—be they Indians, mestizos, or Spaniards—into a universal framework, precisely the opposite tactic that he used to criticize Ulloa's overarching statements about the Indians earlier in his account.

More generally, the critic's assessment of Ulloa's technical predilections was an admonishment of the use of instruments to describe traits that could only be determined through a more subjective, cultural approach. Just as the critic impugned Ulloa's use of geographical language to place "the brutes" within a particular climatic regime, he also criticized Ulloa for relying on devices that removed the observer from direct contact with the culture being described. The critic's own perspective, replete with descriptions drawn from on-the-spot experience and direct "ocular" testimony, was an indictment of an ethnography performed at a physical and emotional remove using technical tools and devices. Following in the mold of Feijóo, the anonymous critic placed instrumentation into the deformative category of distance, arguing instead for an ethnography based on extensive time spent in situ, reliable commerce with balanced and unbiased sources, and constant attention to a moral code that would infuse the text with a higher humanistic purpose. At the very end of his critique, the anonymous critic wrote that he would be "consoled if by improving (*reformando*) Don Antonio Ulloa's description, the reputation of the miserable Indians of Quito, so unjustly and massively defamed, will be restored."[68] There is little indication that the anonymous critic's methodological critique served in any way to counter the Indians' defamation in subsequent editions of the *Relación histórica*; there is less indication still that his robust defense of the Indians of Quito significantly ameliorated their state of collective oppression under colonial Spanish rule.[69]

Conclusion: Useful Lights and Specious Falsehoods

Three decades after Feijóo published his "Intellectual Map and Comparison of Nations," the editor of the first English translation of Juan and Ulloa's *Relación histórica* chose to discuss the positive effects that could emerge from the perusal of travelers' tales from beyond European shores. "Books of voyages and travels," the editor recounted,

have been commended by persons of the greatest sagacity, and in the highest reputation, for superior understanding. Voyages and travels [were] the favorite study of the judicious Locke, who looked upon it as the best method of acquiring those useful and practical lights, that serve most effectually to strengthen and also to enlarge the human understanding.[70]

These "useful and practical lights" garnered through the study of travel accounts were counterposed with those "marvelous and very often in-credible relations" that lacked the verisimilitude expected from eyewit-ness observation. In the words of the editor, such "strange and surpris-ing adventures . . . have served rather to mislead than to instruct men's minds by a display of specious falsehoods." This contrast between useful lights and specious falsehoods in the study of travel narratives could only be resolved by "knowing the character of the authors we peruse, that we may judge of the credit, that is due to their reports." Donning rhetorical vestments of his own, the editor praised "the authors whose writings are now offered to the public in an English dress . . . , men distinguished for their parts and learning, and yet more for their candor and integrity." In his view, reading travel literature required a moral judgment about the author in order to determine the veracity of a given account. "For many writers impose on the world not through any evil intention of deceiv-ing others, but because they have been deceived themselves," the editor continued. "They relate falsehoods, but they believe them."[71]

The English editor's descriptions of Ulloa and Juan's "scrupulous at-tention" stand in stark contrast to the polemical volleys of the "Juicio Imparcial." This manuscript critique also relied on a moral code to deter-mine the value of Ulloa's narrative, but it sought to attack Ulloa's status as a credible narrator rather than praising his "circumspection." The "Juicio Imparcial" claimed to defend the natives of Quito against an im-moral ethnographic observer who had garnered praise from the learned community because of his authority and character. But this handwritten diatribe over ethnographic practice never reached the public eye in the same way as did Ulloa's printed account, which was translated into En-glish, French, German, and Dutch in the decades that followed. Critical judgments against the *Relación histórica*, it seemed, were left for oth-ers to detect and unravel in the hazy blur of eighteenth-century anon-ymity.

For Spain, the publication of the *Relación histórica* represented the opening of Spanish intellectual culture to a global network. Juan and Ulloa's account provided an international community of savants and academicians with a tangible sign of Spain's renewed scientific vigor. In the decades that followed, Madrid became a central pole in botanical ex-changes connecting the Spanish capital with the Parisian Jardin du Roi,

London's Kew Gardens, and other gardens and laboratories in between. Spain appears on equal footing with her European neighbors to the north in the epistolary circles and specimen-sharing networks encouraged by José Quer, first director of the Real Jardín Botánico in Madrid, and his successor, José Antonio Cavanilles. And it was Antonio de Ulloa himself that was responsible for creating the first cabinet of natural history, even if in its earliest years it resembled more of a storehouse than a display case for serious study of the natural world.[72] Pedro Franco Dávila, the native of Guayaquil inspired by Maldonado who spent two-and-a-half decades amassing natural curiosities in Paris, became the Real Gabinete de Historia Natural's first director in 1771. And Spanish expeditions to the Americas at the end of the eighteenth century went a long way toward sealing Spain's status as an important partner in natural historical investigations overseas. But the *Relación histórica*'s overture to a broader European audience through print also opened Spain's nascent scientific culture to increased criticism within the learned periodicals and correspondence of the period. This criticism was neither limited to Spanish critics nor exclusively composed of non-Spaniards. Indeed, one indication of the openness of the eighteenth-century Republic of Letters lay precisely in the ability of those without access to the technologies of print to communicate and discuss their grievances with the author of particular treatises—and, potentially, to the broader public as well.

Apologists for Spain's newly minted scientific culture did not sit idly by as Ulloa's text was maligned by an anonymous critic, however. A response, "Defensa à favor de los Indios de Quito" ("A Defense in favor of the Indians of Quito"), signed by an "impartial politician" who *also* claimed to be a protector of Quito's natives, was soon produced in the wake of the criticisms set out in the "Juicio Imparcial." In the eyes of this anonymous counter-critic, the "Juicio Imparcial" had been baseless and inappropriate. The counter-critic used his title as "defender of the Indians of Quito" less as a marker of cultural affinity than as a rhetorical veil, behind which he intended above all else to vindicate the honor and "great merit" of Ulloa and his text:

[I]t seems to me all the more dangerous when I consider the maliciousness with which the author [of the "Juicio Imparcial"] wishes to offend the good reputation of the travelers, and confuse the truth, with a set (*cumulo*) of ingenious sophisms... [They] are used to carry forward his vain ideas and make himself famous at the expense of a work whose impartiality leaves no doubt as to the accuracy of its information... despite the difficulties that present themselves in wishing to verify these materials because of the great distances that exist between those lands and these.[73]

In order to protect the "good reputation" of the two Spanish travelers, whose account is portrayed in the "Defensa" as representing truth and impartiality in contradistinction to the intellectual sophistry of the anonymous critic, this author used the representation of the Indians as a foil for his own ideological purposes. In the hands of the counter-critic, the indigenous populations of Quito became an instrument to shore up the *Relación histórica*, a monument to Spanish glory that had been put forth for public view only a couple of years before. The Indians were transformed from objects of voyeuristic curiosity into a blunt weapon with which to batter the claims of an anonymous critique. And while a full analysis of the response elicited by the "Juicio Imparcial" lies beyond the purview of this chapter, the riposte of an individual who kept his identity hidden reaffirms the malleability of the Indians of Quito as an instrument of critical discourse. It also underlines the importance of occultation in forging such a discourse within the vigorous, if understudied, manuscript culture of eighteenth-century Spain.[74]

This manuscript response to the attacks of an anonymous critic did fit, however, within a larger pattern of eighteenth-century apologies for Spanish culture. In response to the perceived cultural decadence of the Hispanic tradition, Spanish intellectuals such as Mayans, Feijóo, Burriel, and others sought to make forceful arguments about the capacity of Spanish culture. Feijóo, in his *Glorias de España*, proclaimed that other nations had attacked Spain precisely because of its own brilliant past: "Since the unaccustomed eyes of other nations could not withstand the resplendence of such illustrious glory, they chose instead to obscure it, painting the errors (*desórdenes*) that our [people] committed in their conquests with the darkest shades."[75] But the targets of these apologists for Spanish culture were almost always to be found beyond the Pyrenees. Rarely were the likes of Burriel and his fellow apologists obligated to critique a member of their own national community, either in Spain or in one of Spain's numerous territories overseas.

For this reason, the critical efforts of an anonymous author, who presented himself both as a Spanish patriot (recall his loyalty to the monarchy and his use of the term "our Provinces") and as an in situ observer of colonial affairs, went against the grain of most previous models in which national or colonial rivalries were to be found as the primary motivation for epistemological critiques. The individual that defended the Indians of Quito against the "calumnies" of Ulloa's description wanted his identity and his deeper motivations to remain hidden. But Feijóo's brief evocation of the efforts made by other nations to "obscure" the glories of Spanish conquest certainly did not fit the critical work of the anonymous critic, a seemingly internal foe who was not so easily dismissed by the apologetic rhetoric of Spanish patriots. Instead, the strategic omission of the critic's identity required a different strategy, an anonymous counter-critique that sought to suppress a debate that might have emerged had the text been allowed to circulate widely. While he indicted Ulloa for having consciously omitted certain favorable traits of the indigenous character, the anonymous critic ironically employed a similar strategy of strategic effacement: leaving his own identity to be determined by the power and force of his arguments rather than a knee-jerk reaction to his status or prestige within colonial or metropolitan society.

In the end, whoever the author of the "Juicio Imparcial" might have been—motivated by ecclesiastical goals, Creole patriotism, or bearing a more mundane grudge against Ulloa—he was not particularly sanguine about the utility of criticism within the culture of eighteenth-century Spain. "Even in fictional accounts," the critic wrote in the first paragraph of the critique, "a Momus is always disliked by the Gods," referring to the classical figure of Momus, the Greek god of censure and ridicule who was banished from the towering peaks of Olympus for having dared to criticize the judgment of his fellow deities. The maelstrom of dialogue and invective that took place within both print and manuscript spheres following the publication of the *Relación histórica* demonstrates nonetheless that the genre of critique put forward in defense of the downtrodden of the New World was seen as a justifiable exercise, a project to which significant intellectual and ideological resources were committed, and that merited equally vigorous rebuttals in defense not only of the Indians of Quito but also of a historical relation that was written largely at the Indians' expense.

SIX

Incas in the King's Garden

My mother's uncle . . . gave me a long account of the origin of [the Inca] kings . . . which I have attempted to translate faithfully from my native language, that of the Inca, to a foreign tongue, Castilian, although I have certainly not written it with the magnificence (*majestad*) with which the Inca spoke nor with all the meaning that the words of that language contain . . . Instead, I have shortened [his account] by taking out some of the things that would have made it tedious (*odioso*). In the end, it will be enough to have conveyed the true meaning of [these words], which is what is most important for our story.

INCA GARCILASO DE LA VEGA, *PRIMERA PARTE DE LOS COMENTARIOS REALES* (1609)

Everyone has a copy of [the Incas'] history. Everywhere [in Europe] one can find monuments to the wisdom of their spirit and the solidity of their philosophy.

MADAME DE GRAFFIGNY, *LETTRES D'UNE PÉRUVIENNE* (1748)

Antonio de Ulloa may have offered harsh descriptions of eighteenth-century Amerindians, as we saw in chapter 5, but the *Relación histórica* contained little but praise for the Andes' most celebrated precolonial inhabitants. The Incas, according to Ulloa, were "remarkable in [their] understanding, capacity, talents, and rationality," and despite the "ignorance" and "barbarity" he perceived in their eighteenth-century descendants, it was impossible for Ulloa to "doubt those earliest reports regarding [their] industriousness, good conduct, and laws."[1] In fact, Ulloa's chapter "Intelligence, Customs, and Character of the Indians or Natives of the Province of Quito" was conceived explicitly to reconcile these two highly conflicting visions—the one historical, the other contemporary. What is more, the final section of the *Relación histórica* portrayed the Incas as an important link between indigenous

Figure 46. The "Resumen historico del origen, y succession de los Incas, y demas soberanos del Perù" was published as part of the *Relación histórica del viage a la América meridional* (1748), and was designed not only to recount the history of the Inca empire but also to legitimate the "other sovereigns of Peru," the Spanish kings. The Incas are pictured in this frontispiece alongside their later "successors," and are portrayed as noble precursors to Spanish rule. The sole exception is Atahuallpa, who is shown holding his axe horizontally (unlike the other Incas, who hold their arms upright) as he gazes upward with envy at his half-brother Huascar. The Spaniards portrayed Atahuallpa's usurpation of his brother's rule as treacherous, which further legitimated Francisco Pizarro's conquest of Peru. Ferdinand VI, recently installed on the Spanish throne, is at the center of the image. Courtesy of the John Carter Brown Library at Brown University, Providence, R.I.

rule in the Andes prior to the Conquest and the now-consolidated Spanish sovereignty in the viceroyalty of Peru. Accompanied by an extravagant plate replete with Inca walls and a Palladian cathedral, the "Historical Summary of the Origin and Succession of the Incas" sought to justify the stewardship of the kings of Spain by literally writing Fernando VI (r. 1746–59) and his predecessors into Incan history. In this image, they were portrayed as following the Inca rulers in a natural and ordered monarchical succession. Ulloa papered over the "rustic" and "irrational" Amerindians from his earlier chapter and highlighted instead the enlightened character of the Incas' noble and august rulers, political leaders who in his depiction had yielded their scepters and submitted without contest to Spanish rule (fig. 46).

The source for Ulloa's account was Garcilaso de la Vega's *Primera parte de los Comentarios reales* (1609), a text that set out to revise the earliest accounts of the Inca empire and was also, ironically for Ulloa's purposes, a subtle critique of Spanish colonial administration in the New World. Born in Cuzco, the cultural and political center of the vanquished Inca empire, Garcilaso (1539–1616) became the most renowned and respected author of Incan history in the seventeenth century. Despite having left the Andes at the age of twenty-one to take up residence in Spain, he boastfully asserted in his account that he was the only person capable of reproducing "clearly and distinctively . . . what existed in that [Incan] republic prior to the Spaniards' [arrival]."[2] Garcilaso "the Inca" (used to distinguish him from the homonymous poet of the Spanish Golden Age, to whom he was in fact related) offered a utopian portrayal of the Incas that emphasized the sophisticated system of political and military control that had emerged under their rule. As a mestizo—in this case, the son of a Spanish conquistador and an Incan "princess of the Sun"—Garcilaso had sought to record the traditions of his Andean ancestors for a Spanish audience.[3] His text chronicling Inca expansion began with the mythic birth of two "children of the sun" on the shores of Lake Titicaca and moved through the peaks and valleys of the Andean cordillera, as the numerous followers of the Incas brought an ever-growing number of disparate tribes and peoples under their dominion. Most importantly, the *Comentarios reales* described the extraordinary economic and agricultural structures put into place by Garcilaso's Andean forebears: an imperial organization that provided amply for its citizens while safeguarding their political rights under a prosperous and enlightened hereditary monarchy.

But the seventeenth-century mestizo would never have imagined the extent to which the "true meaning" (*verdadero sentido*) of his account was to change language, form, and content in the editions and translations that followed, nor how it might be appropriated for ends far different from what he or his contemporaries had intended. Nearly one-and-a-half centuries after Garcilaso wrote what he considered to be the definitive history of the Inca empire, a new translation of his work was produced in Paris under dramatically different editorial, typographical, and natural historical circumstances. This latest edition of the history of the Incas was translated and reconfigured by a group of naturalists at the Parisian Jardin du Roi, home to the king's Cabinet d'Histoire naturelle, which had just been augmented by an extraordinary set of objects. During their South American sojourn, Louis Godin and La Condamine had sent the cabinet several Peruvian treasures meant to represent the useful work they had carried out in the name of the French monarch. The bountiful assortment of objects included various indigenous artifacts, a dictionary and grammar of the Inca language, and numerous natural historical specimens, sent in five separate shipments from Portobelo (Panama), Quito, Lima, Cartagena, and Tucumán (this last shipment had traveled with a container that held a set of "monstruous bones" thought to belong to Patagonian giants, but nervous and superstitious sailors later jettisoned its contents at sea). These material objects were folded into the king's collection, but they were not included in any ordinary catalog of the cabinet. Instead, they were placed inside a newly revised and translated history of the Incas, alongside the historical description of one of South America's most captivating pre-Columbian polities.

This was an unprecedented project undertaken by the Jardin du Roi: a new translation and reedition of Garcilaso's classic account that employed Incan history in order to display natural history from an eighteenth-century French collection. Conceived and carried out during the seven-year intendancy of Charles-François de Cisternai du Fay (1732–39) and published in 1744 in a two-volume octavo edition, the *Histoire des Incas, rois du Pérou* heralded a text that was "newly translated from the Spanish" according to "an improved organization" (fig. 47). What made this edition different from the many earlier editions of the text were significant structural changes that allowed freshly acquired and potentially useful knowledge to be displayed prominently within the volumes. During a period in which French society was experiencing social unrest and agricultural crisis, many of the topics highlighted by this new organization reflected contemporary concerns as much as they did aspects of Inca life from before the Spanish conquest. Unlike all previous editions and translations, the 1744 *Histoire des Incas* profoundly modified Garcilaso's textual arrangement of Incan history, signaling a new editorial approach that abridged and compressed materials in order to highlight information of social and political relevance. The naturalists at the Jardin du Roi used typography and textual organization to their society's collective advantage in a period of social and agricultural turmoil.[4]

An analysis of the structural changes in this new edition shows how material and editorial strategies both enabled and constrained naturalists in France as they reconstructed the social, political, and natural history of the precolonial Andes. By examining this particular translation of Garcilaso's history, as well as the academic *mémoires* and travelers' accounts the editors consulted, this chapter reassesses a critical moment in the reception history of the Incas in Europe.[5] Remarkably, the conservation and showcasing of South American flora and fauna in the king's cabinet played a critical role in the representation of the Inca in Europe. Through

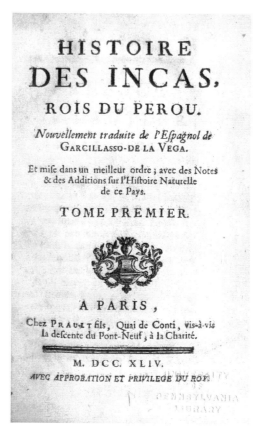

Figure 47. The *Histoire des Incas, rois du Pérou* (Paris, 1744) was edited at the Parisian Jardin du Roi with close attention to the natural historical specimens sent from South America by members of the Quito expedition. As the title page makes clear, the French editors intended this to be a "reordered" history of the Inca empire, using footnotes and other "additions" to showcase these new botanical specimens and to provide greater readability, and thus greater utility, for an eighteenth-century audience. Courtesy of the Rare Book and Manuscript Library, University of Pennsylvania.

the artifice of European editorial and typographical conventions, the editors of the *Histoire des Incas* were able to create a hybrid space within the interstices of a classic text of New World history by bringing specimens of Peruvian natural history into contact with a mestizo's account of indigenous culture. The Jardin du Roi provided a Parisian theater for collecting and displaying, but it did so through text and history rather than a portrayal of exotic specimens within the closed cabinet.[6]

The establishment within the domain of the royal gardens of a "universal meeting place for all the productions of nature," as Diderot referred to the cabinet in the *Encyclopédie*, was a gesture aimed at centralizing power by bringing back to the metropole the disparate pieces of useful natural knowledge from throughout the globe. The production of a new text of Incan history provided an ideal opportunity to showcase the superior abilities of French naturalists in harvesting the hidden fruits of botanical knowledge. Especially treasured was knowledge revealed by indigenous populations in the Americas, since it provided an opportunity to vaunt French capacities vis-à-vis the Spanish, whom the French saw as incapable of harnessing the benefits of indigenous knowledge. This sense of ultra-Pyrenean superiority was admirably expressed by Voltaire in a letter anticipating the Quito expedition, where he wrote that "if in politics we are humble servants to the Spanish, our Academy of Sciences avenges us. The French gain nothing through war, but they do measure (*toisent*) the Americas."[7] The production of a history of the Incas at the Jardin du Roi was thus part of a concerted effort to distinguish French intellectual aspirations from Spain's territorial conquests: the Jardin du Roi did not seek in this instance to assert power in a specific colonial domain but rather through an intellectual dominion over the natural world. And the Incas became a convenient conduit on a number of levels.

The "Marvelous Order" of Inca Agriculture and the Editorial Disorder of Garcilaso's History

When Garcilaso's *Comentarios reales* was first published in 1609, it was widely praised as an insider's view of the Andean civilization that had been decimated by sixteenth-century Spanish invaders, and his text quickly became the authoritative history of the Incas in both the New World and the Old. Having grown up listening to the oral histories of his Incan ancestors in their native tongue, Garcilaso portrayed his *Comentarios reales* as compensating for earlier Spanish chroniclers who wrote about topics "in such a terse fashion that [he] had difficulty understanding even those [customs with which he was] quite familiar."[8] He proclaimed that his intention was not to attack or criticize these historians, which included such classic sixteenth-century authors as José de Acosta, Francisco Lopez de Gómara, and Pedro Cieza de León, but rather to "provide commentary [to them] and serve as an interpreter for many of the Indian terms that, as foreigners to that tongue, they interpreted outside their appropriate meaning (*propriedad*)."⁹ Garcilaso's bilingual abilities in Quechua and Castilian allowed him to write the history of his people with a confidence and authority that, as he saw it, were uniquely his own.

In the intervening years between this first edition and the 1744 text produced at the Jardin du Roi, a host of editions and translations had grown out of Garcilaso's original text. The first was Jean Baudoin's 1633 French translation, *Le commentaire royal, ou L'histoire des Yncas, roys du Peru*, which provided a more or less literal translation of the original Spanish text published twenty-four years earlier.¹⁰ Baudoin's translation was followed by an English edition in 1688 and two nearly identical French-language editions, published in Amsterdam in 1704 and 1715. These Amsterdam editions were the first to point out the disorder of Garcilaso's original text. In their respective prefaces, they emphasized both the positive traits of Garcilaso's history as well as the numerous improvements that had been implemented for their own revised editions:

While Garcilasso [*sic*] de la Vega, who is the author of this text, deserves credit as a faithful historian, one can also say that his style contains many vestiges of the land in which he was born. There are many annoying repetitions in his text, and Baudoin, to whom we owe this French translation, had the courage to maintain them.¹¹

Bemoaning the "scrupulousness" of Garcilaso's first translator in providing such a literal translation, the editor(s) of the 1704 and 1715 editions went on to claim that he had not left a single sentence of Garcilaso's original text untouched by his revisions, assuring the readers that they would find "an infinite number of spots clarified and [many] doubts removed."¹²

In the two decades that followed the publication of these Amsterdam editions, there was an explosion of European interest in Incan history. In 1723 in Spain, Andrés Gonzalez de Barcia produced the first authoritative Spanish edition of the *Comentarios reales* since Garcilaso's original, under the auspices of his project to recuperate important narratives of New World history.¹³ The decade that followed saw a flowering of dramatic productions related to the history of the Inca in France as well, including the first productions of Jean-Philippe Rameau and Louis Fuzelier's *Les Indes galantes* (1735) and Voltaire's *Alzire* (1736). This enthusiasm was not wholly unrelated to the expedition to Quito; Voltaire famously wrote to Algarotti that his muse had been together with La Condamine's compass in Peru, and Voltaire and La Condamine were in intermittent contact while the latter was in South America.[14] A parody of Voltaire's drama soon followed from the pen of Jean-Antoine Romagnesi: *Les sauvages: Parodie de la tragédie d'Alzire* was performed by the *Comédiens italiens* in March 1736. And capping off this three-decade-long Incan fervor, an even more elegant French edition of Garcilaso's text was published in Amsterdam in 1737. Although it hewed closely to the two previous Amsterdam editions of 1704 and 1715, the 1737 edition added several attractive and evocative engravings to the text, portraying the Incas as an ancient polity with overtones of classical civilizations (fig. 48). Thus, at right around the time of the Quito expedition's departure, several competing editions of Garcilaso's text were circulating in Europe, providing ample opportunity for authors and readers alike to catch the wave of Inca-inspired enthusiasm that appeared to be sweeping the continent.[15]

Françoise de Graffigny was one of those who participated in this Inca frenzy. On May 5, 1743, she attended a performance of Voltaire's *Alzire* and already the very next day was perusing an edition of Garcilaso's Incan history. In a letter to Nicolas Devaux, she explained that she found the history of the Incas far more stimulating than the history of France she was reading concurrently. Several years later, in 1748, Graffigny published a fictional account of an Inca princess named Zilia who had been captured by French soldiers and deposited in France. Through a tale told in thirty-eight letters and entitled *Lettres d'une péruvienne*, Graffigny used a captive woman of Peruvian extraction as a foil for criticizing her own culture, rather than attempting, like Garcilaso, to portray his culture's history through its own ancestral voices. Graffigny's strategy was not unfamiliar in eighteenth-century France. Letter writing either with an invented nom de plume or through a fictional conversation with foreign visitors allowed a sardonic author to express things that he or she could never say directly. The most famous contemporary example was Montesquieu's *The Persian Letters* (1721), in which two young Persians visited Paris and made critical remarks about the habits and moral character of Europeans. But the source for Madame de Graffigny's text was not a work of dramatic fiction, even if her text may have been inspired by both *The Persian Letters* and Voltaire's *Alzire*. Instead, the raw material out of which Graffigny sculpted her Peruvian letters was Garcilaso's *Comentarios reales*, rereleased in its 1744 edition precisely during the period in which she was working on her text. One of eighteenth-century France's most successful literary adventures, inspired indirectly by a scientific expedition to the New World and an outpouring of enthusiasm

Figure 48. This portrait of Manco Capac from the *Histoire des Yncas,
rois du Perou* (Amsterdam, 1737) shows the first Inca "baptizing" the
imperial capital of Cuzco. This translation of Garcilaso's text provided
dramatic images like this one to capture a broader fascination with
Inca culture in Europe during this period, exemplified by Rameau and
Fuzelier's *Les Indes galantes* (1735) and Voltaire's *Alzire* (1736).
Courtesy of the Special Collections Library, University of Michigan.

toward one of South America's indigenous polities, was thus also connected to one of the French monarchy's oldest and most esteemed botanical institutions.[16]

Since its inception as the Royal Physic Garden in 1626, the Jardin du
Roi was always a place where a deep interest in medicine and agriculture
merged with the study of exotic plant species. Guy de la Brosse himself
commented that the role of the Jardin Royal would be to recognize "all
plants by their diverse names and shapes" and to teach "their qualities,
faculties, and properties, and their simple and compound usages."[17] But
the idea of using a classic history of the Incas as the conduit for expressing new ideas about botanical and agricultural organization was unprecedented. The seeds and saplings shipped by Godin and La Condamine served

as the material muse for this new edition of Garcilaso's text, just as La Condamine's compass in Peru inspired Voltaire in the creation of *Alzire*. Once these crates arrived at the Jardin du Roi, the New World specimens they contained caught the attention of Thomas-François Dalibard, an enthusiastic apprentice to the Comte de Buffon who meticulously compared the plants, animals, insects, and minerals with the travel narratives that had been gathering dust for decades on the bookshelves of the king's natural history cabinet. It was a perfect opportunity to demonstrate the potential for social improvement that a botanical garden held within its walls. By comparing these texts with live specimens from the Jardin du Roi, Dalibard sought to contribute to a program for social and agricultural renovation in the mid-eighteenth century. And the *Histoire des Incas* became the showcase—a botanical theater—for Dalibard's demonstration.[18]

In the sections of the *Histoire des Incas* for which he bore direct responsibility, Dalibard showed a keen familiarity with New World travel narratives, and he closely examined the natural historical specimens sent from the Americas so that he could find new ways of putting old knowledge to good use. In part owing to their increased access to natural objects brought in from around the world, Dalibard and his colleagues at the Jardin du Roi came to see their role not merely as translators of older accounts—changing terms from one language to another—but rather as active compilers licensed to transform the editorial configuration of previous texts, especially those works that dealt with topics about which they now claimed to know more than their original authors. The epistemological framework within which the editors of the 1744 edition were to interpret, assess, and edit Garcilaso's text had been transformed not only by the arrival of new material specimens from South America but also by stylistic changes in typography that allowed for the hierarchical placement of new information through cross references and more elaborate bibliographical techniques. In addition to the reconfiguration of older material, massive abridgements were effected as well: in some places, entire chapters of Garcilaso's original 1609 text were compressed into two or three lines, and often entire paragraphs were transmogrified into one-line footnotes. Indeed, this last feature of the 1744 edition—the addition of footnotes—was entirely foreign to Garcilaso's original text and had been utilized only rarely in translations prior to this new edition. While few details were added to the historical portion of the text (which was also condensed, reordered, and compacted into the *Histoire's* first volume), some new observations—such as pointing out the literary source of *Robinson Crusoe*, first published in 1719 and translated into

French in 1720–21—clearly extended the temporal reach of Garcilaso's text into the eighteenth century. The use of typographical features such as footnotes and brackets facilitated other kinds of additions in the second volume, allowing the editors of the *Histoire des Incas* to give a privileged place within the corpus of the work to details from reference texts, more up-to-date sources, and botanical treasures recently arrived from overseas.[19]

The organizational strategies used by those who cut, arranged, and collated materials to bring Garcilaso's text "up-to-date" for an eighteenth-century audience were revealed to readers in a sixteen-page translator's preface. In it, the editor explained the historiographical motivations and methods that were followed throughout the *Histoire des Incas*. The individual or individuals responsible for writing this and a second preface, to whom I will refer collectively as "editor" throughout this chapter, presumably maintained some semblance of editorial control over the translation and abridgement of the *Histoire des Incas*. It is impossible to say with certainty who may have composed or translated particular sections, with the exception of Dalibard's contribution to the chapters on natural history. But the term "editor" not only emphasizes the structural changes that were made to the text but also stands in for the multiple tasks and possibly multiple identities of the individual(s) who carried forth this project.[20] Beginning with a brief biographical sketch of "Garcillasso de la Ve'ga, Author of this History," this translator's preface highlighted particular features of Incan social, cultural, and economic life, including sections on the enlightened principles of Inca rule, agricultural practices, and Inca scientific achievements as well, especially in astronomy. The editor described Garcilaso's mixed birth, his education in a Peruvian school designed especially for Inca youths, and the young mestizo's encounter with early Spanish chronicles, which he had found so woefully inadequate.

But the editor of the 1744 edition also leveled a series of criticisms at Garcilaso's *Comentarios reales* that was even more sustained than earlier editors, especially concerning the "disorder" and "confusion" of the original text. According to the editor, Garcilaso had failed to produce a "methodical text," one that evinced a particular historical logic and a prefabricated and recognizable structure. The editor described instead the "heaping up" of facts, a disorderly accumulation of details that drowned the reader and "[made] it impossible to follow [the text's] progression."[21] The preface argued for the perfectibility of historical texts, that the organizational methods employed by eighteenth-century authors, even those who produced what the editor called "the most mediocre works," allowed for greater readability and ultimately greater utility for the public than many previous authors, including Garcilaso. The twin pillars of ever-improving readability and utility were the epistemological underpinnings of this new editorial impulse, which was reflected in the explosive increase in the number of natural historical compendia, geographical dictionaries, and compilations of travel literature that were becoming available in Europe at midcentury. The celebrated publication of Diderot and d'Alembert's *Encyclopédie* was, in itself, merely a conspicuous example of an emerging tendency: to parse knowledge into its component parts and reconstruct the pieces in an abridged form suitable for easy digestion by a broad, learned public (as we will see in the following chapter).

The translator's preface represented a wholly original editorial contribution to the *Histoire des Incas*. As such, it should be seen not as a summary of Garcilaso's original text but rather as an explication of the specific reasons why this reedition was produced for its intended public at precisely this moment. In his prefatory text, the editor discussed the structural transformations undertaken to "remedy" the disorder of the original history and the useful purposes to which, he hoped, the translation would eventually be put. The subjects discussed, the organizational changes, the editorial conventions, and the stylistic and typographical strategies used to facilitate the expression of particular ideas constituted what the editor called the "principal merit" of this new translation: its "precision" and "exactitude." But in what context was this new edition more "precise" and "exact" than before? For the editor, the utility of his "remedied" text seemed to lie in its didactic lessons, revealed by certain "great examples" of Incan history: "We thought it could be extremely useful to society because of the great examples that [Garcilaso] presents, of goodness, gentleness, justice, and moderation on the part of the Sovereign, of gentleness, submission, attachment, and respect on the part of the Subjects. The utility of these subjects made us undertake this new Translation." But why would these subjects seem particularly useful at this moment? In Garcilaso's day, these subjects were ideologically useful for him as well. For this reason, it is important to see how the editor infused his preface with examples of great moral and political beneficence on the part of the Inca kings and emphasized those aspects of Incan culture that may have resonated with a French audience. For instance, he praised the kings for their generosity, moderation, and benevolence toward their subjects. The editor portrayed Manco Capac, the first Inca, as the most noble of monarchs, explaining that "he steered all the rules he considered toward the general good of society." Describing in turn the close bonds that linked the Incas to their sovereign, he went on to explain that the people came to look upon Manco Capac "not only as the direct descendent of the sun but as the shared father of all peoples," employing a paternal metaphor about royalty familiar to an Old Regime society accustomed to positive comparisons between politics and family life. The editor also drew attention to aspects of Incan religious practices that demonstrated the strong similarities that existed between Inca religion and Christianity.[22]

But most significantly, the focus of the preface was on the agricultural innovations of the Incas and their methods for ensuring that all members of society received adequate provisions through "state" intervention. The editor emphasized how Manco Capac had taught his people "the art of cultivating the Earth, and the means to take from the bosom of this common mother all the assistance they might ever need to preserve their lives and their health." While this aspect of Incan life also featured prominently in Garcilaso's original account, the editor gave special attention to the issue in the brief space allotted to him by focusing on the "wise" rules imposed by the Inca for the betterment of his society: "[E]ach individual had sufficient provisions each year for himself and his family. If, despite these precautions, it so happened that some [members of the society] were wanting, they provided them with free supplies from the public storehouses with considerable order, since they knew precisely how many needy there were and the degree of their needs."[23]

This keen attention to Incan agricultural policies likely stemmed from conditions just outside the editor's window. Between 1737 and 1741 it would have been difficult to ignore the growing sense of social crisis throughout France, especially among the urban poor, whose pangs of hunger, the result of bread shortages, were increasingly becoming a regular subject for monarchical critics and other observers. On the street as well as in the correspondence of the bourgeois elite, the plight of the destitute was coming to be recognized as a universal social concern, and the agricultural situation of 1738–41 was bringing this issue to the fore. In January 1739 a *mémoire* on agricultural subsidies was published, "On the political and financial situation in France on the 1st of January, 1739, with [a report on] the way to bring a prompt solution," indicating that the problem of grain distribution was urgent enough to attract the attention of the provincial consuls and the intendants. In May of that year, reports circulated that women carrying bread along provincial roads had been murdered for their victuals, leading the Marquis d'Argenson to comment that a piece of bread "is more coveted today than a bag of gold was in times past." D'Argenson (often an alarmist, it must be said) went on to recount a story about the Duc d'Orléans having offered the king a sample slice of bread brought in from the provinces, presumably not of good quality, and exclaiming, "Sire, look at the kind of bread with which your subjects feed themselves!"[24]

To understand the editor's effusive praise of the Incas' agricultural, political, and scientific pursuits, then, we need to turn away from the Andes and back to the social and political environment of eighteenth-century France. Rather than compare Louis XV's reign positively with the leadership of the Incan kings, the editor chose to muffle what would have been a self-evident strategy of patronage by doing precisely the opposite. In a setting like the Jardin du Roi, unctuous nods to royal support were frequent and natural components of prefaces and forewords. And yet the editor emphasized the singularity of the Incas seemingly at the expense of his own monarch. This kind of foil had several well-known contemporary antecedents, as discussed above. Voltaire set his political satire *Alzire* in the mountains of Peru, far from the palaces of Paris and Versailles, and Montesquieu chose for his *Persian Letters* another distant venue. The translator's preface concluded with a description of a tyrant, the "last king" of the Incas, who had subverted one of the Incas' fundamental laws by naming one of his youngest sons to the throne. His fate was sealed in a bloody coda by the arrival of the Spanish, "who seem to have been sent to Peru for nothing other than to make this unfortunate [king] suffer the just punishment for his misdeeds."[25] This allegory certainly arrived in print too early to prefigure Louis XVI's violent end on the guillotine, but it may have echoed the frustration felt by many at the monarchy's seeming indifference to the suffering of its people. In the market and on the street, the popular classes had already begun to demand changes and threaten collective action if the king, largely understood to be responsible for the dearth of bread and the price fixing that went along with it, did nothing to ease the crisis. "Unless the government changed its system of dealing with grain," the Duc d'Orléans commented to Louis XV that same year, the monarch would end up alienating "the hearts of all his subjects." In yet another example of the rising tide of urban unrest, the king was greeted in the street with jeers, insults, and threats.

Instead of "Long live the king," the crowd that gathered in the Faubourg Saint-Victor cried, "Misery! Famine! Give us bread!"[26]

The frequency of these bread shortages likely gave an added degree of urgency to eighteenth-century French naturalists' interest in and experiments with new non-European botanical specimens.[27] Of course, European interest in the salutary medicinal properties of exotic herbs and spices stretched back to the earliest Iberian writings on Asia and the Americas, including the treatises of Garcia da Orta and Francisco Hernández, among others.[28] Extra-European excursions under French auspices, including the transatlantic expeditions of Jean de Léry and André Thévet to Brazil, also set out in the sixteenth century with an eye toward finding new plants and minerals that would bring benefit to the European metropole. But it was not until the early seventeenth century that religious missionaries were sent as part of a coherent program under the auspices of the French state to explore colonial outposts for potential scientific exploitation. The Jesuit Raymond Breton (1609–79) and the Dominican Jacques Du Tertre (1610–87) were sent by Cardinal Richelieu to Guadeloupe in the 1630s and 1640s, and another Dominican, J. B. Labatt (1666–1738), traveled to Martinique and other islands of the French Antilles at the turn of the eighteenth century. But Guy-Crescent Fagon (1638–1718), superintendent of the Jardin du Roi from 1699 to 1718, linked botanical exploration most explicitly with the burgeoning interest in medical research and made the Jardin du Roi a new center for overseas experimentation. It was Fagon who dispatched two emissaries to the Antilles to carry out important long-term research programs: Charles Plumier (1646–1704) and Louis Feuillée (1660–1732). Both returned with detailed descriptions and drawings of the plants of the Americas. And while the work of these religious missionaries was largely aimed at resolving medical problems, they nevertheless provided important details about species that could be transplanted into European soil for both medical and agricultural advantage. Their work was not immediately recognized, however, because the correlation between exotic specimens, herbal pharmaceuticals, and domestic subsistence still remained in their day largely unexplored. A certain degree of dissemination and experimentation would be required before the "fruits" of their labors could eventually reach a wider audience and be applied broadly throughout the pharmacies and fields of France. This experimentation would eventually take place under the direction of Charles-Cisternai du Fay, and later Buffon, at the Jardin du Roi.[29]

Du Fay's Hothouses and the Footnoted
Florilegium of the King's Garden

Louis Feuillée, a priest of the mendicant Minim order, was a prescient man. In 1714, twenty-five years before d'Argenson exclaimed that bread had come to be more valuable than gold in France, Feuillée was already conscious of the figurative (and perhaps literal) riches that might be accrued were Frenchmen to collect, transport, and transplant in France some portion of the botanical offerings growing in abundance on American shores. During his five-year voyage to Chile, Peru, and the Caribbean (1707–12), Feuillée observed, collected, described, and sketched a multitude of native plants, cataloging their uses, and he later published these detailed descriptions as addenda to his *Journal des observations physiques, mathématiques et botaniques*. In the introductory pages of his *Journal*, Feuillée outlined the strategies he had used to acquire his information and included a list of instruments and a précis of his overriding scientific concerns. He emphasized that the Comte de Pontchartrain had recommended he "put everything to good use during [his] journey, & . . . send him all the observations and experiments that [he] might accomplish at every possible occasion."[30] Feuillée also drew sketches of the plants and trees whose fruits were unknown in Europe and attempted to learn about the properties of these specimens through close attention to indigenous knowledge and practices. He took special note of the Incas, whose ninth king, Inca Pachacutec, had refused to "give the name of Doctor to any of his subjects, unless [the individual] were fully knowledgeable in all the qualities of Plants, both harmful and salutary." Continuing in his description of the *Tithymalus perennis*, Feuillée lauded the extensive applied knowledge of an entire people: "As this Law was rigorously observed during the reigns of the Incas, all the people of this vast Empire applied themselves diligently to this admirable [field of] knowledge, which being passed from father to son after the destruction of the Incan Empire . . . has been preserved up until today."[31] In a supplement to his *Journal* written in 1725, Feuillée asked rhetorically if it wouldn't be possible to "find similarly effective remedies for other diseases, which for the good that they would provide to the Public, could be regarded as far more valuable treasures than those that are extracted from the mines of Peru," with more than a hint of superciliousness toward his Spanish predecessors.[32] While Feuillée's primary object of study was the many diseases (and botanical cures) he had come across while traveling through the West Indies, his attention to the agricultural properties of non-European specimens that could be transplanted in France was not lost on subsequent natural historians, who carefully examined his work back in Europe. Neither was his praise for the Incas' botanical knowledge.

One individual who closely examined Feuillée's printed accounts was Dalibard, who we know spent at least a portion of his career as a naturalist at the Jardin du Roi. In the early 1750s Buffon chose Dalibard to translate Benjamin Franklin's treatise on electricity, and it is for this work—and the electrical experiments Dalibard performed both in Marly-la-Ville and before Louis XV with Buffon and Delor—that he was, and still is today, most widely recognized. Earlier in his career, Dalibard had published the *Florae Parisiensis Prodromus,* a catalog that confirmed his detailed knowledge of Parisian plant species and was reputed to be one of the first texts to employ Linnaean terminology in France. Unfortunately, Dalibard's role at the Jardin during the period in question is unclear; we do not know whether he was an official charge of the king or merely a knowledgeable or curious visitor. What is clear, however, is that the preface to the *Histoire des Incas* refers to Dalibard specifically as having collected previous accounts by travelers to South America and having compared them with more recent "discoveries" made in the realm of natural history. For this reason, the form and content of the revised work probably owe more to Dalibard than any other individual who participated in the project. Dalibard's collation of travel accounts from South America, including Feuillée's *Journal,* with descriptions of plant species contained at the Jardin du Roi, swelled the second volume of the *Histoire des Incas* and transformed the text into a treasure trove of useful botanical data.[33]

The period in which the new translation of the *Histoire des Incas* was being prepared was a dynamic moment in the history of the Jardin du Roi, a period in which the opportunities were multiplying for the cultivation of exotic plant species within the king's Cabinet d'Histoire naturelle, which was inaugurated as such in 1729. Just prior to Du Fay's intendancy, Pierre Chirac had paved the way for the acclimatization of non-European species by incorporating the Jardin des apothicaires de Nantes under the Jardin's jurisdiction in 1719, privileging thereby "the care and cultivation of plants from foreign countries" through an ambitious program of overseas seed exchange.[34] In 1714, a hothouse had been built especially for a coffee plant imported from Amsterdam, and at some point after Du Fay took over the intendancy in 1732, he ordered the construction of at least two other hothouses. Hothouses had existed at the Jardin du Roi prior to 1732, but according to Fontenelle, they were not in particularly good shape: "Foreign plants became thin within the poorly maintained hothouses, and were discarded; when these plants perished, it was permanent, [and] they were not replaced . . . [T]he breaches in the enclosing walls were left unrepaired, [and] large plots of land lay fallow.[35] In this way, the great bulbous structures of Du Fay's new hothouses transformed the Jardin from a desolate location into a rich space "consecrated to plants of the pharmacopoeia, into an experimental botanical garden open to all species."[36] They also linked the royal metropolitan institution with the larger French colonial enterprise, a connection that was emphasized in Fontenelle's 1739 "Éloge de M. du Fay," in which the sécretaire perpetuel de l'Académie explained that "doctors or surgeons . . . received their instruction at the Garden . . . [and] from there spread themselves throughout the colonies."[37] With the number of foreign specimens growing at an exponential rate thanks to the overseas expeditions of Feuillée, Pierre Barrère, Jean-François Artur, and others still, the hothouses served as artificial but essential repositories for the increasing botanical store: the hothouses were "constructed in such a way as to be able to represent different climates, since the idea is to make these diverse plants forget their natural climates. Changes in temperature are accomplished in small degrees, from the strongest to the most temperate, and all of the advances that modern physics has been able to teach us in this regard have been put into practice."[38] Du Fay thus masterminded the production of physical spaces into which these newly arrived non-European botanical specimens would arrive and saw to their acclimatization in the diverse climatic zones of the Jardin's hothouses. Godin's natural historical specimens, which included insects as well as plant species, may have been some of the first shipments to profit directly from the thick panes of Du Fay's hothouses, and the newly arrived collection may have been given special attention by Dalibard precisely on account of these new technologies of protection, preservation, and presentation.

Whoever the anonymous editor of the *Histoire des Incas* may have been, it is clear that his activities were tightly interwoven with those of the Jardin. In volume 2, the editor observed that one of the banana trees kept in the Jardin du Roi "flowered and brought forth fruit in 1741," and he was certainly familiar with the materials Godin had sent to the cabinet in 1737. The *Histoire des Incas* was probably begun at some point between 1737 and 1739, augmented over the course of subsequent years, and finally printed and distributed in 1744. This seven-year period not only coincided with the series of social crises discussed above; it also corresponded to a moment of great transition in the history of the Jardin du Roi, Buffon having been named intendant of the Cabinet d'Histoire naturelle late in 1739. Between the period of Buffon's ascension to the post of intendant général and the publication of the first volume of his *Histoire naturelle* in 1749, however, few details about the day-to-day functioning of the king's cabinet are known. Even less is known about the pre-Buffon period. For these reasons, information gleaned from the preface to volume 2 of the *Histoire des Incas* about a shipment of more than 120 "rarities" to the "Cabinet d'Histoire Naturelle du Jardin Royal" from the academicians in Peru deserves especially close attention.

As noted above, the hothouses or "serres chaudes" had been constructed at the Jardin du Roi just prior to the arrival of Godin's shipment of Peruvian specimens, making the items from his shipment one of the first collections to be acclimatized within the new setting conceived by Du Fay. In many ways, chapter 28 of the *Histoire des Incas,* which appears as the first complete chapter on Peruvian botany, resembles a virtual tour through these newly built hothouses, since Dalibard meticulously recorded the presence at the Jardin of the species sent by Godin, as well as other forms of plant life from the Americas that had preceded his Peruvian rarities. Not surprisingly, the focus of this chapter, "Des Grains," is on those trees and plants that appeared in Garcilaso's text, and Dalibard culled information about these specimens from several chapters of the original edition. But he also augmented Garcilaso's sometimes naïve commentaries with observations made during previous naturalist missions for which easily accessible descriptions existed in the travel narratives and natural historical treatises of Feuillée, Frézier, Marggraf, and Clusius. And, most of all, he described in detail the treasures contained within the Parisian hothouses.

If we were to accompany Dalibard on a tour of the hothouses by peering into chapter 28 ("Des Grains") of the *Histoire des Incas* as if it were a set of display cases within the hothouses themselves, we would find an exuberant assortment of plants, shrubs, and fruit trees. From corn and quinoa to the potato and the palm tree, Dalibard described these items in great detail, always emphasizing whether a given species was to be found on the grounds of the Jardin du Roi, within the collection of the king's cabinet, or in neither location. The typographical strategies used to construct this virtual hothouse in print will be discussed in greater detail in the following section. For our immediate purpose— which is to illuminate the virtual hothouse from within and understand its internal logic—we can divide chapter 28 into two discrete sections. The first section of the chapter hewed closely to Garcilaso's original descriptions in the *Comentarios reales,* loosely following the order of the descriptions provided in book 8, chapters 9–10 of the pre-1744 editions. Only some of these initial species could actually be located within the

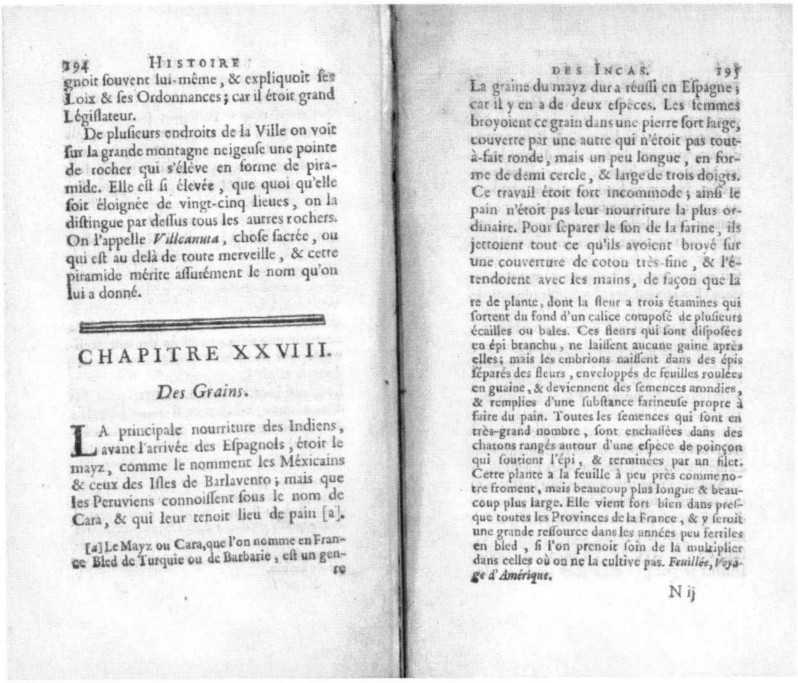

194 HISTOIRE
gnoit souvent lui-même, & expliquoit ses Loix & ses Ordonnances; car il étoit grand Législateur.

De plusieurs endroits de la Ville on voit sur la grande montagne neigeuse une pointe de rocher qui s'élève en forme de pyramide. Elle est si élevée, que quoi qu'elle soit éloignée de vingt-cinq lieues, on la distingue par dessus tous les autres rochers. On l'appelle *Villcanuta*, chose sacrée, ou qui est au delà de toute merveille, & cette pyramide mérite assurément le nom qu'on lui a donné.

CHAPITRE XXVIII.

Des Grains.

LA principale nourriture des Indiens, avant l'arrivée des Espagnols, étoit le mayz, comme le nomment les Méxicains & ceux des Isles de Barlavento; mais que les Peruviens connoissent sous le nom de Cara, & qui leur tenoit lieu de pain [a].

[a] Le Mayz ou Cara, que l'on nomme en France Bled de Turquie ou de Barbarie, est un genre

DES INCAS. 195
La graine du mayz dura réussi en Espagne; car il y en a de deux espèces. Les femmes broyoient ce grain dans une pierre fort large, couverte par une autre qui n'étoit pas tout-à-fait ronde, mais un peu longue, en forme de demi cercle, & large de trois doigts. Ce travail étoit fort incommode; ainsi le pain n'étoit pas leur nourriture la plus ordinaire. Pour séparer le son de la farine, ils jettoient tout ce qu'ils avoient broyé sur une couverture de coton très-fine, & l'étendoient avec les mains, de façon que la

re de plante, dont la fleur a trois étamines qui sortent du fond d'un calice composé de plusieurs écailles ou bales. Ces fleurs qui sont disposées en épi branchu, ne laissent aucune gaine après elles; mais les embrions naissent dans des épis séparés des fleurs, enveloppés de feuilles roulées en gaine, & deviennent des semences arrondies, & remplies d'une substance farineuse propre à faire du pain. Toutes les semences qui sont en très-grand nombre, sont enchaissées dans des chatons rangés autour d'une espèce de poinçon qui soutient l'épi, & terminées par un filet. Cette plante a la feuille à peu près comme notre froment, mais beaucoup plus longue & beaucoup plus large. Elle vient fort bien dans presque toutes les Provinces de la France, & y seroit une grande ressource dans les années peu fertiles en bled, si l'on prenoit soin de la multiplier dans celles où on ne la cultive pas. *Feuillée, Voyage d'Amérique.*

N ij

Figure 49. Chapter 28 of the *Histoire des Incas*, "Des Grains," shows the division of the text into primary and secondary sections using bracketed footnotes. The description of corn in footnote [a] makes reference to the potential benefits of cultivating American maize in France during the years when wheat is scarce. Courtesy of the Rare Book and Manuscript Library, University of Pennsylvania.

Jardin du Roi. Dalibard began with a description of corn, quinoa, the *papa* (*solanum tuberosum esculentum,* a rootlike tubercule closely related to the potato), the potato, and the *inchi,* a sinuous plant that contained bean-like grains "filled with a white interior, which taste like hazelnuts, and are cooked and eaten as dessert." Each of these five species, described both in Garcilaso's text and in Dalibard's footnotes, served either as a substitute for bread ("The *papa* . . . replaces bread [in their diet]") or could be used in some way as a comestible grain or root. In other words, the entrance to Dalibard's virtual hothouse was first dedicated to specimens of an edible variety, all of which were contained within Garcilaso's account but most of which were not to be found at the Jardin du Roi (fig. 49).[39]

Let us take the first object, corn, as an example: it is described as the "principal food of the Indians." After the initial description of the Spanish and Peruvian names for corn, the editor commented that corn replaced bread in the Peruvian diet. It is at this point that the first brack-eted footnote mark, "[a]," appears. The main text discusses the success of transplanting corn in Spain and the manner in which Amerindian women prepared the grain to be cooked—all of which was an abridged version of Garcilaso's description in Book 8, chapter 9, entitled "On corn and what they call rice. And other seeds." The vast portion of the footnote, however, which takes up nearly two-thirds of the following page, offers a detailed description of corn's physical characteristics and was more or less lifted directly from Feuillée's *Journal des observations physiques, mathématiques et botaniques*. But at the very bottom of the text, prior to assigning Feuillée authorial credit for these details, there is an editorial addendum: the corn seed "grows very well in almost all the provinces of France, and if we were careful to bring it in to those provinces where it is not cultivated, it could serve as a great resource in the years that are less fertile in wheat."[40] This citation is an entirely new editorial addition that was not drawn from any of the texts that Dalibard cited. Feuillée, for example, had merely written that "few peo-ple in France are unaware of what corn is," but he stopped short of declaring it a great resource for confronting agricultural crises. This tex-tual commentary inserted by the editor, then, provides proof that those who worked in and around the Jardin du Roi were acutely aware of the potential uses to which Peruvian species such as corn might be put in France. The appearance of this commentary in the first footnote of the first species signals the shifting focus to contemporary concerns, brought into typographic relief through the use of bracketed footnotes within the text.

The next series of specimens in this first section included plants and trees described by Garcilaso as well, but the editor specifically empha-sized those species found in the Jardin. Dalibard included a note to this effect at the bottom of each species's description in the footnotes. Such was the case with the potato, the first item to be listed explicitly as being present at the Jardin: "There are potatoes at the Jardin du Roy." Likewise the guava: "There are some at the Jardin du Roi." Following the guava (*Inga siliquis longissimis: vulgò Pacai*) come the *paltas*, "what are called in our Isles *Avocados*. . . . This tree is at the Jardin du Roy"; the *lucuma*, a large tree bearing a heart-shaped fruit ("The Lucuma is at the Jardin du Roy"); and the *mulli*, a tree whose fruit "turns black as it ripens and tastes like pepper. . . . The *Mollé* with jagged leaf is at the Jardin du Roy." This section ends with descriptions of the *maugei* and the coca tree, two species for which there were no specimens at the Jardin du Roi but which are referred to in book 10, chapters 13 and 15, respectively, of Garcilaso's original edition. Despite these objects not being present at the Jardin, the *maugei* merited placement within the chapter because of its "utility," and the coca tree deserved to be mentioned because of the tree's "important commerce," which transformed it into "one of the greatest treasures of Peru."[41] We therefore see a progression, within the text and footnotes of this first section of chapter 28, from comestible varietals about which few detailed in situ observations had been made—including corn, potatoes, and *inchi*—to those species represented within the actual soil of the Jardin du Roi's hothouses. Both kinds of descriptions relied primarily on details drawn from travel accounts, but the items from this second category suggest that further observation was possible because of their presence within the hothouse. The link between them was the potential use that could be made of certain species within the agricultural or medical economies of eighteenth-century France.

The second section of chapter 28, separated entirely from the first by brackets, described species at the Jardin not listed by Garcilaso in his original text but that nonetheless had been sent from Peru or observed empirically by Dalibard. This section provided specific details of botanical "events" that had occurred at the Jardin and was meant to highlight the "perfectibility" of the eighteenth-century natural sciences made possible by Europeans' role in the accumulation of knowledge about Peruvian species: "Whether motivated by their own self-interest or their curiosity, [these individuals] have made discoveries . . . that their ancestors never would have suspected." Travel narratives, in particular, were exhibited to cast doubt on Garcilaso's knowledge and authority; after perusing these travel accounts at the Jardin, it became evident to the editor that "some travel accounts bear witness to the fact that there are many other plants than those that our historian [i.e., Garcilaso] has spoken of," thus diminishing the credibility of his account.[42] The first specimen described in this section, in fact, was the quinquina tree, which had been described in a detailed "mémoire académique" published by La Condamine in 1738. La Condamine's "Mémoire sur l'arbre du Quinquina" will be discussed in the following chapter, but the manner in which the *Histoire des Incas* paraphrased its conclusions largely mirrored the way in which the *Encyclopédie* abridged his account by suppressing the source of much of La Condamine's information, a local informant who lived near where he had carried out his research and from whom La Condamine garnered much of his in situ knowledge.

Following the eight-page discussion of the quinquina tree, this second section included descriptions of the papaya, cotton, and guava trees, the "Floripondio," the banana and palm trees, the "Sang-Dragon," the

"Cierge," and many others. These specimens had all been observed empirically at the Jardin and were subsequently described in the text. The
first of these species, for example, was the papaya tree, and the description of this species was based primarily on Willem's Piso account in the
Historia naturalis Brasiliae (1648). In the paragraph that followed this description, however, more recent observations about the specific holdings
of the Jardin came to the fore: "The papaya trees that have blossomed in
the hothouses of the Jardin du Roi are all females, and have not borne
fruit, due to the lack of males. We hope that some [male trees] will be
found among those that we keep there."[43] The flowering of the papaya
trees brought a degree of temporal currency to the description offered
by Piso and showed that the hothouse could permit useful research far
from the papayas' native soil in the Americas.[44] The material presence of
these species and the ability to naturalize them within the hothouse gave
natural historical observers the opportunity to place the on-site study of
botanical materials on the same level as those decades-old descriptions
provided by travelers to the New World. These recent observations, based
on the behavior of material specimens, in turn transformed the identity
of the *Histoire des Incas* from a static 150-year-old chronicle into a dynamic vessel for empirically based experiments.

What is perhaps most telling about the transformation of this natural
historical material from the pre-1744 editions to the 1744 *Histoire des
Incas* is the suppression of Garcilaso's descriptive chapter headings in
order to accommodate the more capacious chapter 28, which contained
descriptions ranging from corn and quinoa to palm trees and peppers:
sixty-six pages of seeds, legumes, fruits, trees, medical plants, and nar-
cotics, all subsumed within a single chapter. What was lost in this universalizing schema that pretended to describe the vast collection of species
at the Jardin du Roi were Garcilaso's own organizational principles: book
8, chapter 10, for instance, described the "vegetables (*legumbres*) that are
grown beneath the earth," including the potato and the *ínchi*; book 8,
chapter 11 of Garcilaso's original edition referred to the "fruits of large
trees," and was followed in turn by chapters dedicated to "the *mulli*
tree and the pepper," "the *maguey* tree and its uses," and "the banana,
pineapple, and other fruits." In the editorial reconfiguration of the 1744
Histoire des Incas, the editor combined all of these various objects into
an oddly titled single chapter, "On Grains," papering over what he had
called the "confused" organization of Garcilaso's text with footnotes,
brackets, and notes from travelers' accounts that made this new rendi-
tion of Peruvian natural history appear to the enlightened observer more
ordered, up-to-date, and "exact."

There is no doubt that the epistemological expansion of the translator's charge and the shifting notion of what a translated edition of a classic work should contain were due in large part to the burgeoning specimens sprouting and blossoming within the artificial conditions of Du Fay's hothouses. The blooming trees and ripening fruits allowed the editor of the *Histoire des Incas* to put the empirical observations at the Jardin du Roi on the same plane as the "day-to-day research" undertaken in Peru following the arrival of the Spaniards, thereby shifting the figurative center of botanical experimentation from Peru back toward Paris. But the clean typographical mechanisms and editorial layering that highlighted useful pieces of information for an eighteenth-century audience belie a deeper and more insidious effacement, an editorial epistemology that allowed an editor preoccupied with the shape and function of the king's garden to dismiss entirely a logic that conceivably derived from more local sources. In the end, Garcilaso's description of Peruvian specimens, their uses by the Incas, and the manner in which they were cultivated became subordinated to a new editorial epistemology that distinguished the *Histoire des Incas* from previous translations of Garcilaso's work. Like the physical hothouse, the editor constructed a virtual hothouse in print and guided the reader through an organizational scheme of his own making, highlighting the name, provenance, and agricultural or botanical properties of the various species according to a logic that he had devised and configured. Keenly aware of the structural and typographical stakes of this new translation, the editor of the *Histoire des Incas* revised and reformulated natural knowledge on the printed page using travel narratives, natural historical compendia, and verdant specimens found within the "elegant edifices" Du Fay had built a decade before.[45]

A Peruvian Garden Enclosed by Parisian Brackets

Conscious of the need to guide the reader through this transformed version of Garcilaso's text, and ever mindful of the augmented readership it might attract as a virtual hothouse and a history of the Incas, the editor included a second preface that outlined the criteria he had employed for expanding the range of materials included in this second volume. In addition to a discussion of Godin's shipment of specimens, the revelation of quinquina's properties by La Condamine, a special section on Spanish "indifference" to learning about the features of American natural history, and a detailed account of volume 2, the preface to the second volume explained that Garcilaso had been ill equipped to comment on the natural historical features of his native country. For this reason, the text had been revised "so that the reader who is knowledgeable about the past of this vast country can also be so about [its] present state."[46] The editor went on to describe the process by which the text was to be transformed: "[W]e will add to what [Garcilaso] says some notes to better introduce the vegetables, animals, and minerals he mentions. At the end of the chapters, we will add short descriptions of things he does not speak about."[47] This license to add and abridge according to the interests of the "educated reader" was an explicit and central feature of the editorial strategy as revealed in the preface to the second volume. That the articles on Inca religion, culture, ceremonies, customs, science, geography, clothing, and industry had been "mixed together in the original with historical facts, and were often confused between the two" added further weight to the editor's call for a new and revised edition. By separating these topics from the historical materials, bringing them together "under the same point of view," and organizing them according to chapters, the editor hoped to "add more order to this work," even though he admitted that much more attention was needed in order to bring the work to the "perfection" it truly deserved.

One of the typographical additions was the strategic deployment of footnotes to portions of the text. But in the second volume another typographical strategy enhanced the text's readability even further: the use of a system of brackets to separate material not originally part of Garcilaso's *Comentarios reales*. The first twenty-seven chapters of this second volume, which dealt with the religious, cultural, literary, agricultural, and "scientific" features of Inca life, were largely free of bracketed material. Indeed, there were only two uses of brackets in the first twenty-seven chapters, both in the section on Inca astronomy, to which we shall return shortly. But on page 194, which corresponded to the beginning of chapter 28, "On Grains," a series of brackets appeared. Like a trapdoor leading to a series of underground chambers, a new sublayer of information emerged from within the original text, splitting the page into two separate but interrelated sources. The typographical mechanism that linked these upper and lower portions of the text was the bracketed footnote, which began to appear with regularity throughout this section (fig. 49 above).

A paragraph in the second preface devoted to typographical technologies suggests the extent to which printing practices—in this case, the insertion of brackets—were considered germane matter for a preface or foreword during this period. As the editor explained, "In the printing, we have not distinguished additional materials with the use of different characters, which is almost always incommodious, but rather we have marked them with an Asterism, and have added to each article the name of the Author."[48] "Different characters," in this case, would be italics, which were employed sporadically in the first and second volumes. But what did the editor mean by marking items with an "Asterism"?[49] The reader finds nothing resembling asterisks that would signal or adorn inserted materials. What one does find are brackets, an alternative form perhaps of "asterism" and part of a multilayered system of typographical marks used to alert the reader to new material that had entered the second volume.[50]

Typographically speaking, what did two brackets typically separate in the mid-eighteenth century? This arcane question merited a response in Martin-Dominique Fertel's typographical manual *La science pratique de l'imprimerie,* published in 1723. Fertel explained that the bracket, a variation on the omnipresent parenthetical mark, had three possible uses:

Brackets are occasionally used to mark some discourse that should be in italic character or that may be transposed, which often happens in alphabetical tables or the contents of a Book.

They are also used to enclose things to which the reader should pay close attention, and this principally in church books like prayerbooks (*missels*), rituals, &c...

They can also be used to enclose the figures of running titles at the top of the page, where they are placed at the center of the line; this is done when there is no textual material to be placed there; as well as to enclose alphabetical letters of a cross-reference (*renvoy*) of additions in the margins, or at the bottom of the page.[51]

The third definition comes closest to describing the typographical mechanisms on display in the *Histoire des Incas*. What is peculiar about the *Histoire des Incas,* however, is not the insertion of these "additions" at the bottom of the page but rather the complex system of footnotes, bracketed footnotes, and bracketed text, all interwoven and layered throughout the second half of the second volume. The struggle between a codified system of typographical rules as described by Fertel and the demands of a text that had become a hybrid container for disparate materials—animal, vegetable, mineral, and historical—became evident as the editor attempted to fold information drawn from 150 years' worth of natural historical exegesis in various languages into an ordered narrative structure.

At the same time, the *Histoire des Incas* was an ideal candidate for the kind of typographical instructions that Fertel proposed. Print allowed societies to preserve knowledge that would otherwise have been lost, according to Fertel's preface. Without print, "the richest talents would remain buried, the most interesting research would remain unknown, and the most felicitous discoveries would still be unheard of."[52] As such, the *Science pratique de l'imprimerie* served as a technical manual to help control and standardize the use of typographical strategies for organizing printed texts. We do not know whether the editor of the *Histoire des Incas* was aware of Fertel's treatise, but these themes resonated strongly in certain sections of the *Histoire des Incas*. Indeed, the opening lines of the section on astronomy could have served very well as an epigraph for Fertel's text: "Without the use of the alphabet, the Incas were not able to extend their knowledge: and, consequently, their astrology and natural philosophy were extremely mediocre.... In general, their understanding perished along with its inventors."[53] It is probably mere coincidence that the editor of the *Histoire des Incas* specifically cited the section from which this quotation was drawn in order to point out the improvements made since Baudoin's translation of 1633. Nevertheless, in the translator's preface to volume 1, the editor encouraged the reader to compare these two sections on astronomy in order to observe "the simplicity to which we have attempted to reduce each object" and the manner in which each section was designed to be "more pleasant for the Public." For the editor of the *Histoire des Incas,* however, it was not print itself (for, of course, Garcilaso's original text was printed as well) but rather the *configuration* of the printed text on the page that allowed knowledge to be passed on efficiently from one generation to the next. And this was also Fertel's point in writing his treatise. It is perhaps no wonder, then, that the first section to employ brackets to facilitate comprehension and clarity was this very section on astronomy, meant to showcase a new commentary inserted into the 1744 translation by "a man versed in [Astronomy], and well known in the Republic of Letters."[54]

Notwithstanding the two instances where brackets were used in the section on Inca astronomy, the first section to employ the bracket as an extended typographical feature was chapter 28, "On Grains," dedicated to Peruvian natural history. This section, as we have seen, recovered a series of chapters originally published by Garcilaso in the middle of book 8 of the *Comentarios reales.*[55] There are several planes of editorial organization within this extended—one might say distended—section, each represented by a different typographical stratum used by the editor to create order in a system where information from a multiplicity of sources was being collated, abridged, transformed, and displayed at once. The first level of textual organization could be called the principal text—that

is, the (more or less direct) translation of Garcilaso's *Comentarios reales*. As with the first volume, this text was dramatically abridged and reshaped but by and large still followed the narrative thrust of Garcilaso's original. This primary level also contained footnotes denoted alphabetically and which began renumbering on each page instead of numbering consecutively throughout. These footnotes, like those of the first volume, oscillated between commentaries inserted *ex post translatio* by the editor and those that were condensations or transplantations of information provided by Garcilaso. An example of the former would be the editor's comment in a footnote on page 207 that the *Huchu*, a kind of piquant pepper, "is our long, or Guinea, pepper" (fig. 50).[56] The other category of footnote included page references to classic texts such as Acosta's *Historia natural y moral de las Indias,* lifted from the body of Garcilaso's text, as well as longer commentaries about particular Amerindian customs also extracted from Garcilaso's text and reinserted using the organizational mechanism of the footnote.[57]

Beyond the material culled from Garcilaso's original text and placed discretely within the parenthetical footnote form "(a)," the second level of editorial reconfiguration consisted of bracketed footnotes injected into Garcilaso's text by the editor, typographical references that pointed to natural historical material based largely on post-Garcilasan travel narratives to South America. The editor employed several texts to make these editorial annotations to Garcilaso's text, including Louis Feuillée's *Journal des observations* (1714–25), Amédée François Frézier's *Voyage de la mer du sud* (1717), Georg Marggraf and Willem Piso's *Historia naturalis Brasiliae* (1648), and Carolus Clusius's *Histoire des plantes* (1557).[58] But the text contained within these bracketed footnotes can also be further subdivided, since the commentaries contained not only descriptions based on travel narratives but also occasional indications that certain specimens of these plants could be found at the Jardin du Roi, as we saw above. Additionally, these descriptions sometimes provided details on the recurrent attempts to "naturalize" exotic species into French soil.

A third and final level of editorial organization took the shape of an extended, multipage description of the Peruvian natural historical specimens found at the Jardin du Roi, separated from the principal text and (once again) enclosed within brackets. This section bore no relation to Garcilaso's original text; indeed, it began with a statement from the editor that it "should come as no surprise that Garcillasso [*sic*] de la Vega did not know all the plants of his country."[59] Interestingly, this text was separated not only by a line break but also by two closed brackets placed side by side, which seemed to symbolize a contained but ultimately

qu'à l'amer ; ils gardent cette décoction trois ou quatre jours auparavant que d'en faire usage ; elle est très-bonne pour la colique , les maux de vessie & la gravelle ; elle est meilleure & plus délicate en la mêlant avec le breuvage fait de mayz. Cette même eau bouillie jusqu'à s'épaissir se change en bon miel , & en vinaigre quand elle est exposée au Soleil avec des drogues que les Indiens connoissent. La graine de *Mulli* & la raisine qu'il produit font admirables pour les blessures. L'eau dans laquelle on a fait bouillir ses feuilles est très-bonne pour se laver, elle guérit la galle & les ulcères , & son bois sert à faire des curedens. J'ai vû dans la vallée de Cozco un nombre infini de ces arbres ; mais on les a presque tous abattus pour faire du charbon meilleur que tous ceux des autres bois , car il garde toujours sa chaleur ; & ne s'éteint que lorsqu'il est réduit en cendre [i].

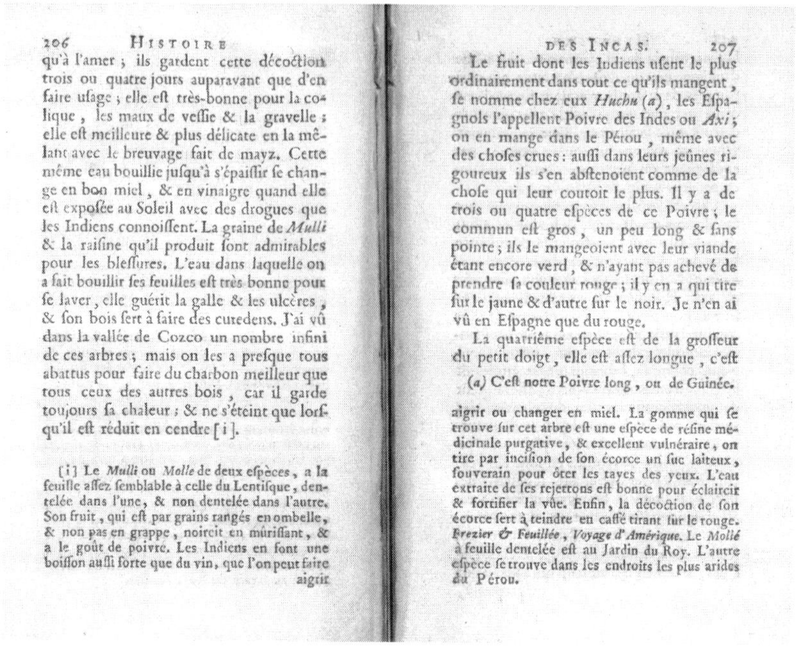

[i] Le *Mulli* ou *Molle* de deux espèces , a la feuille assez semblable à celle du Lentisque , dentelée dans l'une, & non dentelée dans l'autre. Son fruit , qui est par grains rangés en ombelle, & non pas en grappe , noircit en mûrissant, & a le goût de poivre. Les Indiens en font une boisson aussi forte que du vin, que l'on peut faire aigrir

Le fruit dont les Indiens usent le plus ordinairement dans tout ce qu'ils mangent , se nomme chez eux *Huchu* (a) , les Espagnols l'appellent Poivre des Indes ou *Axi* ; on en mange dans le Pérou , même avec des choses crues : aussi dans leurs jeûnes rigoureux ils s'en abstenoient comme de la chose qui leur coutoit le plus. Il y a de trois ou quatre espèces de ce Poivre ; le commun est gros , un peu long & sans pointe ; ils le mangeoient avec leur viande étant encore verd , & n'ayant pas achevé de prendre sa couleur rouge ; il y en a qui tire sur le jaune & d'autre sur le noir. Je n'en ai vû en Espagne que du rouge.

La quatrième espèce est de la grosseur du petit doigt , elle est assez longue , c'est

(a) C'est notre Poivre long , ou de Guinée.

aigrir ou changer en miel. La gomme qui se trouve sur cet arbre est une espèce de résine médicinale purgative , & excellent vulnéraire , on tire par incision de son écorce un suc laiteux , souverain pour ôter les tayes des yeux. L'eau extraite de ses rejettons est bonne pour éclaircir & fortifier la vûe. Enfin , la décoction de son écorce sert à teindre en caffé tirant sur le rouge. *Frezier & Feuillée , Voyage d'Amérique.* Le Mollé à feuille dentelée est au Jardin du Roy. L'autre espèce se trouve dans les endroits les plus arides du Pérou.

Figure 50. The first two levels of the *Histoire des Incas*'s editorial annotation are shown here on both pages. The first level is represented on page 207 by the text beginning "Le fruit dont les Indiens usent le plus ordinairement," with the alphabetical footnote marker (a) shown underneath. The second level appears in the lower lefthand corner of page 206, represented by the bracketed footnote marker [i] and its accompanying text. Note the reference to the presence of the "Mollé à feuille dentelée" at the Jardin du Roi at bottom right. Courtesy of the Rare Book and Manuscript Library, University of Pennsylvania.

open-ended display case within which additional material was allowed to extend for several pages (fig. 51). The section elaborated within the brackets, then, brought together information gleaned from travelers' accounts as well as details gathered by observing the natural specimens in situ at the Jardin. This fusion of textual reference and firsthand observation was meant to compensate for Garcilaso's unstudied approach to classifying the natural historical materials of his native land. The editor (probably Dalibard in this case) drew upon several reference texts, citing the appropriate author at the end of each paragraph and extracting various kinds of information to complement Garcilaso's descriptions with more technical observations.

The editor justified these changes to Garcilaso's text by linking his overall editorial project to a positivistic conception of scientific expansion and "commerce" in the New World. According to the preface to

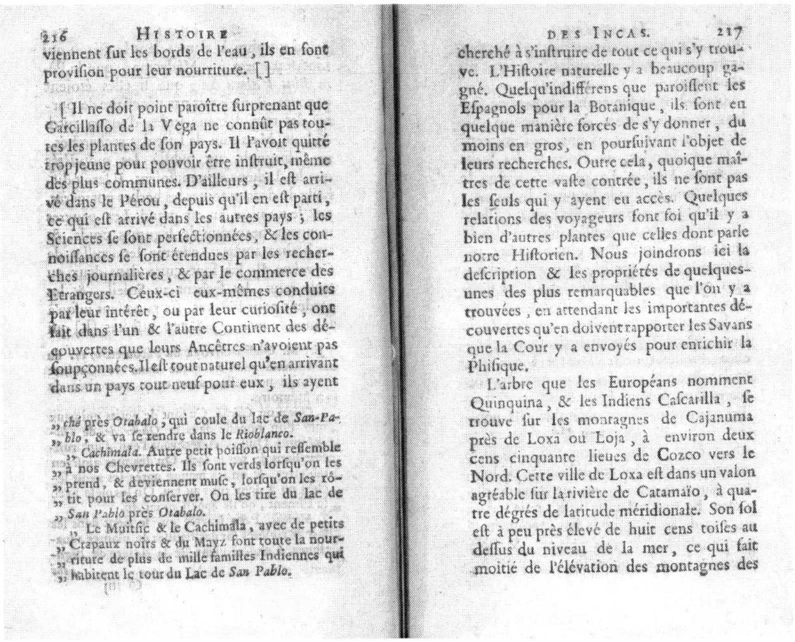

216 HISTOIRE
viennent fur les bords de l'eau, ils en font
provifion pour leur nourriture. [ ]

[Il ne doit point paroître furprenant que
Garcillaffo de la Vega ne connût pas toutes les plantes de fon pays. Il l'avoit quitté
trop jeune pour pouvoir être inftruit, même
des plus communes. D'ailleurs, il eft arrivé dans le Pérou, depuis qu'il en eft parti,
ce qui eft arrivé dans les autres pays; les
Sciences fe font perfectionnées, & les con-
noiffances fe font étendues par les recherches journalières, & par le commerce des
Étrangers. Ceux-ci eux-mêmes conduits
par leur intérêt, ou par leur curiofité, ont
fait dans l'un & l'autre Continent des découvertes que leurs Ancêtres n'avoient pas
foupçonnées. Il eft tout naturel qu'en arrivant
dans un pays tout neuf pour eux, ils ayent

„ thé près *Otabalo*, qui coule du lac de *San-Pa-*
„ *blo*, & va fe rendre dans le *Rioblanco*.
„ *Cachimala*. Autre petit poiffon qui reffemble
„ à nos Chevrettes. Ils font verds lorfqu'on les
„ prend, & deviennent mufc, lorfqu'on les rô-
„ tit pour les conferver. On les tire du lac de
„ *San Pablo* près *Otabalo*.
„ Le Muftic & le Cachimala, avec de petits
„ Crapauds noirs & du Mayz font toute la nour-
„ riture de plus de mille familles Indiennes qui
„ habitent le tour du Lac de *San Pablo*.

DES INCAS. 217
cherché à s'inftruire de tout ce qui s'y trouve. L'Hiftoire naturelle y a beaucoup ga-
gné. Quelqu'indifférens que paroiffent les
Efpagnols pour la Botanique, ils font en
quelque manière forcés de s'y donner, du
moins en gros, en pourfuivant l'objet de
leurs recherches. Outre cela, quoique maî-
tres de cette vafte contrée, ils ne font pas
les feuls qui y ayent eu accès. Quelques
relations des voyageurs font foi qu'il y a
bien d'autres plantes que celles dont parle
notre Hiftorien. Nous joindrons ici la
defcription & les propriétés de quelques-
unes des plus remarquables que l'on y a
trouvées, en attendant les importantes découvertes qu'en doivent rapporter les Savans
que la Cour y a envoyés pour enrichir la
Phifique.
L'arbre que les Européans nomment
Quinquina, & les Indiens Cafcarilla, fe
trouve fur les montagnes de Cajanuma
près de Loxa ou Loja, à environ deux
cens cinquante lieues de Cozco vers le
Nord. Cette ville de Loxa eft dans un valon
agréable fur la rivière de Catamaïo, à quatre dégrés de latitude méridionale. Son fol
eft à peu près élevé de huit cens toifes au
deffus du niveau de la mer, ce qui fait
moitié de l'élévation des montagnes des

Figure 51. The third level of editorial organization in the *Histoire des Incas* is revealed in the first two pages of the extended bracketed section of volume two. Note the two brackets placed side by side in the second line, at upper left on page 216. This eighteenth-century equivalent of a hyperlink connected the previous section to an additional textual division, which extends for several pages in the bracketed text that follows. Text from the bracketed footnote [k], at bottom left, carries over from the previous page. This particular text is a "note tirée [du] mémoire" of M. Godin that describes a collection of insects he sent to the Cabinet d'Histoire naturelle in 1737. One of the supposed "insects," the "Cachimala," is a fish. Courtesy of the Rare Book and Manuscript Library, University of Pennsylvania.

the second volume, Garcilaso's sixteenth-century understanding of natural history had been entirely superseded by "day-to-day research and
the commerce of foreigners" in Peru, including but not limited to the
three academicians sent to measure the shape of the Earth: Louis Godin,
Charles-Marie de La Condamine, and Pierre Bouguer. This traveling tri-
umvirate makes several cameo appearances within the *Histoire des Incas*.
They are referred to regularly as "new Argonauts" who would augment
"the collections of the Jardin du Roy, & the *mémoires* of the Academy [of
Sciences]." The eight-page exposition of the quinquina tree from Loja
mentioned earlier, written by La Condamine and published in the 1738
Mémoires de l'Académie Royale des Sciences, is but one example of the kind
of material that the bracketed section in chapter 28 included.[60] As the

editor made clear, much of the impetus behind the editorial reorgani-
zation of the *Histoire des Incas* was the arrival of the new materials sent
to France by Godin and his colleagues; these "additions," using Fertel's
terminology, required the deployment of brackets as figurative display
cases. In this way, typographical markers served as the structural girding
for mounting these material specimens within the text, one of the many
tools employed by European printers and editors to naturalize exotic
flora within the expanded bounds of a historical treatise.

Clearly, the tripartite typographical division of chapter 28 overlaps
considerably with the material considerations discussed earlier in this
chapter, paralleling and facilitating the shifting focus to contemporary
concerns reflected in the nature of the information given about the various specimens. The progression from specimens of agricultural interest
to those about which material evidence could be drawn from empirical
observation was made possible, in fact, by the use of these typographical strategies. The multiple strata visible through typographical divisions
were windows onto the state of natural historical knowledge at the Jardin
du Roi. Indeed, the text went on to describe other Peruvian species as
well, with chapters on wild animals, aquatic birds, emeralds and other
precious stones, horses, fruits, grains, and gold-mining techniques. In
each of these sections, pride of place was given to the recent additions to
the king's cabinet, including the "tail of a rattlesnake (*Serpent à sonette*)
that M. Godin sent from Peru in 1737 to the Cabinet d'Histoire naturelle
at the Jardin du Roy."[61] Descriptions of snakes such as this were drawn
largely from Marggraf and Piso's *Historia naturalis Brasiliae,* but the ending to chapter 30 became a serpentine coda for the politics of collecting
and an overdue *éloge* to both the divine "master" and his terrestrial "protector":

One can see such well-preserved skins of many of these serpents in the Cabinet
d'histoire naturelle of the Jardin du Roy. This treasure, which is not at all sufficiently
known to the public, contains an unusual abundance of natural products of all genres,
assembled with care and great effort from every corner of the world. If these treasures speak to the grandeur of their august master, the careful attention with which
they have been augmented and the order and elegance with which they have been
arranged make us admire no less the exquisite taste of their illustrious and zealous
protector, as well as the enlightened precision of [the individual] to whose supervision
they have been confided.[62]

This archaeology of typographical strategies demonstrates that the tripartite structure of the footnote, the bracketed footnote, and bracketed

text were understood as bringing "order and elegance" to the *Histoire des Incas* as well. But it was not merely a neutral, value-free order these devices brought. Rather, this new epistemology, represented by brackets and material specimens, sought to express the "progress" of natural historical knowledge and, indeed, its ability to confront contemporary problems through botanical solutions imported into Europe. Whether in the physical form of the hothouse or the typographical form of the bracket, these structures provided stability for hierarchically conceived layers of knowledge. In the *Histoire des Incas* as in the Jardin du Roi, these units of storage operated as mechanisms for centralization and consolidation. When we consider the relationship between natural knowledge and political power, the brackets served a function similar to that of Du Fay's hothouses: they represented large enclosed spaces where exotic specimens could be planted, cultivated, and made visible while being exploited for social and political purposes.

Several years before Louis-Jean-Marie Daubenton composed his description of the "Cabinet d'Histoire naturelle" for Buffon's multivolume *Histoire naturelle,* Dalibard had already performed an initial walkthrough of this burgeoning collection. Godin's shipment of Peruvian specimens likely served as the catalyst for Dalibard's interest in describing the botanical species of a distant land. If so, then the *Histoire des Incas* became the figurative soil into which he transplanted his enthusiasm, combining details gleaned from the previous century's travel accounts with the material pleasures of the objects themselves: items that were collected, cultivated, and displayed in the glittering hothouses that were of such recent vintage. The technical and typographical strategies he later employed to transform these physical specimens into print enabled their hierarchical arrangement through an elaborate system of texts and references. In conjunction with the editor, or perhaps *as* the editor, Dalibard gave us an interconnected chain of natural specimens, linked in print by footnotes and brackets that transformed a seventeenth-century chronicle into a revitalized treatise on nature and history.

Whether consciously or not, the three levels of editorial hierarchy that divided the *Histoire des Incas* also represented a stadial hierarchy of natural historical progress. The first level, an abridged and translated version of text originally written by Garcilaso, presented pre-Columbian "savage" knowledge without reference to the effects of European civilization, deleterious or salutary, and attempted to reconstruct indigenous patterns of culture "before the arrival of the Spaniards." The second level focused on observations culled from native peoples after European contact, employing scientific descriptions of plants, seeds, and fruits in tandem with native explanations of remedies drawn from postcontact observation. But the third level represented most explicitly the eighteenth-century aspiration to empirically based science, in which on-the-spot discoveries and new systems of natural classification allowed for the determination of universal laws of nature, observable through botanical "events" and live specimens drawn from throughout the known world.

This theory of natural historical progress was represented typographically within the pages of the *Histoire des Incas*. The typographical strategies employed by the editor throughout both volumes sought to emphasize the currency and utility of a people's history, even though the Incas and their empire had already been decimated some three centuries before. The material specimens sent by Godin from the ancient homeland of this rich if largely diminished culture made the use of these bracketed footnotes necessary. The brackets, in turn, offered the specimens a unique environment in which to be displayed: a *typographical* space within the margins of a pre-Columbian history where social and material interests merged together and an *ideological* space where useful objects were put forth as triumphant novelties to adorn the natural historical collections of the king's cabinet.

Conclusion: Staking Claims to Incan History

The republication of Garcilaso's history in 1744 shows that translation and appropriation in a natural historical context were dependent on many factors: material acquisition of new species; the construction of structures in which to house these specimens; an increased reliance on typographical mechanisms to organize and structure printed accounts; and access to books and reports that recounted prior natural historical knowledge, including indigenous practices that frequently formed the basis of European understanding of a given plant's value or utility. In the case of the *Histoire des Incas*, social and cultural conditions appear to have played a role as well, since references to agricultural crisis provided hints as to why the translation came to be produced when and in the form that it did. The two prefaces—the only portions of the *Histoire des Incas* that represent wholly original contributions—open a window through which to understand the interrelated stakes of historical and natural historical knowledge in a period of increasing botanical exchange both within and across imperial lines.

At the opening of the Jardin Royal in 1640, Guy de la Brosse had already lauded the "variety of species of its plants, brought from the two

Indies and from all the provinces of the Earth where French intelligence has been able to extend, which is what gives it a large and rich advantage over all the others."[63] Even then, bushes from Peru stood alongside other plants and trees that were growing in French soil, blurring the line between the "natural" environments of each and making it possible to think in global terms about science and applied botany within an institutional setting. But the construction of hothouses in the early eighteenth century provided a structural impetus for the collection of new species and the publication of new texts. The specimens sent by Godin passed unsuspected through the tight cordon of Spanish colonial administration that sought to prevent others from profiting from the riches of the Indies. Once in France, however, these seeds, travel logs, and academic *mémoires* provided the material ingredients for a potent brew from which to formulate new configurations of botanical knowledge. This knowledge, gleaned from native practitioners of the botanical arts, had been collected by European naturalists, compared with specimens sent by scientific explorers, and then molded together into a sourcebook that could be used to address contemporary social problems through agricultural experimentation, all under the protective guise of a historical treatise on a faraway region. Assisted by typographical methods that allowed a broader range of potential readers to profit from this knowledge, the texts that resulted from this process became hothouses in print, shifting possession from the field to the domestic and imperial institutions these individuals served.

The transfiguration of seeds, fruits, and flowers into textual descriptions on the printed page in the *Histoire des Incas* mirrored an earlier set of metamorphoses that had taken place in the gardens of the Incas. In the midst of the bracketed section on natural history, between the *Tara* and the *Onagra laurifolia flore amplo, pentapetalo,* the editor of the *Histoire des Incas* placed the description of a plant that "because of the beauty of its flower deserved a place in the admirable gardens of the Incas": the *Emérocallis floribus purpurascentibus maculatis,* known more commonly as the "Pelegrina," or wanderer. According to this account, when the "perpetual Springtime" of the Peruvian season finally started to fade and the "Pelegrina" itself began to wilt and decay, the natives took an extraordinary measure to maintain the sun-gilt appearance of their gardens: "[T]hey substituted in their place new plants created with gold and silver, which their artistry had imitated with perfection. Trees made from these precious metals forming long walkways, fields entirely filled with artificial corn whose flowers and ears were of gold, and whose stalks and leaves were of silver, served as convincing proof of the wealth of this land, the skill of the Indians, and the magnificence of their sovereigns."[64] These shiny shrubs of silver and gold took their place as monuments to the Incas' artistic prowess. With more than an air of nostalgia, La Condamine had described similar objects at the end of his account of the Incan ruins at Cañar, alluding to their presence in the royal houses of Cuzco and to descriptions in the accounts of Cieza de León, Lopez de Gómara, Father Andrés de Zarate, and others. "During the time that I was in Quito," La Condamine wrote in the concluding paragraph of his *mémoire*, "I had always heard about such figures of animals, insects, and other works of solid gold, safeguarded for more than a century out of curiosity." But when La Condamine finally decided to visit the royal treasury, shortly before his departure, he discovered that these extraordinary pieces of Incan artistry had been melted down and sent to Cartagena as gold bars. Lamenting the politics that led to the destruction of such prized works of art, he wrote that "the unique products of nature and artistry would cease to be rare, if there were responsible people to collect them."[65]

As La Condamine observed, this botanical El Dorado perched among the Andes' jagged peaks was not to last. The transcription by Dalibard recapitulated in print the importance of horticulture to the original residents of the Andes, but their golden treasures had long since been taken out of their possession. Feuillée's reference to plants as objects that were "far more valuable treasures than those extracted from the mines of Peru" was more prescient than he himself may have realized. Likewise, the editor of the *Histoire des Incas* criticized the Spanish for not being more interested in acquiring native knowledge; if they had, he claimed, they would have quickly realized that the mines of Potosí were not "the only sources from which one could extract riches in Peru." Seeking to profit from these founts of native knowledge, French naturalists at the Jardin du Roi studied the material botany brought back by their monarch's minions from overseas. In turn, they employed editorial and typographical techniques to transform this knowledge into a shape they could employ and promote. The impulse to parse, deconstruct, and reassemble transformed Garcilaso's *Comentarios reales* as well, and the descriptions and concatenations of new materials offered by Dalibard and others eventually served as a source for new riches, extracted this time from the botanical mines and floral lodes of Peru. But their legacy as building blocks was mixed. The *Histoire des Incas, rois du Pérou* (1744) was used by Madame de Graffigny to describe the history of the Incan people in her *Lettres d'une péruvienne*, but it was also deployed by Buffon and Cornelius de Pauw later in the century to denigrate the American continent

and the culture of its native inhabitants. In the end, the "important discoveries" brought back to France by the king's "new Argonauts" depended not only on the material conditions in which they traveled and
the typographic conditions their work was to encounter in print. Their
fate also depended upon the uses to which this reconfigured knowledge
was eventually put. The bounty of naturalists and the texts that described what specimens they had acquired made their itinerant way
within wooden crates and printed brackets. But neither vessel offered sufficient protection from loss or erasure from the historical record, a danger
that was as real for melted Inca artifacts and spoiled natural specimens as
it was for the abridged portions of Garcilaso's history of the Incas. This
phenomenon of abridgement and suppression would even be repeated
in some of the eighteenth century's most iconic and capacious texts.

SEVEN

The Golden Monkey
and the Monkey-Worm

[Travelers] almost always exaggerate what they have seen, or what they might have seen; and so as not to leave their travel narratives in an imperfect state, they report on what they have read in other authors, since they were fooled at first [by these authors] in the same way they now fool their readers . . . [T]he assurances they offer of having verified innumerable falsities that were written before them have the singular effect of rendering the sincerity of every traveler suspect, since the censors of good faith and the accuracy of authors never give sufficient warnings about their own [good faith and accuracy]. CHEVALIER DE JAUCOURT, "VOYAGEUR," IN *ENCYCLOPÉDIE, OU DICTIONNAIRE RAISONNÉ* (1765)

Denis Diderot had few words of praise for travelers or their tales. Especially in his later years, he portrayed the peripatetic philosopher as a partial and selective observer, one "born with the taste for the marvelous," exaggerating for personal gain and self-servingly justifying the routes he had taken. His indictment of the "unfortunate, errant, and dissipated" nature of the traveler was often couched as a critique of the behavior of Europeans overseas, a position most clearly associated with his 1771 *Supplément au voyage de Bougainville*.[1] Diderot's fictional *Supplément*, which purported to reveal a "suppressed" addition to Bougainville's wildly popular account of his journey to the Pacific, criticized the treatment the Tahitians had received from Bougainville and his crew. After reading this celebrated narrative, Diderot responded with an apostrophic command: "Monsieur de Bougainville, move your boat away from the banks of these innocent and fortunate Tahitians; they are happy and you can do nothing but damage their contentment." Travelers such as Bougain-ville, Diderot seemed to imply, were at odds with the very spirit of an enlightened age, debasing the fundamental precept of the eighteenth-century philosophe: "The contemplative man is sedentary," wrote Diderot of the static and immobile thinker; "the traveler," by contrast, "is an ignorant liar."[2]

But problems arose when Diderot and his fellow philosophes set out to compose their "reasoned dictionary" at the midpoint of the eighteenth century. Travelers may have been untrustworthy, but the editors of the *Encyclopédie, ou Dictionnaire raisonné* needed these "ignorant liars" in order to write authoritatively about the world beyond Europe's shores. Voyages of exploration were written into the fabric of the institutions from which the *Encyclopédistes* (as the collective authorial pool of the *Encyclopédie* was known) drew much of their material, including the scientific societies of London, Paris, and elsewhere. While many philosophes expressed an antipathy toward philosophical endeavors undertaken outside of Europe, they did recognize, like the Chevalier de Jaucourt, that the Argonauts of antiquity had "enlightened their nation through wisdom (*lumières*) acquired in visiting foreign countries."[3] For these reasons, Diderot and his associate Jean le Rond d'Alembert found themselves in an epistemological paradox: how were they to use these important yet unverifiable observations from abroad to construct the solid foundations of a universal philosophy at home?

Despite being reluctant to include such accounts in his encyclopedic project, Diderot had already agreed to accept contributions from at least one explorer: Charles-Marie de La Condamine. Only one month had passed since the June 1751 publication of the first volume of the *Encyclopédie*, and while several years had elapsed since La Condamine's dramatic presentation before the Academy in 1745, Diderot seems to have remained familiar with what his friend had published in the intervening years. The just-completed initial volume contained an article on the Amazon River that referred explicitly to La Condamine's voyage. Diderot, in fact, may have been in the audience for La Condamine's initial exposition of his journey, since he contributed regularly to the *Mercure de France* during this period and is thought to have composed the *Mercure*'s review of this event. The review noted that "the public will benefit, in time, from the numerous discoveries that M. de la Condamine has produced during this long and difficult journey, and we will see that the new world, which is today the source of all the gold in Europe, is no less wealthy in literary treasures."[4] Now, six years later, Diderot had asked La Condamine to contribute an article specifically on the pyramid controversy (discussed in chapter 1), implying that in order to seal the episode's definitive passage into posterity, the errant academician would most certainly want to record the authoritative version in the pages of the *Encyclopédie*.[5]

While enthusiastic at the idea of an article on the destroyed pyramids, Diderot may have seen significantly less literary promise in one of the more colorful objects La Condamine brought back with him from his journey. At the picturesque Brazilian port of Pará, near the mouth of the Amazon River, the city's governor had given La Condamine a dazzling departing gift: a golden-haired monkey reputed to be the sole representative of his species. The monkey's tail, in contrast to a body whose sheen radiated like "the color of the most beautiful blond hair," was of a dark and lustrous brown, almost black. But most impressive of all were the bright red tints with which his ears, his cheeks, and his muzzle were adorned, "a vermilion so vibrant," in the words of La Condamine, "that one could hardly be persuaded that the color was natural."[6] Eager to return to Paris accompanied by a live specimen that would entertain his friends, impress his colleagues, and serve as tangible proof of his account, La Condamine attempted to keep the monkey alive while returning to Europe by sea. But his efforts were in vain:

Despite the constant precautions I took in order to preserve him from the cold, the rigors of the season apparently caused his death. Since I had no facilities on board to dry him in the oven, following M. de Réaumur's suggestions for preserving birds, all I was able to do was to preserve him in alcohol, which will suffice perhaps to show that I have in no way exaggerated in my description.[7]

This monkey, which would be classified by naturalists as the "little monkey from Para," merited no more than three short lines in Diderot's *Encyclopédie*, folded into the article "Monkey" (*singe*) along with baboons, apes, and other simians from around the world.[8] What was the classificatory logic that gave the golden monkey such terse acknowledgement while the mysterious "monkey-worm," a vermin found in Asia and America that lodged itself between the skin and the muscle of its prey, merited its own entry in the *Encyclopédie*'s final volume? La Condamine returned from South America with both objects preserved in glass containers. But the one, adapted as it was to wending its way into the flesh of its host, seems to have been more adept at garnering an extended life in print as well. If we view these two objects as eighteenth-century metaphors for textual tenacity, how might we use them and other examples drawn from the Quito expedition to understand compilation and abridgement more broadly in the context of the *Encyclopédie*?[9]

La Condamine's golden monkey did not survive the climatic transition from the banks of the Amazon to European shores. And neither was the article Diderot commissioned on the pyramids of Quito included in the *Encyclopédie*. The objects from the Quito expedition that did survive the transatlantic voyage from the equatorial regions of South America were abridged and transformed according to principles of inclusion that subtly reinforced hierarchies of European geographical knowledge. And their incorporation into the *Encyclopédie* often had the collateral effect of effacing the original sources of that knowledge. While many material specimens found their way to physical homes within the burgeoning collections of the Jardin du Roi, the sheer diversity of the Americas presented the *Encyclopédie*'s editors with a challenge of an intellectual, rather than a logistical, nature: how to condense and encapsulate discoveries from outside Europe without losing the thoroughness they felt had been lacking in previous encyclopedic works. Diderot wrote in his prospectus that the *Encyclopédie* represented an opportunity to "give each entry its appropriate length, to insist on the essential, neglect the minutiae, and to avoid . . . treading heavily where only a [single] word is needed."[10] To carry out this project according to the proposed logic, travel narratives had to be parsed, abridged, and reassembled so that their utility would be readily apparent to potential readers.

But silences inhered in the transcription of knowledge from extra-European spaces back to the metropole. Writers and editors who engaged in the translation and interpretation of these phenomena did not seamlessly divide knowledge into its Baconian branches for insertion into the *Encyclopédie* and other texts. Instead, narrative constraints from the geographic periphery to the metropolitan center transformed complex source texts into equally problematic distilled articles, effacing their origins in order to lay rhetorical claim to omniscience and universal knowledge. These erasures and imposed silences dramatically altered the form of this knowledge when it was compiled within the sheaves of a single bound volume. How, then, could "travel make truth" if the texts that brought these "truths" to a European audience—including encyclopedias, dictionaries, maps, and natural historical treatises—compressed travelers' observations to such a degree that they siphoned off the contributions of those individuals whose experiences formed the very foundation of European knowledge? In order to expand the frontiers of human understanding within an encyclopedic text, the editors of the *Encyclopédie* diminished many extra-European zones of geographic interest to near-infinitesimal proportions. This encyclopedic impulse reinforced the geographical exclusion of objects and spaces that were seen to have less significance for the rationalizing principles of the philosophes, just as it silenced the roots by which this knowledge took hold in their philo-sophical journeys in the first place.[11]

From Display Cases to Dictionary Definitions

Even if his Amazonian monkey had lived, La Condamine did not neces-sarily intend to carry it alongside him as he strolled down the Quai de la Seine following his return from South America. But had he so wished, many were the venues in which he could have done so in a Parisian land-scape increasingly accustomed to overseas oddities on display. Regular commerce with the non-European world had blurred the line between the everyday and the exotic in eighteenth-century France, and spaces for interaction between the Parisian public and these non-European objects and peoples abounded.[12] The ubiquity of non-European animals as pets and curiosities was only one manifestation of this broader phenomenon. The French capital was also assailed during this period with many awe-inducing spectacles, including lightning-rod experiments, balloon launc-hes, mesmeric séances, and public lectures on the thermal qualities of light rays. Against this backdrop, the material bounty brought back from voyages of exploration took its place comfortably within an ever-expand-ing sphere of public science.[13]

The Cabinet d'Histoire naturelle in Paris's Faubourg Saint-Victor was one of the sites in which exotic specimens and public spectacle came together. Twelve to fifteen hundred people were thought to frequent the species within its spacious halls each week. As Diderot wrote in the *Encyclopédie*, visitors came "from every state, from every nation, and in such large num-bers that during the pleasant seasons of the year, when the bad weather did not prevent them from staying in the halls of the cabinet, its space was barely sufficient."[14] The Cabinet, along with *musées*, cafés, public lec-tures, and outdoor events, brought the Parisian public into direct contact with the latest scientific spectacles of the day, and the leading periodi-cals followed their course and riveted attention on these activities with a vigor that outshone many other topics of contemporary interest.[15]

But printed texts also came to be one of the emerging public spaces in which mixing between exotica and eager audiences took place. The frequent publication of periodicals and academic *mémoires* placed the natural world within the reach of a literate audience. With the wide avail-ability of printed matter in the eighteenth century and a voracious pub-lic appetite for new forms of scientific knowledge, massive encyclopedic works like Buffon's *Histoire naturelle*, Brisson's *Le regne animal*, Réaumur's

Mémoires pour servir à l'histoire des insectes (1734–42) and Diderot and d'Alembert's *Encyclopédie* created surrogate communities that substituted reading for in situ observation. These texts provided a forum in which collaborative discussion and shared learning took place among individuals from diverse backgrounds; compilations such as the *Encyclopédie* even came to supplement the role that various academies played within Parisian society. This transition from display to print allowed a wider public to become familiar with new knowledge coming in from overseas.[16]

The challenge of how to reduce natural knowledge into parsed, alphabetically organized units on the printed page, however, remained unresolved. Textual spaces abided by a different kind of organizational logic than the institutions in which physical objects were stored and managed. This transposition of specimens from the cabinet to the catalog represented a shift between two radically different representational modes, one that operated according to the logic of visual display and the other that functioned according to the alphabetical logic of the dictionary. Unlike the Cabinet d'Histoire naturelle, which was heir to the chaotic displays of the *wunderkammer*, textual compilations such as the *Encyclopédie* used alphabetical ordering to vindicate an epistemological program placing knowledge explicitly at the service of humankind.[17] This complex interface between materiality and textuality was one of the epistemological problems confronting the *Encyclopédistes* as they sought to incorporate non-European objects into their reasoned dictionaries.

Discussing this challenge in his description of the Cabinet d'Histoire naturelle, Louis-Jean-Marie Daubenton explained that the order and classificatory structure of the natural history cabinet's material displays depended upon "a certain artistry in [its] arranging [and] continuous care, and a kind of industry to place everything in its proper order and maintain it in a good condition."[18] Daubenton's conception of the cabinet centered around the notion of elasticity. Expansion was seen as the clearest sign of success, which in turn required the constant rearrangement and renovation of a collection's internal structure:

Since the Cabinet du Roy has been augmented considerably over the past several years, it is easy to see that the arrangement [of the materials] has also changed several times, and I fervently wish for it to be regularly the case: it is the clearest proof of the progress made by this establishment.[19]

It was thus mobility, rather than fixity, that defined the progress of the cabinet as an institution. Like an accordion, the cabinet could expand and change shape without ever losing its established function.

For Diderot and d'Alembert, the natural history cabinet was a highly cogent metaphor for the *Encyclopédie* project. The compartmentalization of knowledge within the encyclopedic form mirrored in important ways the form and structure of the cabinet, which was designed to encapsulate knowledge and allow the broad expanse of that knowledge to be viewed from a single location. In the *Encyclopédie*, the laudatory description of the "Cabinet d'Histoire naturelle," penned jointly by Daubenton and Diderot, functioned as an argument on behalf of the order and content of their encyclopedic project as a whole. The cabinet, like the *Encyclopédie*, was a space where one could survey the entire range of human knowledge: "[H]ow can we possibly provide an appropriate sense of the spectacle that would exist if every kind of animal, vegetable, and mineral were assembled in the same place and could be seen, so to speak, from a single vantage point? . . . a *cabinet d'Histoire naturelle* is a reduced version (*un abrégé*) of nature in its entirety.[20]

But the *Encyclopédie* was also an abridged and compressed version of that natural world; the extremes of breadth and compression were contradictory tendencies that nevertheless exemplified the entire project. Diderot even recognized this paradox in his "Encyclopédie" article: on the one hand, the purpose of the *Encyclopédie* was to amass as much material as possible in order to provide an ample catalog of nature from which to augment human understanding; on the other hand, its goal was to create order within a massive database of knowledge, an activity that Diderot acknowledged would require abridgement and compression to achieve its goals. By enclosing the treasures of enlightened knowledge in a vaulted edifice, Diderot sought to execute a project he felt the ancients had been capable of carrying out but had fallen short of accomplishing: the production of an encyclopedic corpus of knowledge, one that, according to Diderot, would have been more valuable than the Library of Alexandria itself. This grandiose project got its start the very month that an errant academician who had been absent from Europe for nearly a decade made his triumphant return to the City of Lights.

South America in the *Encyclopédie*:
Quito without Pyramids, Empires without Incas

La Condamine returned to Paris in February 1745, the same month that the publisher André Le Breton agreed to produce a French translation of Chambers's *Cyclopaedia*, in collaboration with his German colleague Gottfried Sellius and the wealthy Englishman John Mills. Several months later, Diderot was recruited to participate in the translation effort and was named along with d'Alembert to be joint principal editors of the entire project. The two editors had come to realize that a mere translation of Chamber's *Cyclopaedia* would be woefully inadequate to represent the wealth of new knowledge that had been unearthed since the English dictionary's publication. In the words of Diderot, "we examined in full the translation [of Chambers's *Cyclopaedia*], and we found it lacking in many items, both in the sciences and in the liberal arts, one word where pages were needed."[21] What was truly required in their minds was a renovated structure and significant additions to its content.

But Diderot also acknowledged the difficulties he and d'Alembert would confront in bringing their encyclopedic project into print. If on the one hand the philosophes would benefit from "the discoveries of great men and learned associations," they would also have to contend with "the prodigious increase in the amount of material." Diderot resorted to metaphors of saturation: the Republic of Letters was "flooded" with treatises, and the eighteenth century was surrounded by this "sea" of texts and objects. Only by "giving each subject an appropriate size, insisting on what is essential, [and] neglecting the minutiae" were the editors to succeed in stimulating "the genius to reveal unknown paths and to advance toward new discoveries." The articulation of these challenges also provided a frame in which the *Encyclopédistes* could forestall the criticisms they knew would most certainly follow them throughout their project.[22]

The Quito expedition exemplified this flood of new material. New artifacts and specimens accompanied the numerous written accounts of the academicians' quarrels and controversies, augmenting the libraries and natural historical collections of European monarchs and their academies. While many of these objects were lost along their transatlantic journey—recall the monstrous bones from chapter 6 dumped overboard on their return voyage to Europe—there were many items that *did* of course arrive intact, as well as numerous texts that described topics ranging from the customs of native peoples to South American geography. This rich vein of material brought back from South America by members of the Quito expedition would in principle be easily assimilated into the *Encyclopédie*, since their accounts were exceptionally fresh, almost contemporaneous, and the specimens were only recently arrived from across the seas. The *Encyclopédistes* could base their articles and observations on reports from members of their own order rather than merely rehashing old myths and cribbing from texts that had circulated for decades, a practice they themselves frequently criticized in the context of travel literature. And the temporal proximity of the academicians' reports would necessarily yield thorough accounts in the body of the *Encyclopédie*. Or would it?

When Diderot asked La Condamine to compose an article on the Quito pyramid controversy, for instance, he explained the importance of hearing directly from the author of the *Histoire des pyramides*, despite the account's having only recently been published and circulated among its eighteenth-century readers:

> There is above all one important article that I would ask you for; it is the history of the pyramids. This [story] does not need to be recorded by us in order to pass into posterity: but it would be unfortunate (*on nous saurait très mauvais gré*) not to speak of it with all the knowledge that we are capable of providing.[23]

Diderot implied that he and his contemporaries had ample knowledge of the events surrounding the construction of pyramids on the Yaruquí plain. But while the pyramid controversy may have been noticed by those within the Republic of Letters, readers of Diderot and d'Alembert's "reasoned dictionary" would have missed the episode entirely, for La Condamine never actually composed the requested article on the Quito pyramids for the *Encyclopédie*. All articles related to pyramids, from the "Pyramide de Porsena" to the "Pyramides d'Égypte," were written by Jaucourt, and none of these articles mentioned Quito. The article Diderot had commissioned La Condamine to write on the pyramids' destruction, even more completely than the pyramids themselves, had disappeared without a trace.

Whatever the specific cause may have been, the controversy over the Quito pyramids was one of many South American topics given short shrift by the editors of the *Encyclopédie*. South American geography, history, and natural history received very little attention in the multivolume work, either relegated to articles that ridiculed strange Amerindian customs or, most often, folded into articles on "America" more broadly. (Even the article "America" was shockingly brief in the *Encyclopédie*'s first volume, and it was not until 1776 that a more thorough article on America, written by Samuel Engel, was added in the *Supplément à l'Encyclopédie*.[24]) Given the interest in the region demonstrated by authors and writers from Voltaire to Buffon, this dearth of detail is striking. Diderot appears to have been far less interested in having articles on Quito's pyramids written with "all the knowledge that we are capable of providing" than his rhetoric initially suggested.

The number of La Condamine's direct authorial contributions to the *Encyclopédie* project was actually quite meager, in fact, especially when considering the output of Daubenton, Jaucourt, and d'Alembert, or even some of the more occasional contributors of articles on natural history, such as the baron d'Holbach, Antoine-Joseph Dezallier d'Argenville, or Jean-Baptiste-Pierre Le Romain.[25] Only four articles, including "Chirimoya," "Couronne," "Guayaquil," and "Guiane," could be explicitly attributed to La Condamine. All of these articles appear in the first third of the alphabet, adding credence to the idea that one of the editors may have banned La Condamine's subsequent participation in the project. D'Alembert, for one, had apparently become frustrated with La Condamine's "carelessness." In a letter to Le Breton, d'Alembert claimed that La Condamine had made him "more furious that I had ever been in my life" by not returning Pierre-Charles Le Monnier's article "Boussole" to him in a timely fashion.[26] The article on "inoculation," suspected by many to be La Condamine's work, was printed as an anonymous contribution. And the four articles that La Condamine did write may have been of his own volition, since it appears he had more or less invited himself to contribute to the *Encyclopédie*.[27]

Like the pyramid article never written by La Condamine, the article "Quito" was also diminished by editorial exigencies. When we compare the articles "Guayaquil" and "Quito," we notice that La Condamine's "Guayaquil" was more than twice as long as Jaucourt's "Quito," even though Quito was the seat of the Audiencia, name of the administrative province, and a city of far greater political significance during the colonial period than its coastal cousin Guayaquil. In fact, Jaucourt published an earlier and shorter version of the entry in the *Encyclopédie*'s seventh tome, and La Condamine's entry on "Guayaquil" was only published at the end of the seventeenth volume, under the section entitled "Omitted Articles." Whereas La Condamine dedicated the entire final paragraph of his entry to the observations and measurements used to determine Guayaquil's precise location, Jaucourt merely included the longitude and latitude measurements for Quito in italicized text at the end of his much briefer entry. Jaucourt did not take advantage of the broad circulation of the narrative accounts from the expedition to Quito. But neither did Diderot or d'Alembert ask La Condamine to compose the Quito article in his stead.

When Jaucourt described Quito, he also failed to fold in any of the dramatic detail that was expressed in the recently published *Carta de la Provincia de Quito*, a map mentioned in Vaugondy's article on "Geography" as one of the most admirable works of eighteenth-century geography and "a . . . work to which we should pay close attention."[28] Unlike the article on the "Rivière des Amazones," in which d'Alembert praised La Condamine's "new map of the river [as] more accurate than any that had preceded it," there was no mention of Maldonado's map in Jaucourt's brief article. Instead, his geographic description tersely expressed the meandering frontiers of the Audiencia according to border distinctions drawn from Guillaume Delisle's *Carte de la Terre Ferme, du Perou, du Bresil, et du pays des Amazones*, originally published in 1703 and reproduced on several occasions in the 1710s and 1720s: "[The Audiencia of Quito's] borders are Popayan to the north, the Audiencia of Lima to the south, the country of the Amazons to the east, and the South Sea to the west."[29] But Jaucourt did not incorporate any of the detail expressed in the *Carta de la Provincia de Quito*, over which so much editorial ink had been spilled between Maldonado, d'Anville, and La Condamine.

A similar fate awaited the Incas in the pages of the *Encyclopédie*. The most recent translation of Garcilaso's history of the Incas (the *Histoire des Incas*, 1744) brought the French reader a wealth of information about Quito's preconquest empire, and readers from Buffon to Madame de Graffigny benefited from this new edition. The *Histoire des Incas* would have been an ideal source for what d'Alembert expressed in the "discours préliminaire" as one of the "principal fruits" that could be gained from the study of past societies:

One of the principal fruits of the study of empires and their revolutions is to examine how human beings, separated so to speak into several large families, have formed diverse societies: how these different societies have given rise to different kinds of governments, [and] how they have sought to distinguish themselves from one another, either by establishing laws or through particular signs that each [society] has invented so that its members could more easily communicate with one another.[30]

The editor of the revised *Histoire des Incas* (1744) undertook the translation precisely because he imagined that the "great examples" of Incan civilization would be "extremely useful" to eighteenth-century French society (see chapter 6). But the history of the Incas was vastly compressed when transferred to the *Encyclopédie*. Rather than highlighting their profound agricultural and political achievements, the articles focused on their risible singularities; these texts poked fun at the people who believed their leaders had descended from the sun and who claimed that a certain stone near Cuzco (the so-called "Pierre des Incas") had mystical qualities:

[T]hey attributed [to this stone] a large number of virtues. In Spanish America, they still make buttons and stones for rings out of these pyrites, and they believe that [the stones] change color when the bearer is threatened by illness . . . We have throughout

Europe a large number of these stones that we could use for the same purpose, if we found it appropriate.[31]

The Incas were thus portrayed as quaint vestiges of a bygone civilization, hardly the image offered either by the most recent translation of Garcilaso or by the editors of the *Encyclopédie* for the historical study of empires past.

The most extended article related to the Incas, in fact, was the article "Amautas" by the abbé Mallet, published in the *Encyclopédie*'s very first volume and, coincidentally, on the same page as d'Alembert's article on the Amazon River. Mallet described the Amautas as "Peruvian philosophers during the reign of the Incas," focusing on their central role in instructing the nobility in history, art, music, and the sciences. He also discussed the theatrical representations they made "before their kings and the court nobility during solemn feasts," comedies in which "[the Amautas] discussed agriculture, domestic affairs, and other events drawn from human life."[32] The abbé Mallet depended on a French translation of Garcilaso's Incan history, but which edition of Garcilaso was he using? His own bibliographical references are ambiguous on this account. The abbé modernized the spelling according to the most recent French edition of Garcilaso (the 1744 *Histoire des Incas*, using an "I" rather than a "Y" for "Inca"), but referred the reader to "liv. II & IV," a citation that corresponded to the book and chapter ordering of the pre-1744 editions of Garcilaso's *Comentarios reales*. This oscillation between "Ynca" and "Inca" persisted throughout other tomes of the *Encyclopédie* as well, due in large part to the plurality of authors and the diversity of their sources. In fact, because of the multiple authors working on the project and the alphabetical order to which the editors had decided to adhere, the Incas were discussed in two articles, in volume 8 ("Incas") and volume 17 ("Yncas"). In the final volume, Jaucourt tersely accounted for "the people considered [to be] descendents of the sun," who had "beautiful palaces, superb gardens, magnificent temples, and compliant subjects."[33] Jaucourt's compressed account of the Incas in volume 17 was far from the programmatic mandate to account for how "different societies gave birth to different forms of government." Indeed, the reference work to which Jaucourt sent his readers was the *"histoire des yncas,"* the 1737 version, an account whose facts according to Jaucourt's contemporaries at the Jardin du Roi had been "heaped together" in such a way as to make them unintelligible.[34]

This encyclopedic dissonance on account of a shifting transliteration from Quechua to Spanish to French was not uncommon when referring

une espece de casque verd; de ce creux sortent des fleurs bleues semblables aux premieres.

* AGUAS, (*Géog.*) peuple considérable de l'Amérique méridionale, sur le bord du fleuve des Amazones. Ce sont, dit-on dans l'excellent Dictionnaire portatif de M. Vosgien, les plus raisonnables des Indiens: ils serrent la tête entre deux planches à leurs enfans aussi-tôt qu'ils sont nés.

* AGUATULCO *ou* AQUATULCO *ou* GUATULCO, ville & port de la nouvelle Espagne, en Amérique, sur la mer du Sud. *Longit. 279. latit. 15. 10.*

Figure 52. These three articles—"Aguas," "Homagues," and "Omaguas," shown in figures 52–54—are from Diderot and d'Alembert's *Encyclopédie, ou Dictionnaire raisonné* (1751–72) and represent the same indigenous group. Note the irony with which the article "Aguas" discusses this group: "They are the most rational of Indians: they squeeze the heads of their infants between two pieces of wood as soon as they are born." This article relies on Vosgien's *Dictionnaire géographique portatif* (The Hague, 1748), which drew on La Condamine's account. Courtesy of the Special Collections Library, University of Michigan.

to South America: other transliterations sowed double or triple visions in their wake as well. The indigenous group of Omaguas, for example, a native people described by La Condamine in his *Relation abrégée* based on Jean Magnin's manuscript, was described three separate times in the *Encyclopédie*: in the articles "Aguas," "Omaguas," and "Homagues" (figs. 52–54). But what these various examples demonstrate is the striking diversity of sources and outright dissonance between articles on South American history and geography within the *Encyclopédie*. The representations of topics that had been studied and assessed by members of the Quito expedition were not based on the texts that had been produced by the academicians just a few years before. It is not that the *Encyclopédie* was merely "imperfect" or "random" or that the editors did not live up to their lofty ambitions. Rather, when condensing Quito, its pyramids, and the Incas into their reasoned dictionary, the authors of the *Encyclopédie*'s entries seem to have preferred textual accessibility to depth of research, employing whatever texts happened to be at hand in their attempt to distill complex subjects into an abridged form. The *Encyclopédistes* made do with the tools at their disposal. There was little interest in carrying out extensive research on a geographical region that lay so far from the quotidian preoccupations of the enlightened elite, despite the public attention bestowed on South America more broadly in the periodical press.

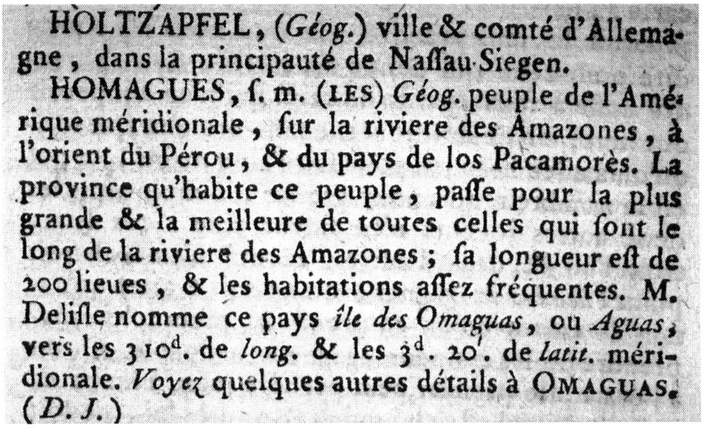

HOLTZAPFEL, (*Géog.*) ville & comté d'Allemagne, dans la principauté de Naſſau-Siegen.

HOMAGUES, ſ. m. (LES) *Géog.* peuple de l'Amérique méridionale, ſur la riviere des Amazones, à l'orient du Pérou, & du pays de los Pacamorès. La province qu'habite ce peuple, paſſe pour la plus grande & la meilleure de toutes celles qui ſont le long de la riviere des Amazones ; ſa longueur eſt de 200 lieues, & les habitations aſſez fréquentes. M. Deliſle nomme ce pays *île des Omaguas*, ou *Aguas*, vers les 310ᵈ. de *long.* & les 3ᵈ. 20ʹ. de *latit.* méridionale. *Voyez* quelques autres détails à OMAGUAS. (*D. J.*)

Figure 53. The "Homagues" tribe, including their latitudinal and longitudinal coordinates from one of Guillaume Delisle's maps. Courtesy of the Special Collections Library, University of Michigan.

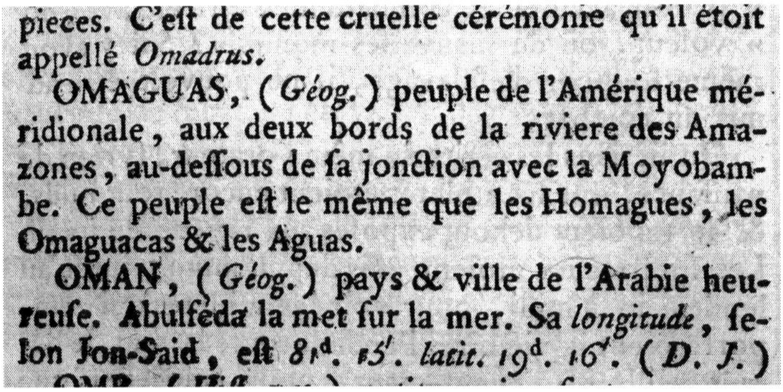

pieces. C'eſt de cette cruelle cérémonie qu'il étoit appellé *Omadrus*.

OMAGUAS, (*Géog.*) peuple de l'Amérique méridionale, aux deux bords de la riviere des Amazones, au-deſſous de ſa jonĉtion avec la Moyobambe. Ce peuple eſt le même que les Homagues, les Omaguacas & les Aguas.

OMAN, (*Géog.*) pays & ville de l'Arabie heureuſe. Abulféda la met ſur la mer. Sa *longitude*, ſelon Jon-Said, eſt 81ᵈ. 15ʹ. *latit.* 19ᵈ. 16ʹ. (*D. J.*)

Figure 54. The "Omaguas" in the *Encyclopédie*. This short entry points out that the "Omaguas" are the same as the "Homagues," "Omaguacas," and the "Aguas" tribes, although "Omaguacas" are not mentioned elsewhere in the *Encyclopédie*. Courtesy of the Special Collections Library, University of Michigan.

From Women Warriors to *Quinquina*: The *Encyclopédie* and Its Amazons

The representation of the Amazon River in the *Encyclopédie* is strikingly different from these previous examples, however. The extensive circulation of La Condamine's *Relation abrégée* within erudite circles and its publication as an academic *mémoire* in the 1745 *Mémoires de l'Académie Royale des Sciences* seem to have made his account more widely accessible when it came time to prepare the *Encyclopédie*. While La Condamine may not have authored a statistically significant number of articles, the account of his journey down the Amazon River featured prominently among texts used by those who wrote articles on the geography and natural history of South America. When conceived in broader terms, La Condamine's contribution takes on greater significance. Indeed, the results from his travels and traces of his academic *mémoires* can be found dispersed throughout the *Encyclopédie*. While it would have been quite possible to live out one's life in London, Paris, or Amsterdam without any knowledge of the massive river system at the heart of South America, a faint but increasingly perceptible crescendo of literature since the middle of the seventeenth century would have made it far easier for the curious reader to learn about the Amazon, its mythology, and the early history of its exploration.[35]

Between 1717, when Samuel Fritz's "Description abrégée du fleuve Maragnon, & des missions establies aux environs de ce fleuve" was published in the *Lettres édifiantes et curieuses*, and the publication of La Condamine's *Relation abrégée* in 1745, there were few texts that referred to the Amazon exclusively in geographic terms. More prevalent in eighteenth-century libraries were discussions relating to Amazon women warriors, some of which began to contain specific remarks on the "modern" Amazons of America and the river that bore their name. Pierre Petit's *Traité historique sur les Amazones*, originally published in 1685, was republished in 1718, and contained a section that corroborated the existence of the Ancient Amazons based on "similar women, living in the manner of Amazons, in the New World, near a river of great breadth and length that has been named the river of the Amazons after these courageous women."[36] In 1737, Louis le Maingre de Bouciquault released in Rotterdam a work of fiction entitled *Les Amazones revoltées*, in which Amazons in Asia Minor fought for their liberty against Turks who set out to subdue them and their land in a region abundant with minerals and agricultural potential. This text contained a separate section on the discovery of the New World as well. And finally, the abbé Guyon combined the ancient history of the Amazons of Asia Minor with those of the American Amazons in his *Histoire des Amazones anciennes et modernes* (1740), relying on "esteemed authors" to report on the "surprising [fact] of having found in America a group of Amazons whose customs were almost identical to those of Thermodon." Indeed, Guyon ends his text with the supposition that it was by way of the island of Atlantis that the American continents were populated by "Amazons from Africa [who] were able to pass to the other Hemisphere, where they inspired other women to live like them . . . near the river that bears their name." The Amazon River region in South America was thus slowly folded into the Old World myth of Amazon warriors, and it became increasingly rare that the one was mentioned without the other being invoked in at least a cursory way.[37]

But while the myth of women warriors continued to hold a tenacious grip on the imagination of Europeans, knowledge based on classic texts slowly began to be supplanted by the increasing number of travel accounts and other materials arriving from South America. La Condamine's *Relation abrégée* was certainly primary among these textual sources, as evidenced by d'Alembert's article on the Amazon River in the *Encyclopédie* that dedicates the bulk of its text to describing La Condamine's account and makes only the briefest allusion to the fact that "Orellana claims to have seen several armed women in his descent of the river." But the history and representation of two commodities in the *Encyclopédie*, the rubber tree and cinchona bark, bear noting as well.[38] It is especially worth considering how La Condamine was able to transform his and others' early notes about these two specimens into printed *mémoires*, which in turn served to increase the impact of his initial observations and allowed his name to resound in tandem with these objects through learned periodicals and natural historical texts.

La Condamine first described the "elastic resin" of the rubber tree, or "caoutchou," in his *Relation abrégée*. He had seen some objects produced with this material by Amazonian natives during his journey. But he did not publish an account of this material until receiving the *mémoire* written by François Fresneau, an engineer and inventor he had met in Cayenne. La Condamine began his "Mémoire sur une résine élastique, nouvellement découverte a Cayenne par M. Fresneau" with these words: "In 1736, I sent the Academy [of Sciences] . . . several rolls of a blackish and resinous mass."[39] While La Condamine gave ample credit to Fresneau for his work, his *mémoire* demonstrated his own early role in bringing European attention to this exotic substance. The title of his *mémoire* heralded a species "newly discovered" and implied that Fresneau had "discovered" the rubber tree only after its having been seen, described, and dispatched by La Condamine. In the articles "Résine" and "Cahuchu," both composed by the Chevalier de Jaucourt, La Condamine's commentaries were the only ones mentioned, even though these texts were printed well after the "Mémoire sur le résine élastique" was published in the *Mémoires de l'Académie*.[40] The compression of the article to conform to the standards of the *Encyclopédie* suppressed the source of much of La Condamine's information—in this case, Fresneau's *mémoire*—even though La Condamine made reference to Fresneau in his longer article.

The case of the quinquina or cinchona tree, producer of the febrifugal Peruvian or Jesuit's bark, sheds light on how the presence of indigenous or local informants could be suppressed both in situ and through the practices of European editors.[41] At the behest of Joseph de Jussieu, who had given him a "*mémoire* on various historical and physical aspects of this tree," La Condamine stopped on his way to Lima at the mountain of Cajanuma (near Loja, south of Quito) in order to observe what was considered to be the best specimens of quinquina in the region. Having read and benefited in advance from Jussieu's account, La Condamine chose to spend the night in a dramatic spot: atop the very mountain where the quinquina tree was harvested, at the house of a "local man" from whom La Condamine learned much if not most of what he later published. This local man was Fernando de la Vega, whose "Virtudes de la cascarilla de hojas, cogollos, cortezas, polvos y corteza de la raiz" was composed at the behest of Miguel de Santisteban, who had traveled through this region several years after La Condamine's and Jussieu's respective visits.[42]

While Jussieu arrived in Loja only several months after La Condamine, the scientific community had to wait more than two centuries to be informed of his discoveries, since Jussieu's *mémoire* was never published in his lifetime. In the meantime, La Condamine published an academic paper that he admitted was inferior to what Jussieu would have been capable of writing; nevertheless, by combining his own observations with those of a local, sifting through myths and "ancient traditions" as recounted by the indigenous populations, and compiling and condensing a series of historical works into a twenty-page account, La Condamine successfully became the spokesperson for the quinquina tree in Europe.[43]

Not surprisingly, he was credited with this knowledge in the pages of the *Encyclopédie* as well.[44] In fact, the *Encyclopédie* to a certain degree conflated La Condamine's personal itinerary down the Amazon River with the biogeography of the quinquina tree. The quinquina tree was the only specimen deriving specifically from the Quito expedition that was graphically illustrated in the published plates of the *Encyclopédie* (fig. 55). In the description that accompanied the drawing, the quinquina tree was linked to the Amazon River region, even though its "discovery" and subsequent description owed more to La Condamine and Jussieu's research in Loja than it did to La Condamine's journey down the Amazon. This may be a further indication of the influence of personality and prestige on the organizational logic of the *Encyclopédie*. The association

Figure 55. According to the *Encyclopédie*, "Quinquina . . . is a tree that grows in America, in the lands near the Equator (*pays voisins de la ligne*), & principally along the Amazon River . . . " Here (at right) it is shown alongside the "Casse" or Cassia tree, a relative of cinnamon that is native to China. Courtesy of the Rare Book and Manuscript Library, University of Pennsylvania.

between La Condamine and the quinquina tree, on the one hand, and La Condamine and the Amazon River, on the other, led the editors of the *Encyclopédie* to consider the quinquina tree primarily as a specimen of Amazonian flora, even though its provenance was more likely the hills near Loja, south of Quito. The itinerary of La Condamine's downstream voyage seems to have spilled over into the geographical logic with which exotic flora was described in the pages of the *Encyclopédie*.

What is more, the transcription of the "Quinquina" article in the *Encyclopédie* reveals a suppression of the sources of local knowledge from which La Condamine derived his own account. The man who served as La Condamine's mountain host, Fernando de la Vega, was described in Jaucourt's article as an individual for whom "the collection of the quinquina bark... was [his] sole occupation." When Jaucourt transcribed this episode in the *Encyclopédie*, he transposed the agency for having selected this location from the local man to La Condamine. La Condamine had written that the man had "chosen to place his domicile [atop the mountain] ... in order to be closer to the quinquina trees, the harvest of which provides his sole occupation and livelihood." What Jaucourt wrote, however, was that "La Condamine... spent the night at the summit, in the home of a local man, in order to be closer to the quinquina trees." This subtle shift illustrates the transformative power of transcription and abridgement, where a small transposition of the *subject* of an action could have a deep impact on the way in which the production of knowledge was narrated and portrayed.[45]

More illustrative still of the coercive power of editorial abridgement was how Jaucourt transcribed La Condamine's description of the *transfer* of knowledge from his local informant to the European traveler. La Condamine had initially written in his *mémoire*: "My host from Cajanuma, who spends his life on this mountain examining these [quinquina] trees, assured me—and this was later confirmed by the testimony of the most learned individuals—that the yellow and red [trees] show no significant difference in their flowers, their leaves, their fruit, or their external bark." The manner in which Jaucourt transcribed this phrase in the *Encyclopédie* transformed the entire first half of the sentence into a universal affirmation: "*It is true that* the yellow and red [trees] show no significant difference in their flower, their leaf, their fruit, or their external bark." Rather than illustrating La Condamine's reliance on local knowledge, Jaucourt merely affirmed that knowledge as a universal truth, removing its source from his text. In yet another example, La Condamine had written that "as for the white quinquina, this same man assured me that its leaves were rounder and smoother than the other two." Jaucourt again removed La Condamine's source: "As for the white quinquina, its leaves are rounder and smoother than the other two." What has been quite explicitly evacuated from the *Encyclopédie* article is the local individual who provided this information. While in his narrative La Condamine showcased the role of this man as an informant—as a knowledge mediator between the natural environment of Cajanuma and a European visitor— the compilatory practices of a "reasoned dictionary" suppressed this information in the abbreviated article. These editorial practices carried out by Jaucourt and the *Encyclopédistes* shaped and channeled knowledge into—and out of—the *Encyclopédie*. In attempting to portray this knowledge as universal, Jaucourt evacuated the sources of La Condamine's information from the text. An omniscient third-person voice was the ultimate authority on the quinquina tree: not "this man had assured me" but rather "it is true." The source of that truth, narrated out by the rhetoric of universality, made his silent exit from the encyclopedic stage.[46]

Monkeys, Manatees, and Bibliographic Dissonance in the *Encyclopédie*

Narration was an important editorial tool for European travelers, a form of power they asserted over events they were otherwise powerless to control. As we saw in chapter 2, narrative priorities often played a crucial role in how La Condamine structured his *Relation abrégée*. La Condamine's journey between Curupá and Pará along several channels and tributaries of the Amazon was another such example. During this voyage, La Condamine shuddered at having been given no choice as to which direction they were to follow, which in turn made it impossible for him to complete his map in accordance with his stated goals:

What was responsible for my safety, and what would have made any other traveler feel secure, made me extremely uncomfortable, since my primary goal was the construction of my map. I had to pay extremely close attention so as not to lose the thread of my route through the labyrinth of islands and innumerable canals.[47]

La Condamine was nonetheless able to reassert control by "redoubling his attention" to the task at hand. This strategy served as an antidote to the endless waterways and patches of land that quickly filed past him. But La Condamine's invocation of a "labyrinth of islands" and "innumerable canals" also served as an elegant lead-in to the disquisition that followed, which related to the natural history of the Amazon River region and the "singularity" of its many birds and animals. This narrative structuring allowed La Condamine to include an extended description of tropical creatures, yet another effort on his part to account for the astounding complexity of Amazonian nature:

I have not spoken at all of the rare fish that are found in the Amazon, nor of the different kinds of rare animals that are seen along its riverbanks. This subject alone would provide material for an entire book, and such a study would require a voyage made for that very purpose and a traveler who had no other duties. I will only mention some of the most rare [objects].[48]

Despite describing his catalog as a selective account of the many animal species and plant types he found along the Amazon's shores, La Condamine's *Relation abrégée* did provide a rich panoply of new names and descriptions that were of interest to naturalists as well as the public at large. From *curupa* seeds to rubber trees, from crocodiles to *chauves-souris*, La Condamine presented a vivid image of a forested riverbank teeming with life, and many of the observations he made were recorded and transcribed as articles in the *Encyclopédie*. As such, the seventeen tomes of the "reasoned dictionary" became adorned with Amazonian monkeys and manatees, jade gemstones and green parakeets, condors as well as cocoa beans. The Amazon River region, in particular, was represented within the *Encyclopédie* by an extreme multivalence of meaning: geographic, natural historical, ethnographic, and mythological. In all, the Amazon River wended its way through eighty-five separate articles, from descriptions of contiguous geographic regions in South America such as "Guiana" and "Brazil" to more peculiar articles such as "Skin" and "Head," with references to features of human anatomy peculiar to the Indians of the Amazon region. There were also descriptions of several species of monkey that lived near the Amazon: the "Kajou," to give one example, was a hairy monkey that resembled an old man with a gray beard and black eyes. There were other descriptions of animals endemic to America and/or the Amazon region, such as the "Lamantin," more commonly known as the manatee or sea cow; the "Danta," the "largest of the quadrupeds of South America"; and the "Condor," all of which, according to the *Encyclopédie*, inhabited regions within close proximity of the Amazon River.[49]

Not surprisingly, many of the *Encyclopédie*'s authors, including d'Alembert and the Chevalier de Jaucourt, employed La Condamine's text as a primary source for these articles. Prime among these entries was d'Alembert's "Amazones, *Rivière des Amazones*" article, which exalted La Condamine's "long, painful, and dangerous" journey down "the world's longest river." Praising his important literary and cartographic contributions to the history and geography of the Amazon, the text remarked that in addition to "a new map of this river more accurate (*exacte*) than any that has preceded it, the celebrated academician . . . has published an extremely interesting (*très-curieuse*) and very well-written account of his voyage, which has been inserted in the Royal Academy of Science's 1745

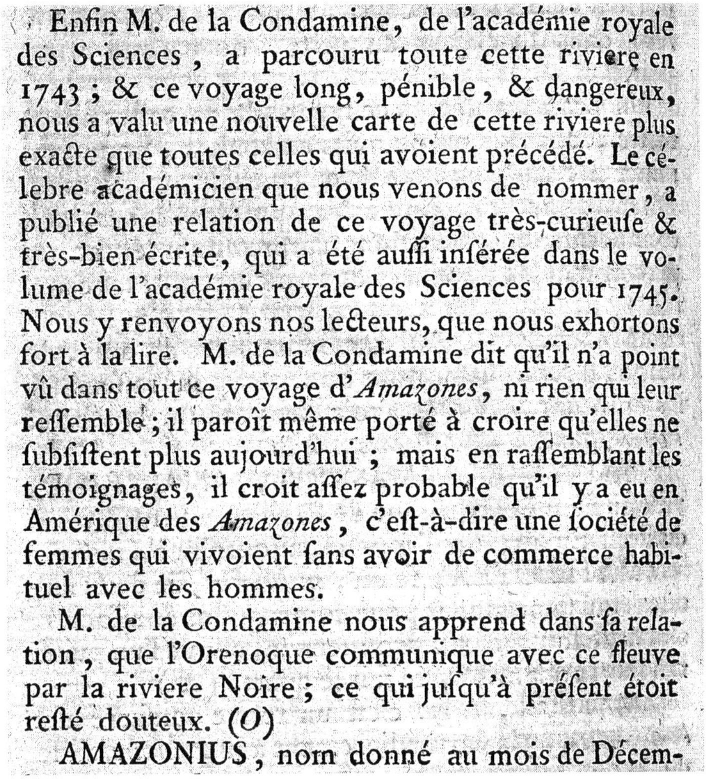

> Enfin M. de la Condamine, de l'académie royale des Sciences, a parcouru toute cette riviere en 1743 ; & ce voyage long, pénible, & dangereux, nous a valu une nouvelle carte de cette riviere plus exacte que toutes celles qui avoient précédé. Le célebre académicien que nous venons de nommer, a publié une relation de ce voyage très-curieuse & très-bien écrite, qui a été aussi insérée dans le volume de l'académie royale des Sciences pour 1745. Nous y renvoyons nos lecteurs, que nous exhortons fort à la lire. M. de la Condamine dit qu'il n'a point vû dans tout ce voyage d'*Amazones*, ni rien qui leur ressemble ; il paroît même porté à croire qu'elles ne subsistent plus aujourd'hui ; mais en rassemblant les témoignages, il croit assez probable qu'il y a eu en Amérique des *Amazones*, c'est-à-dire une société de femmes qui vivoient sans avoir de commerce habituel avec les hommes.
>
> M. de la Condamine nous apprend dans sa relation, que l'Orenoque communique avec ce fleuve par la riviere Noire ; ce qui jusqu'à présent étoit resté douteux. (*O*)
> AMAZONIUS, nom donné au mois de Décem-

Figure 56. An excerpt from the "Amazones, Rivière des Amazones" article in the *Encyclopédie, ou Dictionnaire raisonné*. D'Alembert's article draws attention to La Condamine's narrative account of his voyage (published in 1749 in the *Mémoires de l'Académie Royale des Sciences*), the map that accompanied his account, and La Condamine's speculation that a "society of women" that lived separately from men had once inhabited South America but had since disappeared. Courtesy of the Rare Book and Manuscript Library, University of Pennsylvania.

volume. We send our readers there, and strongly exhort you to read it"[50] (fig. 56). Contributions also included descriptions of Amazonian tributaries (the articles on the "Chingou," the "Maranon," the "Orenoque," and the "Purus" rivers all cited La Condamine's text) and other entries like "Résine," "Jade," and "Pierres des Amazones." These near-verbatim descriptions of Amazonian features were pulled primarily from La Condamine's narrative in the 1745 edition of the *Mémoires de l'Académie* (published in 1749), and they added sharp details to some of the more matter-of-fact geographic descriptions within the pages of the *Dictionnaire raisonné*. One such example was the article entitled "Perroquet vert varié," which became a rainforest menagerie and a colorful showcase for Amazonian birds:

> The species of parakeets and macaws, different in size, color, and shape, are innumerable. The most ordinary parakeets from Pará, which are known in Cayenne as the tahouas or parakeet of the Amazon, are green, with the upper part of their heads, the bottoms and edges of their wings a beautiful yellow... But the rarest of them all are those that are entirely yellow, lemon-colored on the outside, with the bottoms of their wings, and two or three feathers of their tail, a very beautiful green.[51]

What is perhaps more surprising than the prominent use of La Condamine's texts are the *other* texts used to adorn the pages of the *Encyclopédie* with animals, plants, and Indian groups from the Amazon region. These texts included the *Dictionnaire géographique portatif* of M. Vosgien; the *Histoire naturelle des animaux*, written by MM. Arnauld de Nobleville and Salerne; De Laet's *Histoire des Indes Occidentales* from 1622; the Comte de Pagan's *Relation historique et géographique du grand pays et rivière des Amazones en Amérique*, from 1655; and the *Histoire naturelle* of Buffon and Daubenton, to list only those that were cited explicitly within the text. Jaucourt's article on the torrid zone, that burning land of antiquity supposed to be unfit for human habitation, employed poetry from the recently translated French version of James Thomson's *The Seasons* (1748).[52] In this account, Jaucourt echoed Thomson in presenting the Amazon as one of numerous "ocean-like rivers [traversing] . . . unknown realms," timeless and desolate in comparison with La Condamine's descriptions of an abundant land populated by human, animal, and botanical life.[53]

Vosgien's *Dictionnaire géographique portatif* served as the source book for at least seven of the twenty-four articles relating to the geography of Amazonian regions, provinces, islands, and countries. Diderot pulled his description of the "Aguas" tribe explicitly from Vosgien's text, calling them "the smartest of Indians" and adding that they "tighten the heads of their infants between two boards as soon as they are born"[54] (fig. 52 above). Most would read this description as a typically sardonic commentary by Diderot on the cultural norms of a barbaric tribe. But by examining the definition in Vosgien's text, we see that Diderot lifted this remark almost verbatim from the *Dictionnaire géographique portatif*: "They are the smartest and most civilized (*mieux policée*) nation of all the Indians. No sooner are their children born than they tighten their heads between two boards, one of which presses against the forehead, the other against the back."[55] This editorial practice, that is, lifting from certain texts while culling and parsing from others, created blurred and occasionally dissonant representations of diverse geographic phenomena. The source of the original observation ended up submerged beneath layers of cutting and pasting, thus allowing for the perpetuation of stereotypical descriptions by subsequent compilers who were not familiar with the original context.

What these examples suggest is the extent to which diverse and often divergent sources were combined in the construction of separate articles in the *Encyclopédie*. As the editors of the *Encyclopédie* attempted to reduce source texts into article-sized chunks, their materials became pasted together with an out-of-date adhesive. The distillation process by which natural historical information was extracted from La Condamine's travel narrative and grafted into eighty-five short articles necessarily eliminated much of the richness and "abundance" that La Condamine had worked so hard to articulate in his text. Naturalists and travelers who consulted the *Encyclopédie* were likely incapable of seeing the cognitive connection between products of the natural world and their geographic provenance, a set of complex relationships that Alexander von Humboldt would articulate as a "geography of plants" some fifty years later. This dual articulation of scientific catalog and exotic space as reflected within the *Encyclopédie* helps to address one of the curious contradictions of eighteenth-century naturalist practice. One scholar has written that as naturalists "traveled the world and cataloged its diversity in material terms, [explorers] for a long time were ignorant of the cognitive implications of the spatial and geographical relationship between the Earth's natural phenomena."[56] Paradoxically, the encyclopedic impulse of eighteenth-century natural history seems to have diminished this interconnection of natural phenomena at the same time that its alphabetical structure allowed naturalists to catalog and contain its global expanse.

The Golden Monkey and the Monkey-Worm

Of course, Quito and the Amazon were not the only sites that underwent this kind of compression in the *Encyclopédie*, nor should the abridgement of the results of the expedition to South America be seen as a totally isolated phenomenon. Certain European cities, such as Ravenna, were given short shrift merely on account of their being the birthplaces of enemies of the *Encyclopédistes* and their project.[57] The image of China offered by the *Encyclopédie* was far from coherent, even though interest in China and the East more broadly was at its peak during the period in which it was written. Voltaire, for instance, was an avowed Sinophile, but he only invoked China in three of the forty-three *Encyclopédie* articles that he published under his name.[58] Jaucourt's article on India was what one scholar called "an extraordinarily brief summarization of the spirit of a whole subcontinent... [that lacked] the recognition of regional subdivisions that de Jaucourt could bring to other Encyclopedia subjects."[59]

So what is distinctive about the Quito expedition's representation within the *Encyclopédie* in light of these other compressed non-European geographical spaces? Despite the contemporary fascination for India, China, and the "exotic East," interest in the Americas and in regions connected through Atlantic networks to the French maritime world was on the rise in the eighteenth century. Sugar, tobacco, and indigo production in the West Indies was becoming increasingly integrated into the French mainland economy, and French philosophes and monarchical strategists alike came to see such institutional sites as the Jardin du Roi as important storehouses for practical knowledge. Against this backdrop of increased French interest in commercial expansion was a historical and philosophical interest in the relationship of the New World to the Old, a debate over human origins and the stages of societal development that attempted to integrate new information about the Americas into what had previously and exclusively been seen as Old World accounts. Buffon, Rousseau, and Montesquieu all participated in attempts to create a conjectural history of humankind that was universal in its scope. South America, as a site of ancient polities as well as aboriginal cultures, played a prominent role as a stage for the phases of human development that were integral to this history. And the *Encyclopédie* became one repository for this new knowledge, bringing together in a single text the profusion of new information about the non-European world. The Quito expedition, as the hallmark expedition of the mid-eighteenth century, thus serves as a yardstick for understanding the extent to which the most recent examinations into cultural, historical, and natural historical phenomena found a place in the most iconic compilation of an "enlightened" age.

This chapter has challenged the nature of the transition from the closed universe of the *savants* to a broader public fascination with the possibilities inherent in the study of natural history, a process that Daniel Mornet described as being brought about by the explosion of new materials collected through exploration in Europe and beyond. "From the first half of the eighteenth century," wrote Mornet, "[natural history] held the curiosity of the *gens du monde*, the prestige of the moment, and the attention of the pedagogues. These successes increased dramatically after 1750."[60] This chapter has shown that Mornet's "triumph of natural history" in the second half of the eighteenth century was less triumphant than he might have led us to believe. True, there was a dramatic and verifiable increase in the number of "manuals, dictionaries, [and] compilations" during this period, but the nature of what was contained within these tomes neither directly nor transparently translated the experiments in the field or the language used to describe such experiments. And even when they did, that very language was shaped and often compromised by the expectations inherent in the audience for which it had been originally conceived. Piquant examples were often chosen over more mundane descriptions to appeal to a reading public surrounded by public displays of exotica and scientific rarities within the European capitals. Those editors who abridged and distilled natural historical lessons brought back to Europe from the four corners of the globe often did so obliquely, if at all. Less frequently still did they recognize those actors that participated in the transfer of that knowledge from the field to the naturalist's notebook.

European editors and authors were confronted with an immense and often overwhelming quantity of materials from which to compile encyclopedic texts. Maps, geographical texts, travel accounts, and missionary reports were assembled and arrayed in a format that was sometimes clear, sometimes oversimplified, and often highly selective. Poetic language was used to appeal to a broader audience and, it may be imagined, to relieve the author of painstaking and sometimes tedious research. Portable geographic dictionaries were consulted and copied to describe faraway regions whose characteristics could not be easily surmised by comparing contemporary sources. Outdated sources created overlapping regions that often elided eighteenth-century "fact" with seventeenth-century "fable" in the same paragraph. In short, the condensing and compacting process corroborated what Henri Gabriel Duchesne wrote in the preface to his *Manuel du naturaliste* at the end of the eighteenth century—that the "excess of nomenclature" and "frivolous and ridiculous observations" were major obstacles to the writing of natural history in an age of expanding geographic frontiers.[61]

Nevertheless, the encyclopedic impulse as represented by d'Alembert's "Discours préliminaire" captured and incorporated materials from the Quito expedition at precisely the moment that the *Encyclopédie* itself was coming into existence. As such, there is an implicit, if perhaps coincidental, alliance forged between these two projects: to catalog the rivers, forests, mountains, and valleys of the equatorial regions of South America; and the mapping of human knowledge in the *Encyclopédie*. Sitting at opposite ends of the Enlightenment spectrum of human knowledge, the one exotic, the other eminently continental, these two projects nonetheless found common ground. Aside from the Portuguese and Spanish empires, which vied for political control of the Amazon River and its tributaries during the eighteenth century, there were two other empires for which South America came to have particular significance: the empire of science and the empire of letters, explicitly referred to by Diderot in numerous citations.[62] These were not empires in the traditional territorial sense, but their rhetorical invocation as such by Diderot points to their similarities even when knowledge was procured and filtered outside of an explicitly controlling imperial power. The triumphant proclamation of universal knowledge by the philosophes came at the expense of voices that were closer to the specific circumstances in which much of this knowledge was gathered and understood. This knowledge was carried via channels that were facilitated by imperial networks, but its ultimate destination was not always recognizably "imperial" in any overriding political or hegemonic way. What is particularly useful about examining how natural and historical knowledge from Quito and the Amazon made its way into French compendia is precisely the cross-fertilization of knowledge: not a monolithic *imperial* knowledge gathered and deployed by agents of a particular territorial empire, such as Spanish explorers in New Spain, but rather knowledge garnered in an era of exchange within and across imperial territories, outside the aegis of a specific political regime. La Condamine's desire to leave several seeds of the quinquina tree in Cayenne for experimentation by local Crown officials is only one example of the potential ramifications of cross-border knowledge circulation in an age of increasing imperial conflict and competition.[63]

One of most decidedly peculiar articles in the *Encyclopédie* that was related to La Condamine's sojourn in South America described the "Vermacaque" or monkey-worm. On the basis of an account given by La Condamine of the single specimen he had seen, as well as several other corroborating accounts from the East Indies, the *Encyclopédistes* chose to describe in gory detail this long thin worm that burrowed its way into the space between one's skin and muscle, creating a tumor the size of a large bean. To rid oneself of this parasite, Jaucourt explained, it was necessary to extract it with the artful use of an emollient, some string, and a small piece of wood around which one would roll it as it was being pulled from the tumor. La Condamine sketched this worm, deposited it in alcohol to preserve it, and likely returned it to Paris along with the other specimens.

While one can only speculate as to why the monkey-worm would have been deemed fit to be included as an entry in the *Encyclopédie*, it may in the end serve as an appropriate symbol for the challenges of extracting meaning from an entangled corpus of natural knowledge. This worm, which La Condamine explained was called *suglacuru* by the Maynas Indians, and *ver macaque* by the residents of Cayenne, was thought by Jaucourt to be the same as a worm from the East Indies, the *culebrilla*. Integrating these descriptions together in a single article was a prime example of how the *Encyclopédie* was conceived to work. But the merging of these three names into a single species was also indicative of "how difficult it was to be certain that the same objects were in fact the same." Diderot seems to have been conscious of the complications imbedded within their larger project and the challenges and dangers faced when confronting the boundless diversity of an ever-expanding world:

As we worked, we watched as the materials expanded before our eyes, the nomenclature became obfuscated, objects were brought in with numerous names, instruments, machines, and maneuvers multiplied beyond measure, and innumerable detours of an inextricable labyrinth became increasingly complex. We saw how difficult it was to be certain that the same objects were in fact the same, and likewise how hard it was to confirm that things that appeared very different were not actually different. We saw that the alphabetical format, which offered us peace of mind at every instant, allowed us tremendous variety in our work, and from a certain perspective seemed highly advantageous for such a long work, [also] brought with it certain difficulties that at every moment needed to be overcome.[64]

Diderot and d'Alembert's navigation within this "inextricable labyrinth" managed nonetheless to earn them the adulation and respect of most of their eighteenth-century compatriots within the Republic of Letters, if not the Jansenists and those who fiercely opposed their project from the start. And while the monkey-worm managed to burrow its way into the structure of the *Encyclopédie* and remain there, another object described several paragraphs earlier in La Condamine's *Relation abrégée* was given a less prominent post within the pages of the *Dictionnaire raisonné*.

The golden-haired monkey, referred to at the beginning of this chapter, seems to have fallen victim to this inextricable textual labyrinth described by Diderot. Despite having been dedicated an attractive portrait many years later by Buffon in his *Histoire naturelle* (fig. 57), the monkey's

Figure 57. La Condamine's golden monkey, as portrayed in Georges Louis Leclerc, comte de Buffon, *Histoire naturelle, générale et particulière*, 15: 124, plate 18. Courtesy of the Harlan Hatcher Graduate Library, University of Michigan.

experience largely mirrored the way in which the Amazon River, its myriad natural treasures, and other products from the Quito expedition were received within the scientific annals of eighteenth-century Europe. Rather than creating a stir within the zoological community or a controversy such as Bougainville spurred when he brought his young Tahitian to be examined and displayed in the salons and royal courts of Paris, La Condamine's monkey was preserved in alcohol and packaged in a glass container, to be seen and admired, but without the dynamic impact and the accoutrements of celebrity that accompanied great scientific achievements. Squished and sealed within its transparent cage, this monkey is referred to in the *Encyclopédie* only as the "little monkey from Para," hardly the kind of dramatic reception its captor might have initially hoped for.[65] Like a magic genie in a broken lamp, the golden monkey remained trapped and hidden amid the labyrinthine minutiae of the *Encyclopédie*, a tangible reminder of what could happen when ideas about nature were conveyed from the New World and compressed in the Old.

Figure 58. This copperplate engraving shows Minerva, symbol of the Paris Academy of Sciences, surrounded by a group of putti engaged in various experimental pursuits. Other than the palm trees, the rugged landscape, and the Spanish colonial structure at right, the entire scene seems eminently European. This image, reproduced in La Condamine's *Journal du voyage* (1751), was based on a drawing provided by a Jesuit in Quito and later engraved by Morainville to be offered to the Academy in Paris. Courtesy of the John Carter Brown Library at Brown University, Providence, R.I.

Conclusion:
Cartographers, Concubines,
and Fugitive Slaves

The observations collected by M. de La Condamine are quite remarkable; he has published them in the greatest of detail. And I should add that if he passed for someone of the most ardent curiosity in France and England, he is considered in Quito, the very country that he described, as the most sincere and truthful of men. HUMBOLDT, *RELATION HISTORIQUE DU VOYAGE AUX RÉGIONS ÉQUINOXIALES DU NOUVEAU CONTINENT*

Upon returning to Paris, Sr. de la Condamine published a thick historical treatise on [the uprising at Cuenca] . . . But he did not write an account of the famous joke that was played on him, which deserves to be known by everyone.

JUAN DE VELASCO, *HISTORIA DEL REINO DE QUITO*

Facts are not created equal: the production of traces is always also the creation of silences. MICHEL-ROLPH TROUILLOT, *SILENCING THE PAST*

Among the myriad novelties he encountered during a decade-long voyage to South America, La Condamine found the humble peanut to be the most enchanting of all. He marveled at the attractive white flowers of the plant and its long shapely pods. Wherever he went, he would stuff his pockets full of them, ripping open their outer covering to reveal small round "fruits" covered in a fine red film. As he walked the streets of the colonial capital, he would pop peanuts into his mouth and savor their peculiar texture. According to one resident of Quito, La Condamine considered these native legumes to be "the greatest treasure he had seen in the Americas," an extraordinary opinion to have formed in a continent where far more lucrative prizes typically awaited the zealous explorer from overseas.[1]

But however much La Condamine loved peanuts, he cherished his scientific papers further still. Throughout his journey, La Condamine went to great expense to preserve the "fruits" of his empirical investigations. And he often had good reason to worry about their safety: late in the expedition, while preparing for his final departure from Quito, he returned to his house one afternoon to find that all of his papers had been stolen from an open box left on his writing table. "No doubt," La Condamine later explained, "someone expected to find clues about gold mines, which many believed was the secret goal of our mountain excursions."[2] Near desperate at his papers' disappearance, he managed to stay calm long enough for local officials to arrange for their return. And within two days, most of his notes had been restored to his possession, left for him in a bound stack in the inner courtyard of his Quito residence.

So it is easy to understand why a year and a half later, when La Condamine arrived in Cayenne in February 1744, he was still quite nervous about leaving his papers unprotected. After two months of travel from the Portuguese city of Pará, after nearly a decade of observations across the equatorial regions of South America, and having only recently lost his treasured quinquina plants to an unexpected wave off Cape Orange, he told a rather shocked officer welcoming him to French territory that he preferred to eat and sleep in his canoe rather than resting comfortably in the home of Cayenne's military captain. The officer and his colleagues were incredulous: "[N]o one could have imagined that the academician had such important papers [with him]."[3] But after his long absence from Europe, and as he prepared for the very last phase of his journey home, La Condamine had no intention of losing a decade of labors to the thieving whims of the local population. Instead, the next morning, he quickly set out to collect as many additional maps and manuscript descriptions of the region as he could. If anyone would be absconding with scientific data during his short visit to Cayenne, it would be La Condamine himself.

In fact, La Condamine had carried more than scientific papers as he crossed from Portuguese America to French Guyana. At the behest of João de Abreu de Castelo Branco, then governor of Pará, La Condamine's boat provided conveyance for Louis, a "black slave" originally from Cayenne, who along with several others had fled from the cruelties of bondage to seek refuge in Portuguese territory. While Louis had been condemned to death by the French authorities (his crime being contumacy, or obstinate disobedience), a 1732 treaty signed between Portugal and France forbade this punishment as inhumane.[4] Therefore, it was decided in Pará that the Portuguese would return Louis to Cayenne under the condition that he be spared the gallows, absolved of his crime, and returned to his rightful owner. La Condamine's canoe provided for his secure transport from a Portuguese prison back to the shackles of French servitude.

We learn about the presence of Louis not from La Condamine's narrative, however, but rather from the account provided by Jacques-François Artur, a royal doctor in Cayenne and member of the governing council (*conseil supérieur*) that was to decide the slave's fate. There is no direct reference to Louis in La Condamine's text, nor is there any reaction to Louis's predicament, despite the two having spent eight weeks together in the same canoe (including seven days during which, by La Condamine's own description, the craft was "encased in hardened silt" and entirely immobile). But while Louis's presence was *explicitly* absent from La Condamine's *Relation abrégée*, his predicament as a fugitive slave made an unlikely, and entirely unacknowledged, appearance in the text, a text that La Condamine composed during his journey back to Europe shortly thereafter and presented to a public session of the Paris Academy of Sciences only a year later.

Upon his return to Europe, La Condamine's unabashed support for the idea that Amazons once existed in South America had caused excitement as well as consternation. Indeed, half a century later, few could fathom how a representative of France's most prestigious scientific academy could have ever supported such a fantastic theory.[5] But whether or not La Condamine built his case in order to captivate a credulous audience, his arguments in favor of the existence of American Amazons were based not so much on myth but rather social history: the broader history of fugitive slaves in the New World, a history with which he may have been familiar due to his interactions with Louis. Supporting the idea that there were once Amazons in America, he explained that the behavior of women living separately from men would be no more surprising nor difficult to imagine than "what takes place every day in the European colonies of America":

It is all too common that slaves, whether mistreated or unhappy, flee into the forest either in large groups or occasionally alone when they cannot find anyone to join them, and there they spend many years and sometimes all their lives in utter isolation.[6]

While La Condamine suppressed the fact that he had traveled in the company of Louis, his account of fugitive slaves bears an uncanny resemblance to Louis's own predicament. Why would La Condamine repress Louis's presence in his narrative and yet deploy his story in the context of Amazons? Perhaps the testimony of a fugitive slave was not sufficiently credible to bear the burden of a centuries-old myth. Perhaps allowing Louis to speak, figuratively or otherwise, would have diminished the standing of La Condamine's own testimony. But regardless of the reasons for excluding his presence in print, La Condamine unwittingly demonstrated the inextricability of the colonial context from the European production of "scientific" knowledge. Unbeknownst to his readers, La Condamine merged a mythic tale from classical antiquity with the in situ experience of sharing his canoe with a fugitive slave.

But fugitive slaves were not alone in providing support for the Amazon myth. The social conditions of Amerindian women also had a role to play, and La Condamine buttressed his arguments with a material, if hypothetical, analysis of their lives:

If there ever could have been Amazons in the world, it is in America, where the errant life of women who frequently follow their husbands into war, and who are equally unhappy in their domestic life, would have given them both the idea and ample opportunities to cast off the yoke of their tyrants . . . [B]y seeking out an environment in which they could live in independence [they would] not be reduced to the condition of slaves and beasts of burden.[7]

This apparent awareness of the material circumstances of Amerindian domesticity, notwithstanding the excessive generality of his observations, is striking, not least because La Condamine rarely discussed women in his accounts. Despite the fact that he encountered Amerindian women regularly throughout his voyage, he almost always suppressed his experiences with them in situ. Here, however, he seemed to describe their predicament with uncharacteristic sympathy. Still, with the exception of this single reference to the unhappy domestic life of indigenous women "in America," his discussion of gender and domesticity remained almost exclusively linked to Amazon women and their myth.

It is perhaps less of a surprise that La Condamine never discussed any sexual encounters he may have had during his journey. But passing through Cuenca sixty years after him, Alexander von Humboldt made a fascinating discovery about La Condamine's relationships with women in the Americas. During one of his many extended visits to Cuenca, La Condamine apparently visited a prostitute, which resulted in the birth of two daughters, "natural children," as Humboldt referred to them, "whose mother still practices the gallant trade."[8] Humboldt's revelation cannot be corroborated. The two "daughters" may have been attempting to curry

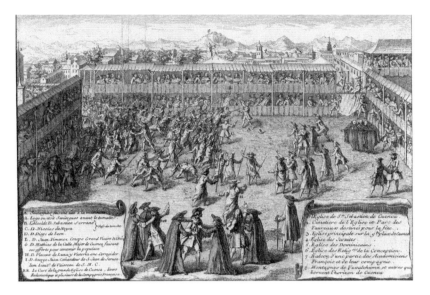

Figure 59. The site of Seniergues' death and the uprising against the academicians, with colonial Cuenca in the background. The balcony from which several members of the expedition viewed the proceedings is at the upper right, placed one level above the other guests. "Vuë d'une Place preparée pour une Course de Taureaux," from La Condamine, "Lettre a Madame ∗∗∗ sur l'emeute populaire excitée en la ville de Cuenca au Perou, le 29 d'aout 1739." Courtesy of the John Carter Brown Library at Brown University, Providence, R.I.

favor with a passing European by claiming in some sense to be Europeans themselves. And yet, the expedition's own history in Cuenca gives us reason to believe that such a liaison between La Condamine and a local woman would not have been unusual. Cuenca was the site of one of the most serious conflicts between the academicians and the local population, a conflict that had its roots in an amorous tryst between Seniergues, the expedition's surgeon, and Manuela Quesada, a member of one of the city's elite families (fig. 59). Like Seniergues, La Condamine was not isolated from the social, physical, or sexual realities of the spaces through which he passed. La Condamine's visitation of a prostitute may have been perceived as inappropriate for or outside the generic conventions of an academic "mémoire," a genre in which, in La Condamine's own words, "everything that did not pertain to geography, astronomy, and physics could not but appear as a digression." Nevertheless, these social encounters transformed the experiences he had, the narratives he wrote, and took on a life of their own beyond his partial and selective representations of them. Like his interactions with Louis the fugitive slave, La Condamine may have sought to maintain his sexual liaisons as an intimate affair, but they inevitably impinged on the nature of his "scientific" conclusions as well as the shape and direction of his empirical itineraries.

In the case of the Amazons, La Condamine painted a portrait that fused the grim realities of African servitude together with the harsh conditions of Amerindian domesticity. This extraordinary conceptual alliance between mistreated fugitive slaves and enslaved native women belies the image of La Condamine as a European living in a figurative scientific bubble, insensitive and impervious to the social circumstances that pervaded his surroundings. It is true that he may not have used this knowledge for anything more than his own aggrandizement; he was certainly no advocate for the rights of Amerindian women or enslaved Africans. But the example underscores an epistemology forged in the crucible of the eighteenth-century Atlantic world: a framework bound as much to the social and material conditions of female subservience, African slavery, and imperial interactions across political and geographic boundaries as it was dependent on empirical observations drawn from telescopes and barometers. Louis's example shows us explicitly that although Europeans may have carried practices of empiricism to the Americas in the eighteenth century, hidden forces from *within* the Americas acted upon those pathways of knowledge and measurement as well. These interactions unsettled empirical itineraries and complicated what have often been portrayed as overly idealized trajectories that converge at a European center. La Condamine was not a solitary and intrepid actor working alone, but rather more like an interpretive conduit of a mobile, characteristically American knowledge. With reference to the Amazons, this knowledge made its way from a classical legend that traversed the Atlantic in the early sixteenth century to a slave-porting canoe off the South American coast, and from the conditions of bondage La Condamine recognized in the roles of indigenous women all the way to the printing presses of the Old Regime. As he made his way through South America and back to Europe, La Condamine attempted to repress these sources, from which he drew his material evidence, his moral and intellectual inspiration, and, occasionally, his sexual sustenance as well.

In print, the Quito expedition presented itself as having the same standards of empiricism and objectivity as it would have had were its members carrying out measurements in Paris rather than in the viceroyalty of Peru. One striking portrayal of this aspiration was an etching included in La Condamine's *Journal du voyage*: an attractive frontispiece to a theological dissertation defended in Quito and dedicated to the academicians who were in attendance (fig. 58 above). Seated at the center is a helmeted Minerva, symbol of the Academy of Sciences, holding a shield emblazoned with the academy's motto: "Invenit et perficit." To the left of the seated figure stands a building oddly reminiscent of the Observatory of Paris, with a large telescope mounted carefully before it. At Minerva's feet is a group of putti at play, carrying out experiments with globes, octants, beakers, and pendulae, and seemingly oblivious to the stage upon which they are seated. Yet on the righthand side one cherubic figure gives the lie to this scene of universal science. He slides down a palm tree clutching a strange object in his hands: a coconut. Behind him, far in the distance, is a small colonial village, the only sign of the peculiar space—half-European, half-American—in which this stately scientific drama was taking place. Otherwise, the putti are protected and focused in their activities, eager to contribute to a scientific program for which location was deemed irrelevant to the larger truths of empirical observation carried out beyond Europe's shores.

As we remember from the introduction, local populations in the colonial Americas did not sit idly by and watch their territory mapped and marked without making their own presence known. Their responses and interactions came to play a central if often unacknowledged role in the knowledge that was carried back across the ocean. One story especially merits retelling in closing, not only because La Condamine suppressed it entirely from his account, as Juan de Velasco explained in the epigraph to this concluding chapter, but also because it encapsulates how knowledge was channeled, diverted, and erased by forces that went beyond the academicians' control. Unlike several cartographic representations of Quito, no map of Cuenca was ever produced by the expedition, despite the fact that this city was an important location for them during their observations. The joke to which Juan de Velasco referred, and which he explained in his *Historia del Reino de Quito*, goes a long toward explaining why this map was never completed:

Having worked on the geographical map of Cuenca, [La Condamine] still needed to measure a few blocks of the city. Not daring to carry out these measurements by day for fear of the anger of the local populace, he went out one moonlit night accompanied by various high-ranking people of the city who could protect him. As he began to carry out [his measurements], an old woman recognized him and proclaimed that he was plotting to use his measurements to carry out some treachery against the city. She incited the neighborhood to such a degree that other women came out with sticks and stones and made them all flee.[9]

This scene of fleeing academicians being chased through the streets by stone-throwing women contrasts profoundly with the image of staid and stolid measurements La Condamine and his associates sought to portray in their published treatises and scientific *mémoires*. La Condamine never included a description of the measurements he made for a map of Cuenca, nor did he write that he had been unable to produce his map because of the ire of the populace. All that he produced, by contrast, was an ornate image of the angry multitude pitted against Seniergues, a portrait in which slaves and concubines are conspicuously absent from the protected balcony where the academicians watched in shock and amazement as an angry mob prepared to lynch a member of their expedition. From their privileged position high above the fray, we have to wonder, were they able to observe the scene as clearly as they suspected?

Notes

PREFACE

1. Simón Bolivar, "Mi delirio sobre el Chimborazo," in *Papeles de Bolívar*, ed. Vicente Lecuna (Madrid: Editorial-America, 1920), 2: 177; Alexander von Humboldt, "Besteigung des Chimborazo," in *Reise auf dem Río Magdalena, durch die Anden und Mexico*, ed. Margot Faak (Berlin: Akademie Verlag, 1990), 2: 100; Charles-Marie de La Condamine, *Journal du voyage fait par ordre du Roi* (Paris: L'Imprimerie Royale, 1751), 68–70. Unless otherwise noted, all translations are my own.

INTRODUCTION

1. For similar accounts from colonial Mexico, see Max Harris, *Aztecs, Moors, and Christians: Festivals of Reconquest in Mexico and Spain* (Austin: University of Texas Press, 2000); and idem, *The Dialogical Theatre: Dramatizations of the Conquest of Mexico and the Question of the Other* (New York: St. Martin's Press, 1993).

2. On the instrumental processes of the expedition's triangulation and astronomical observations, see Antonio Lafuente and Antonio Mazuecos, *Los caballeros del punto fijo: Ciencia, política y aventura en la expedición geodésica hispanofrancesa al virreinato del Perú en el siglo XVIII* (Madrid: CSIC/Serbal, 1987), which is the most thorough and analytically persuasive account of the expedition. See also Michael Rand Hoare, *The Quest for the True Figure of the Earth: Ideas and Expeditions in Four Centuries of Geodesy* (Aldershot: Ashgate, 2005); James R. Smith, *From Plane to Spheroid: Determining the Figure of the Earth from 3000 B.C. to the Eighteenth-Century Lapland and Peruvian Survey Expeditions* (Rancho Cordova: Landmark Enterprises, 1986); Antonio Lafuente and Antonio J. Delgado, *La geometrización de la tierra: Observaciones y resultados de la expedición geodésica hispano-francesa al virreinato del Perú, 1735–1744* (Madrid: CSIC, 1984); Neptalí Zúñiga, *La expedición científica de Francia del siglo XVIII en la Presidencia de Quito* (Quito: Municipio

de Quito, 1986); Jules Loridan, *Voyages des astronomes français à la recherche de la figure de la terre et de ses dimensions* (Lille, 1890); Antonio Lafuente, "Una ciencia para el Estado: La expedición geodésica hispano-francesa al Virreinato del Perú, 1734–1743," *Revista de Indias* 43, no. 172 (July–December 1983): 549–629; Robert Finn Erickson, "The French Academy of Sciences Expedition to Spanish America, 1735–1744" (Ph.D. diss., University of Illinois, 1955); Nelson Gomez, *La misión geodésica y la cultura de Quito* (Quito: Ediguias Ltda., 1987); Georges Perrier, *La mission française de l'équateur* (Paris: F. Leve, 1907); Henri Lacombe and Pierre Costabel, eds., *La figure de la terre du XVIIIe siècle à l'ère spatiale* (Paris: Académie des sciences, 1988); and Roger Mercier, "Les Français en Amérique du Sud au XVIIIe siècle: La mission de l'Académie des sciences (1735–45)," *Revue française d'histoire d'outre mer* 56, no. 205 (1969): 327–74. Contemporary descriptions of geodetic practices abound. See, for example, La Condamine, *Journal du voyage fait par ordre du Roi*, vol. 1, *Mesure des trois premiers degrés du méridien dans l'hémisphere austral* (Paris: Imprimerie royale, 1751). See also La Condamine, "Sur la manière de déterminer astronomiquement la différence en longitude de deux lieux peu éloignés l'un de l'autre," *Mémoires de l'Académie Royale des Sciences* (1735 [1738]): 1–11. For another contemporaneous description of cartographic methods, see François Chevalier, "Sur une manière de lever la carte d'un pays," *Histoire de l'Académie Royale des Sciences* (1707 [1730]): 113–16. For the best exposition in English of the process of geometric triangulation in constructing terrestrial maps, see Josef W. Konvitz, *Cartography in France, 1660–1848: Science, Engineering, and Statecraft* (Chicago: University of Chicago Press, 1987); and Matthew Edney, *Mapping an Empire: The Geographic Construction of British India, 1765–1843* (Chicago: University of Chicago Press, 1997), esp. 19–30. On the relationship of the Quito mission to the Lapland expedition, see Mary Terrall, *The Man Who Flattened the Earth: Maupertuis and the Sciences in the Enlightenment* (Chicago: University of Chicago Press, 2002). By October, 1739, Maupertuis had already notified members of the Quito expedition that their measurements in Lapland confirmed the Newtonian hypothesis, but the academicians in Quito continued with their measurements nonetheless.

3. La Condamine, *Journal du voyage*, 88.

4. La Condamine, *Journal du voyage*, 88.

5. This ethnographic dilemma is posed in Greg Dening, "Theatricality of Observing and Being Observed: 'Eighteenth-Century Europe' 'Discovers' the ?-Century Pacific," in *Implicit Understandings: Observing, Reporting, and Reflecting on the Encounters between Europeans and Other Peoples in the Early Modern Era*, ed. Stuart B. Schwartz (Cambridge: Cambridge University Press, 1994), 475.

6. Thomas A. Abercrombie, *Pathways of Memory and Power: Ethnography and History among an Andean People* (Madison: University of Wisconsin Press, 1998).

7. Amédée-François Frézier, *A Voyage to the South-Sea, along the Coasts of Chili and Peru, in the years 1712, 1713, and 1714* (London: Jonah Boywer, 1717), 263.

8. Homi Bhabha, "Of Mimicry and Man," in *The Location of Culture* (London: Routledge, 1994), 122, 125. See also Edward W. Said, *Orientalism* (New York: Vintage Books, 1994).

9. Robert Darnton, *The Great Cat Massacre and Other Episodes in French Cultural History* (New York: Basic Books, 1984).

10. Michael Taussig, *Mimesis and Alterity: A Particular History of the Senses* (New York: Routledge, 1993), 246–49.

11. For a general overview of these expeditions, see Iris Engstrand, *Spanish Scientists in the New World: The Eighteenth-Century Expeditions* (Seattle: University of Washington Press, 1981).

12. D'Alembert, "Figure de la Terre," in *Encyclopedie, ou Dictionnaire raisonné,* 6: 754.

13. Mary Louise Pratt, *Imperial Eyes: Travel Writing and Transculturation* (New York: Routledge, 1992), 23.

14. Pratt, *Imperial Eyes,* 23. Some scholars have emphasized the national contests the expedition engendered, while others have represented the voyage as a romantic quest. I am indebted to the work of these various authors. My goal, however, is not to rewrite the history of the expedition. Rather than being comprehensive in my approach, I have offered more punctual readings of specific episodes from the Quito expedition and left out other topics that have received ample attention elsewhere (e.g., the popular uprising in Cuenca, Isabel Godin's descent down the Amazon, the conflict between La Condamine and Bouguer). On the uprising against the academicians in Cuenca, see Deborah Poole, *Vision, Race, and Modernity: A Visual Economy of the Andean Image World* (Princeton: Princeton University Press, 1997); on La Condamine's narrative of Isabel Godin's journey down the Amazon, see Robert Whitaker, *The Mapmaker's Wife: A True Tale of Love, Murder, and Survival in the Amazon* (New York: Basic Books, 2004), and Celia Wakefield, *Searching for Isabel Godin* (Berkeley: Creative Arts, 1999); on the conflict over scientific publishing between La Condamine and Bouguer, see James E. McClellan, III, *Specialist Control: The Publications Committee of the Académie Royale des Sciences (Paris), 1700–1793* (Philadelphia: American Philosophical Society, 2003). I would like to thank James McClellan for having shared an early draft of his work with me.

15. Recent studies have shown how the presence of African, indigenous, and female actors were specifically disregarded as participants in the practice of empirical science in a colonial context. See Londa Schiebinger, *Plants and Empire: Colonial Bioprospecting in the Atlantic World* (Cambridge: Harvard University Press, 2004); and Susan Scott Parrish, *American Curiosity: Cultures of Natural History in the Colonial British Atlantic World* (Chapel Hill: University of North Carolina Press, 2006). Likewise, Jorge Cañizares-Esguerra has argued that northern European prejudice led to a similar phenomenon of exclusion in the case of Spanish American Creoles. See Cañizares-Esguerra, *How to Write the History of the New World: Histories, Epistemologies, and Identities in the Eighteenth-Century Atlantic World* (Stanford: Stanford University Press, 2001). Unless otherwise stated, I use the term "Creole" throughout this book to imply those born in the New World of European descent.

16. Marie-Jean-Antoine-Nicolas Caritat, marquis de Condorcet, *Esquisse d'un tableau historique des progrès de l'esprit humain* (Paris: Agasse, year III [1794]).

17. Cornelius de Pauw, "Observations sur les voyageurs," as cited in Michèle

Duchet, *Anthropologie et histoire au siècle des Lumières: Buffon, Voltaire, Rousseau, Helvétius, Diderot* (Paris: Albin Michel, 1995), 99.

18. Denis Diderot to Madame Necker, La Haye, 6 September 1774, in Diderot, *Oeuvres Complètes*, ed. Roger Lewinter (Paris: Club Français du Livre, 1971), 1043.

19. On the place of exotic knowledge in France and throughout Europe, see the classic study by Gilbert Chinard, *L'Amérique et le rêve exotique dans la littérature française au XVIIe et au XVIIIe siècle* (Paris: Hachette, 1913). On the circulation of natural objects, the role of scientific travelers in collecting and transporting these objects, and especially their materiality, see Marie-Noëlle Bourguet, "La collecte du monde: Voyage et histoire naturelle (fin XVIIème siècle–début XIXème siècle," in *Le Muséum au premier siècle de son histoire*, ed. Claude Blanckaert et al. (Paris: Muséum national d'histoire naturelle, 1997), 163–96; Bruno Latour, "Comment redistribuer le Grand Partage?" *Revue de Synthèse* 110 (1983): 202–36; Pamela H. Smith and Paula Findlen, eds., *Merchants and Marvels: Commerce, Science, and Art in Early Modern Europe* (New York: Routledge, 2001); Louise E. Robbins, *Elephant Slaves and Pampered Parrots: Exotic Animals in Eighteenth-Century Paris* (Baltimore: Johns Hopkins University Press, 2002); E. C. Spary, *Utopia's Garden: French Natural History from Old Regime to Revolution* (Chicago: University of Chicago Press, 2000); Ann Secord, "Science in the Pub: Artisan Botanists in Early Nineteenth-Century Lancashire," *History of Science* 32, no. 97 (1994): 269–315; and Jean-Marc Drouin, "La moisson des voyages scientifiques: Les singularités, l'inventaire, la loi et l'histoire," in *Anais do VI Seminário Nacional de Historia da Ciência e da Tecnologia*, ed. Isidoro Alves and Elena Moraes Garcia (Rio de Janeiro: Sociedade Brasileira para a História da Ciência, 1997). On the biological and scientific expansion of Europe, and particularly its ecological consequences, see Alfred W. Crosby, *Ecological Imperialism: The Biological Expansion of Europe, 900–1900* (Cambridge: Cambridge University Press, 1986); Richard Grove, *Green Imperialism: Colonial Expansion, Tropical Island Edens and the Origins of Environmentalism, 1600–1860* (Cambridge: Cambridge University Press, 1995); William K. Storey, ed., *Scientific Aspects of European Expansion* (Aldershot: Variorum, 1996); E. C. Spary, "Of Nutmegs and Botanists: The Colonial Cultivation of Botanical Identity," in *Colonial Botany: Science, Commerce, and Politics in the Early Modern World*, ed. Londa Schiebinger and Claudia Swan (Philadelphia: University of Pennsylvania Press, 2005). On Diderot's and Herder's incorporation of ethnography in their critiques of empire, see Sankar Muthu, *Enlightenment against Empire* (Princeton: Princeton University Press, 2003). In *Émile*, Rousseau criticized those who traveled for no purpose, explaining that "to travel for the sake of traveling is to wander [*errer*], to be a vagabond." Rousseau, *Émile*, 455. On eighteenth-century attitudes toward travel and objectivity more generally, see Juan Pimentel, *Testigos del mundo: Ciencia, literatura y viajes en la Ilustración* (Madrid: Marcial Pons, 2003).

20. Jean-Jacques Rousseau, *Discours sur l'origine et les fondements de l'inégalité parmi les hommes* (Amsterdam: Marc Michel Rey, 1755), 236–37.

21. Roger Chartier, *Inscrire et effacer: Culture écrite et littérature, XIe–XVIIIe siècle* (Paris: Seuil, 2005). On the social world of print in early modern Europe, see

Donald F. McKenzie, *Bibliography and the Sociology of Texts* (Cambridge: Cambridge University Press, 1999); Adrian Johns, *The Nature of the Book: Print and Knowledge in the Making* (Chicago: University of Chicago Press, 1999), 90–91; D. F. McKenzie, "Printers of the Mind: Some Notes on Bibliographical Theories and Printing-House Practices," *Studies in Bibliography* 22 (1969): 1–76; Anthony Grafton, "Correctores corruptores? Notes on the Social History of Editing," in *Editing Texts/Texte Edieren*, ed. Glenn W. Most (Göttingen: Vandenhoeck & Ruprecht, 1998). While the importance of print culture to the elaboration of New World narratives has received a lot of attention for the earliest period of European contact, far less attention has been paid to the material conditions in which knowledge traveled in the eighteenth century, with the significant exception of the work of Marie-Noëlle Bourguet. See her "La collecte du monde," in Blanckaert, *Le Museum*. Classic accounts of Europe's encounter with the Americas include Chinard, *L'Amérique et le rêve exotique*; John Elliott, *The Old World and the New: 1492–1650* (Cambridge: Cambridge University Press, 1970); Anthony Grafton, *New Worlds, Ancient Texts: The Power of Tradition and the Shock of Discovery* (Cambridge: Belknap Press, 1992); Edmundo O'Gorman, *La invención de América* (Mexico City: Fondo de Cultura Económica, 1986); Anthony Pagden, *European Encounters with the New World* (New Haven: Yale University Press, 1994); Rolena Adorno, "The Discursive Encounter of Spain and America: The Authority of Eyewitness Testimony in the Writing of History," *William and Mary Quarterly* third series, 49, no. 2 (1992): 210–28.

22. *Dictionnaire de l'Académie Française* (Paris: J. B. Coignard, 1694), s.v. "commémorer."

23. On changing conceptions of the laboratory in early modern Europe, see Steven Shapin and Simon Schaffer, *Leviathan and the Air-Pump: Hobbes, Boyle, and the Experimental Life* (Princeton: Princeton University Press, 1985); Owen Hannaway, "Laboratory Design and the Aim of Science: Andreas Libavius versus Tycho Brahe," *Isis* 77 (1986): 585–610; and Steven Shapin, "The House of Experiment in Seventeenth-Century England," *Isis* 79 (1988): 373–404. On the scientific laboratory as a place of production, see Bruno Latour and Steve Woolgar, *Laboratory Life: The Social Construction of Scientific Facts* (Princeton: Princeton University Press, 1986); Bruno Latour, *Science in Action: How to Follow Scientists and Engineers through Society* (Cambridge: Harvard University Press, 1987); and Jan Golinski, *Making Natural Knowledge: Constructivism and the History of Science* (Cambridge: Cambridge University Press, 1998), esp. chap. 3. On the idea of the "immutable mobile" and the diverse spaces of scientific production, see Latour, "Visualization and Cognition: Thinking with Eyes and Hands," *Knowledge and Society* 6 (1986): 1–40; and Adi Ophir and Steven Shapin, "The Place of Knowledge: A Methodological Survey," *Science in Context* 4 (1991): 3–21.

24. Such studies include John Gascoigne, *Science in the Service of Empire: Joseph Banks, the British State, and the Uses of Science in the Age of Revolution* (Cambridge: Cambridge University Press, 1998); James E. McClellan III, *Colonialism and Science: Saint Domingue in the Old Regime* (Baltimore: Johns Hopkins University Press, 1993); Michael Osborne, *Nature, the Exotic, and the Science of French Colonialism*

(Bloomington: Indiana University Press, 1994); Richard Drayton, *Nature's Government: Science, Imperial Britain, and the "Improvement" of the World* (New Haven and London: Yale University Press, 2000); D. Graham Burnett, *Masters of All They Surveyed: Exploration, Geography, and a British El Dorado* (Chicago: University of Chicago Press, 2000); and Grove, *Green Imperialism*. Only recently have less institutionally bound studies begun to emerge: Schiebinger, *Plants and Empire*; Judith Carney, *Black Rice: The African Origins of Rice Cultivation in the Americas* (Cambridge: Harvard University Press, 2001); Parrish, *American Curiosity*; and Ralph Bauer, *The Cultural Geography of Colonial American Literatures: Empire, Travel, Modernity* (Cambridge: Cambridge University Press, 2003).

25. *Dictionnaire de l'Académie Française* (Paris: J. B. Coignard, 1694), s.v. "itinéraire"

26. For a review of scholarship in Atlantic history and a critique of its limitations, see Alison Games, "Atlantic History: Definitions, Challenges, and Opportunities," *American Historical Review* 111, no. 3 (June 2006): 741–57. Games's article confirms that while Atlantic history has emerged as its own discipline, "few works exist that have attempted to capture the entire Atlantic across imperial, regional, and national boundaries." For a methodological discussion of how microhistorical research can be compatible with the broader aims of Atlantic history, see Lara Putnam, "To Study the Fragments/Whole: Microhistory and the Atlantic World," *Journal of Social History* 39, no. 3 (2006): 615–30.

27. On the importance of pan-Atlanticism, see Jorge Cañizares-Esguerra, *Puritan Conquistadors: Iberianizing the Atlantic, 1550–1700* (Stanford: Stanford University Press, 2006). Scholars of colonialism have done much to reject bipolar models, but the history of natural knowledge production has yet to catch up. For instructive examples and discussions as to how this may be done, see James Delbourgo, *A Most Amazing Scene of Wonders: Electricity and Enlightenment in Early America* (Cambridge: Harvard University Press, 2006); Maya Jasanoff, *Edge of Empire: Lives, Culture, and Conquest in the East, 1750–1850* (New York: Knopf, 2005); Ann Laura Stoler and Frederick Cooper, "Between Metropole and Colony," in *Tensions of Empire: Colonial Cultures in a Bourgeois World,* ed. Frederick Cooper and Ann Laura Stoler (Berkeley: University of California Press, 1997); Londa Schiebinger, "Forum Introduction: The European Colonial Science Complex," *Isis* 96 (2005): 52–55; Sanjay Subrahmanyam, *Explorations in Connected Histories: Mughals and Franks* (New York: Oxford University Press, 2005).

28. George-Louis Leclerc, comte de Buffon, "Réponse à M. de la Condamine," *Histoire naturelle, générale et particulière: Supplément* (Paris: L'Imprimerie Royale, 1774–76), 4: 24–26.

29. On the relationship between early modern science and theater, see Helmar Schramm, Ludger Schwarte, and Jan Lazardzig, eds., *Collection, Laboratory, Theater: Scenes of Knowledge in the Seventeenth Century* (Berlin: Walter de Gruyter, 2005). I do not argue, as does Ann Blair in the case of Jean Bodin, that the participants in this expedition explicitly used theatrical metaphors to portray their endeavors. For a provocative article on the pitfalls of using theatrical metaphors

to describe scientific practice, see Svante Lindqvist, "The Spectacle of Science: An Experiment in 1744 concerning the Aurora Borealis," *Configurations* 1, no. 1 (1993): 57–94, where Lindqvist argues that the model most appropriate to be invoked in the "science as theater" metaphor is that of the improvisational popular theater found in medieval fairs rather than the closed and rigid form that emerges in the classical Baroque period.

30. Sarah Maza outlines the importance of drama in the social consciousness of the Old Regime, as well as the importance of theatrical images in forging political power, in *Private Lives and Public Affairs: The Causes Célèbres of Prerevolutionary France* (Berkeley: University of California Press, 1993). See also Gregory Brown, "A Field of Honor: The Cultural Politics of Playwriting in Eighteenth-Century France" (Ph.D. diss., Columbia University, 1997); and Charles B. Paul, *Science and Immortality: The Éloges of the Paris Academy of Sciences, 1699–1791* (Berkeley: University of California Press, 1980), which assesses the theatricality of Condorcet's "éloges."

CHAPTER ONE

1. Alexander von Humboldt, "Besuch der Pyramiden von Yaruquí," in *Reise auf dem Río Magdalena, durch die Anden und Mexico*, ed. Margot Faak (Berlin: Akademie-Verlag, 1990), 2: 68.

2. Humboldt, "Besuch," 2: 71.

3. On the relationship between gestures, theatricality, politics, and culture in the Old Regime, see Sophia Rosenfeld, *A Revolution in Language: The Problem of Signs in Late-Eighteenth-Century France* (Stanford: Stanford University Press, 2001); Paul Friedland, *Political Actors: Representative Bodies and Theatricality in the Age of the French Revolution* (Ithaca: Cornell University Press, 2002); and Jeffrey Ravel, *The Contested Parterre: Public Theater and French Political Culture, 1680–1791* (Ithaca: Cornell University Press, 1999).

4. Louis Godin to Cromwell Mortimer, 4 November 1736, British Library, Add. MSS 4433, fols. 299–300b.

5. On scientific authorship in the eighteenth century, see Peter Galison and Mario Biagoli, eds., *Scientific Authorship: Credit and Intellectual Property in Science* (New York: Routledge, 2003), 1–112 (esp. chapters by Chartier, Iliffe, Johns, and Terrall); Mary Terrall, *Man Who Flattened the Earth*, 292–309; James E. McClellan III, *Specialist Control: The Publications Committee of the Academie Royale des Sciences (Paris), 1700–1793* (Philadelphia: American Philosophical Society, 2003).

6. The nexus between science and imperialism has been treated recently in several excellent monographs and edited works. See David Philip Miller and Peter Hanns Reill, eds., *Visions of Empire: Voyages, Botany, and Representations of Nature* (Cambridge: Cambridge University Press, 1996); Drayton, *Nature's Government*; Burnett, *Masters of All They Surveyed*; McClellan, *Colonialism and Science*; Roy MacLeod, ed., "Nature and Empire: Science and the Colonial Enterprise," *Osiris* 15 (Fall 2000); and Felicity Nussbaum, ed., *The Global Eighteenth Century* (Baltimore: Johns Hopkins University Press, 2003). Scholars have largely limited their studies of the pyramid controversy to descriptions of arguments used by either side and

have failed to reflect on its implications in a more symbolic and material context. The most complete treatment is to be found in Luis J. Ramos Gomez, *Epoca, genesis y texto de las «Noticias Secretas de America», de Jorge Juan y Antonio de Ulloa,* part 1, "El viaje a America (1735–1745), de los tenientes de navio Jorge Juan y Antonio de Ulloa, y sus consecuencias literarias" (Madrid: Consejo Superior de Investigaciones Científicas, 1985), 183–202. See also Lafuente and Mazuecos, *Los caballeros del punto fijo,* esp. 206–12.

7. Alexander von Humboldt, *Vue des cordillères, et monumens des peuples indigènes de l'Amérique* (Paris: F. Schoell, 1810), 242 (plate 42).

8. Voltaire to Pierre-Louis Moreau de Maupertuis, 15 December 1732, in *Correspondance,* ed. Theodore Besterman (Paris: Gallimard, 1977), 1: 366.

9. On La Condamine's role in the Pelletier-Desforts lottery scandal, see Jacques Donvez, *De quoi vivait Voltaire?* (Paris: Deux Rives, [1949]), 37–55.

10. Marie Jean Antoine Nicolas Caritat, marquis de Condorcet, "Éloge de M. de la Condamine," in *Éloge des Académiciens de l'Académie royale des sciences, morts depuis l'an 1666 jusqu'en 1790,* vol. 1 (Paris: Frédéric Vieweg, et Fuchs, 1799), 233.

11. Jacques Delille, "Discours de Réception à L'Académie Française," in *Oeuvres de Jacques Delille* (Paris: L. G. Michaud, 1825), cxvi.

12. Condorcet, "Éloge de M. de la Condamine," 188–89.

13. Many historians have followed Pierre Conlon in seeing La Condamine's profound "curiosity" and his inquisitive nature as the driving forces behind his intellectual endeavors. See Pierre Conlon, "La Condamine the Inquisitive," *Studies on Voltaire and the Eighteenth Century* 55 (1967): 361–93. I suggest instead that La Condamine's manipulation of resources formed part of a coherent strategy meant to increase his prestige and augment his right to speak on behalf of the mission as a whole, and that rather than attributing La Condamine's admittedly indefatigable spirit to curiosity alone, his activities should be conceived as commemorative practices and reinserted into their appropriate sociological and cultural contexts. As Neil Kenny reminds us, "curiosity" is a philosophical category that was "always context-bound and therefore endlessly variable" during the period in question. See Kenny, *Curiosity in Early Modern Europe: Word Histories* (Wiesbaden: Harrassowitz Verlag, 1998).

14. [Antoine-François Prévost (abbé)], untitled article, *Le Pour et Contre* 77 (1735): 35.

15. Ibid., 29.

16. On the foundation of the Academy of Inscriptions and Belles Lettres, see Blandine Kriegel, *Les académies de l'histoire* (Paris: Presses universitaires de France, 1988).

17. Archives of the Academy of Inscriptions and Belles Lettres (Paris), Procès-verbaux, 13 May 1735, 238.

18. La Condamine, *Histoire des pyramides de Quito* (Paris: L'Imprimerie Royale, 1751), 4.

19. On eighteenth-century surveying in France, Lapland, and Peru, see Terrall, *Man Who Flattened the Earth*; Rob Iliffe, "'Aplatisseur du monde et de Cassini':

Maupertuis, Precision Measurement and the Shape of the Earth in the 1730s,"
History of Science 31 (1993): 335–75; Konvitz, *Cartography in France*; John L. Greenberg, *The Problem of the Earth's Shape from Newton to Clairaut* (Cambridge: Cambridge University Press, 1995); Tom B. Jones, "The French Expedition to Lapland,
1736–1737," *Terrae Incognitae* 2 (1970): 15–24; Harcourt Brown, "From London
to Lapland and Berlin," in *Science and the Human Comedy: Natural Philosophy in
French Literature from Rabelais to Maupertuis* (Toronto: University of Toronto Press,
1976), 167–206; and Lacombe and Costabel, *La figure de la terre.*

20. La Condamine, *Histoire des pyramides*, 6.

21. José Merino Navarro and Miguel Rodriguez San Vicente made this observation in their introduction to Jorge Juan and Antonio de Ulloa, *Relación histórica
del viaje a la América meridional* (Madrid: Fundación Universitaria Española, 1978),
xx, n. 33.

22. Duguay-Trouin is best known for his surprise attack on Rio de Janeiro
at the tail end of the War of Spanish Succession. For a biography of Duguay-
Trouin, see Yves-Marie Rudel, *Duguay-Trouin, corsaire et chef d'escadre (1673–1763)*
(Paris: Perrin, 1973); and Michel Vergé-Franceschi, "Duguay-Trouin (1673–1736):
un corsaire, un officier général, un mythe" in *Revue historique* 295 (1996): 333–52.

23. La Condamine, "Observations mathématiques et physiques faites dans
un voyage de Levant en 1731 & 1732," *Mémoires de l'Académie Royale des Sciences*
(1732 [1735]): 295–322.

24. La Condamine, *Histoire des pyramides*, 12.

25. La Condamine, *Histoire des pyramides*, 15. The asterisk that follows "Tetrahedron" in the citation is a footnote marker.

26. Scipione Maffei, "Per I Signori Academici delle Scienze spediti al Peru,"
reproduced in La Condamine, *Histoire des pyramides*, 8–10. On Maffei's views of
inscriptions and antiquities more generally, see Paula Findlen, *Possessing Nature:
Museums, Collecting, and Scientific Culture in Early Modern Italy* (Berkeley: University of California Press, 1994), 396–97.

27. Daniel Stolzenberg, "Egyptian Oedipus: Antiquarianism, Oriental Studies
and Occult Philosophy in the Work of Athanasius Kircher" (Ph.D. diss., Stanford
University, 2004).

28. On the symbolic meanings of pyramids prior to the eighteenth century,
see Zur Shalev, "Measurer of All Things: John Greaves (1602–1652), the Great
Pyramid, and Early Modern Metrology," *Journal of the History of Ideas* 64, no. 4
(October 2002): 555–75. For the uses of symbolism in the Freemasonry movement, see Margaret C. Jacob, *The Radical Enlightenment: Pantheists, Freemasons and
Republicans* (London: George Allen, 1981), esp. 115. The system of using pyramids as geographical markers was conceived by the Baron de Bouis in his *Parterre
géographique et historique, ou nouvelle Méthode d'enseigner la géographie et l'histoire*
(Paris, 1753). For more on this symbolic system, and geographical gardens more
generally, see Jean-Marc Besse, *Face au monde. Atlas, jardins, géoramas* (Paris: Desclée de Brouwer, 2003). On Kircher's views of pyramids and hieroglyphs, see
Stolzenberg, "Egyptian Oedipus"; and Daniel Stolzenberg, ed., *The Great Art of*

Knowing: The Baroque Encyclopedia of Athanasius Kircher (Stanford: Stanford University Libraries, 2001). Anthony Grafton briefly discusses Kircher in "The Ancient City Restored: Archaeology, Ecclesiastical History, and Egyptology" in *Rome Reborn: The Vatican Library and Renaissance Culture*, ed. Anthony Grafton (New Haven: Yale University Press, 1993), as does Findlen in her study of early modern Italian museology, *Possessing Nature*. On Kircher's interest in Egyptology, see Erik Iversen, *The Myth of Egypt and Its Hieroglyphs in European Tradition* (Princeton: Princeton University Press, 1993), esp. 89–100.

29. James Stevens Curl, *Egyptomania: The Egyptian Revival; A Recurring Theme in the History of Taste* (Manchester, Manchester University Press, 1994); and Iversen, *Myth of Egypt*, 100.

30. Paul Lucas, *Troisième voyage du sieur Paul Lucas, fait en MDCCXIV, &c. par ordre de Louis XIV, dans La Turquie, L'Asie, La Sourie, La Palestine, La Haute et La Basse Egypte, &c.*, 3 vols. (Rouen: R. Machuel, 1719). La Condamine mentions Lucas's texts as one of the travel narratives he read in preparation for his voyage. See La Condamine, "Observations mathématiques," *Mémoires de l'Académie Royale des Sciences* (1732).

31. Lucas, *Troisième voyage*, 2: 41.

32. Lucas, *Troisième voyage*, 2: 242.

33. Lucas, *Troisième voyage*, 2: 304.

34. Jean Terrasson, abbé, *Sethos, Histoire, ou, Vie tirée des monuments anecdotes de l'ancienne Egypte* (Paris: H. L. Guerin, 1731).

35. La Condamine, *Histoire des pyramides*, 11.

36. La Condamine, *Histoire des pyramides*, 6.

37. Voltaire, *Lettres anglaises*, xv, as cited in Loridan, *Voyages des astronomes français*, 15.

38. La Condamine, *Histoire des pyramides*, 41–42.

39. On the predicament of the philosopher-traveler and the problem of securing reliable knowledge at a distance, see Shapin, *Social History of Truth*, 243–47; and Pimentel, *Testigos del mundo*, chapter 1. On empiricist thinking in the seventeenth and eighteenth centuries, see Lorraine Daston, "The Factual Sensibility," *Isis* 79 (1988): 452–70, idem, "Marvelous Facts and Miraculous Evidence in Early Modern Europe," *Critical Inquiry* 18 (1991): 93–124, and idem, "Baconian Facts, Academic Civility, and the Prehistory of Objectivity," *Annals of Scholarship* 8 (1991) 337–63. On inscription and materiality in the scientific voyage for a slightly later period, see Barbara Stafford, *Voyage into Substance: Art, Science, Nature, and the Illustrated Travel Account, 1760–1840* (Cambridge: MIT Press, 1984).

40. Charles-Marie de La Condamine, *Journal du voyage fait par ordre du Roi, à l'Équateur, servant d'introduction historique a la mesure des trois premiers degrés du Méridien* (Paris: Imprimerie Royale, 1751), 12. The asterisk that follows "inscription" in the citation is a footnote marker provided in the original text. Part of the inscription shown in the image, *Observationibus Astronomicis. . . Hocce Promontorium Æquatori Subjacere Compertum Est. 1736,* appears in this footnote.

41. On science and seafaring at the Academy of Sciences, see Jordan Kellman, "Discovery and Enlightenment at Sea: Maritime Exploration and Observation in the Eighteenth-Century French Scientific Community" (Ph.D. diss., Princeton University, 1998).

42. It is also worth noting that the interior of the continent appears as a dark, unknown landscape overgrown with trees and shrubs of many sizes, in contrast to the open coastal space on the engraving's lefthand side. This contrast between light and dark conforms to a symbolic trope familiar to eighteenth-century readers, reinforcing the need for "wild" spaces to be "illuminated" by tools of Western science.

43. For similar acts of symbolic possession, see Stephen Greenblatt, *Marvelous Possessions: The Wonder of the New World* (Chicago: University of Chicago Press, 1991). This powerful symbolic view can also be read as an episode in the history of littoral representation in European philosophy and art. See Alain Corbin, *Le territoire du vide: L'Occident et le désir du rivage, 1750–1840* (Paris: Flammarion, 1988).

44. La Condamine, *Journal du voyage*, 124.

45. La Condamine, *Histoire des pyramides*, passim. Other laborers included the "various individuals" who accompanied La Condamine in his search to find stones to fill the interior of the base, the "stone-cutters" who were always scarce in relation to the other workers, a day laborer whose sole responsibility was to travel between Yaruquí and Quito to repair broken picks, and the generic "Indians" who, according to La Condamine, only accelerated the pace of their work when a deluge of rain descended upon them.

46. A similar example is the African slave who accompanied La Condamine during part of his voyage down the Amazon River, acknowledged briefly in a moment of emergency and subsequently forgotten. See chapter 2.

47. La Condamine, *Histoire des pyramides*, 15.

48. La Condamine, *Histoire des pyramides*, 16.

49. Ronald Escobedo Mansilla, *Las comunidades indígenas y la economía colonial peruana* ([Bilbao]: Servicio Editorial Universidad del País Vasco, 1997).

50. La Condamine, *Histoire des pyramides*, 47. See also La Condamine, "Mémoire sur quelques anciens monumens du Perou, du tems des Incas . . . ," *Histoire de l'Academie royale des sciences et belles lettres, année 1746* (Berlin: Ambroise Haude, 1748).

51. "Even the ruins themselves were destroyed." Lucanus, *Pharsalia*, book IX, verse 969.

52. La Condamine, *Histoire des pyramides*, 46.

53. La Condamine, *Histoire des pyramides*, 48.

54. La Condamine, *Histoire des pyramides*, 47.

55. La Condamine, *Histoire des pyramides*, 47.

56. In addition to being the name of an administrative district within the larger viceroyalty of Peru, the Audiencia was also the highest juridical body of the region, composed of a president, four *oidores*, a fiscal officer, and a *protector de*

naturales serving a host of administrative and juridical functions within colonial society. While its primary function was to adjudicate cases as a court of second instance (that is, to hear appeals that would come from judges throughout the Audiencia), it could also hear criminal cases that took place within the capital itself. See Tamar Herzog, *Upholding Justice: Society, State, and the Penal System in Quito (1650–1750)* (Ann Arbor: University of Michigan Press, 2004); and Kenneth Andrien, *The Kingdom of Quito, 1690–1830* (Cambridge: Cambridge University Press, 1995), esp. 174–80. On the nature of the Bourbon reforms in the Audiencia of Quito and the impact of the establishment of the viceroyalty of New Granada, see Rosemarie Terán Najar, *Los proyectos del imperio borbónico en la Real Audiencia* (Quito: Abya-Yala, 1988).

57. La Condamine, "Peticion de Don Carlos de la Condamine," in *Documentos para la historia de la Audiencia de Quito*, ed. José Rumazo (Madrid: Afrodisio Aguado, 1948–1950), 5: 412. This volume reproduces the "Autos de las pirámides de Quito," Legajo Quito 374, Archivo General de Indias, Seville, Spain.

58. As agents of the Spanish king, Ulloa and Juan had a similar conception of their own superiority vis-à-vis the local judicial system. See Herzog, *Upholding Justice*, 223.

59. As Ramos Gomez suggests and Herzog confirms, the Audiencia's approval of La Condamine's project likely had to do with an earlier dispute between Ulloa and José Araujo y Río, president of the Audiencia of Quito from 1736 to 1743. The conflict arose when Ulloa refused to address Araujo y Río with proper deference. This battle over protocol and hierarchy, which may have prejudiced the Audiencia against the Spaniards, is discussed in Herzog, *Upholding Justice*, 221–25. See also "El proceso contra Jorge Juan y Antonio de Ulloa en Quito (1737)," *Anuario de estudios americanos* 5 (1948): 747–80.

60. Jorge Juan, "Petizion de los dos officiales de Marina Españoles nombrados para asistir a las observaciones de los Franceses," in *Documentos para la historia de la Audiencia de Quito*, ed. José Rumazo (Madrid: Afrodisio Aguado, 1948–1950), 5: 401. The Spaniards' arguments are discussed in Merino Navarro and Rodriguez San Vicente's introduction to Juan and Ulloa, *Relación histórica*, xx–xxiv; Ramos Gomez, *Epoca, genesis y texto*, 183–202.

61. La Condamine, "Peticion," 5: 415.

62. On the anti-Hispanic prejudice of northern European authors in the eighteenth century, see Cañizares-Esguerra, *How to Write the History of the New World*, chapter 1.

63. La Condamine, "Peticion," 5: 414.

64. La Condamine, "Peticion," 5: 424

65. La Condamine, "Peticion," 5: 424.

66. Denis Diderot, *Sur la liberté de la presse*, ed. Jacques Proust (Paris: Éditions sociales, 1964). On contemporaneous disputes over intellectual property taking place in France and throughout Europe, see Roger Chartier, introduction to Chrétien Guillaume de Lamoignon de Malesherbes, *Mémoires sur la librairie et sur la liberté de la presse* (Paris: Imprimerie nationale, 1994); and Carla Hesse,

"Enlightenment Epistemology and the Laws of Authorship in Revolutionary France, 1777–1793," *Representations* 30 (Spring 1990): 109–37.

67. Simon-Nicolas-Henri Linguet, *Canaux navigables, ou Développement des avantages qui résulteraient de l'execution des plusiers projets en ce genre pour la Picardie, l'Artois, la Bourgogne, la Champagne, la Bretagne, & toute la France en géneral* (Amsterdam: L. Cellot, 1769).

68. Linguet, *Canaux navigables*, 241–42.

69. On his understanding of the rights and privileges of authorship in the Old Regime, see the discussion of Linguet's *Mémoire sur les propriétés et privilèges exclusifs de la librairie: Présenté en 1774* (Paris, n.d.) in Hesse, "Enlightenment Epistemology." On Linguet more generally, see Darline Gay Levy, *The Ideas and Careers of Simon-Nicolas-Henri Linguet* (Urbana: University of Illinois Press, 1980).

70. "Historia de las Pyramides de Quito, corregida, añadida, y con reflexiones," Biblioteca Nacional de España, MS 8428, 46.

71. Ibid., 67, 70.

72. Ibid., 52.

73. Burriel and the *Relación histórica* are discussed at greater length in chapter 5 below.

74. MSS 8428 and 7406 are briefly discussed in Ramos Gomez, *Epoca, genesis y texto*; and Merino Navarro and Rodriguez San Vicente, introduction, *Relación histórica* (1978).

75. Juan, "Petizion," 5: 402.

76. Rosenfeld, *Revolution in Language*, 107. On the question of languages and science in Europe, see Roger Chartier and Pietro Corsi, eds., *Sciences et langues en Europe* (Paris: Centre Koyré/EHESS, 1996).

77. Condorcet, *Esquisse*, 205.

78. On this point, see Rosenfeld, *Revolution in Language*, 107–8. Rosenfeld relates Condorcet's discussion of scientific language to the broader interest in sign language as a model for social and moral reform.

79. Condorcet, *Esquisse*, 234.

80. Hesse, "Enlightenment Epistemology."

81. As cited in Hesse, "Enlightenment Epistemology," 112.

82. Marie-Noëlle Bourguet Christian Licoppe and H. Otto Sibum, introduction, in Marie-Noëlle Bourguet, Christian Licoppe and H. Otto Sibum, *Instruments, Travel and Science: Itineraries of Precision from the Seventeenth to the Twentieth Century* (London and New York: Routledge, 2002), 16.

83. Humboldt, "Besuch," *Reise*, 2: 72.

84. Humboldt, "Besuch," *Reise*, 2: 72.

85. Humboldt, "Besuch," *Reise*, 2: 69.

CHAPTER TWO

1. José Gumilla, *El Orinoco ilustrado, historia natural, civil, y geographica, de este gran rio, y de sus caudalosas vertientes* . . . (Madrid: M. Fernandez, 1741), 4.

2. On the generic conventions of the seventeenth-century *relation*, see Sara E. Melzer, "The French *Relation* and Its 'Hidden' Colonial History," in *Companion to the Literatures of Colonial America*, ed. Susan Castillo and Ivy Schweitzer (Oxford: Blackwell Publishers, 2005), 220–40.

3. Charles-Marie de La Condamine, *Relation abrégée d'un voyage fait dans l'intérieur de l'Amérique méridionale. Depuis la côte de la mer du Sud, jusqu'aux côtes du Brésil & de la Guiane, en descendant la riviere des Amazones; lûe à l'assemblée publique de l'Académie des Sciences, le 28. avril 1745* (Paris: Veuve Pissot, 1745 [1746]). On the mixed genre of popularized scientific account and travel narrative, the stylistic evolution of the *mémoire académique* in the eighteenth century, and a more detailed treatment of the narrative strategies La Condamine used to navigate between these two publics, see Neil Safier, "Unveiling the Amazon to European Science and Society: The Reading and Reception of La Condamine's *Relation abrégée*," *Terrae Incognitae* 33 (2001): 33–47. Texts such as Maupertuis's *Discours sur les différentes figures des astres* (1732) and Voltaire's *Élemens de la philosophie de Newton* (1738) exemplified the way that a broad readership could be attracted to fields that were previously considered limited to the narrow purview of the academic elite.

4. *Carte du cours du Maragnon, ou de la grande rivière des Amazones. Dans sa partie navigable depuis Jaen de Bracamoros jusqu'à son embouchure et qui comprend la Province de Quito, et la côte de la Guiane depuis le Cap de Nord jusqu'à Essequebè. Levée en 1743 et 1744 et assujettie aux observations astronomiques par M. de la Condamine de l'Ac. Rle. des Sc. Augmenté du cours de la rivière Noire et d'autres détails tirés de divers mémoires et routiers manuscrits de voyageurs modernes.* The map was created by d'Anville and engraved by Guillaume Delahaye. Guillaume Sanson (d. 1703) was the son of Nicolas Sanson (1600–67), the king's first official cartographer, who worked closely with Colbert to collect geographic information to produce the first complete survey of France. For what is still the most thorough discussion of early modern cartographic practice in France, see Konvitz, *Cartography in France*. Sanson's son took over the mantle of *géographe du roi* upon Nicolas's death. It was La Condamine who called Sanson's map "very defective," *Relation abrégée*, 13. For more on the Jesuit missions in Brazil, and Jesuit cartographers specifically, see Serafim Leite, *História da Companhia de Jesus no Brasil*, 10 vols. (Lisbon: Livraria Portugalia, 1938–50), esp. 3: 255–58 and 4: 281–87. On Fritz and his maps of the Amazon, see André Ferrand de Almeida, "Samuel Fritz and the Mapping of the Amazon," *Imago Mundi* 55 (2003): 103–12.

5. Alexander von Humboldt, *Relation historique du voyage aux régions équinoxiales du nouveau continent* (Stuttgart: F. A. Brochhaus, 1970), 2: 456.

6. David Woodward and J. B. Harley, eds., *History of Cartography* (Chicago: University of Chicago, 1987), 1: xvi. Other recent works of interest include Christian Jacob, *L'empire des cartes. Approche théorique de la cartographie à travers l'histoire* (Paris: Albin Michel, 1992); Edney, *Mapping an Empire*; Burnett, *Masters of All They Surveyed*; Ricardo Padrón, *The Spacious Word: Cartography, Literature, and Empire in Early Modern Spain* (Chicago: University of Chicago Press, 2003); Mary

Pedley, *The Commerce of Cartography: Making and Marketing Maps in Eighteenth-Century France and England* (Chicago: University of Chicago Press, 2004); and Mark Monmonier, *From Squaw Tit to Whorehouse Meadow: How Maps Name, Claim, and Inflame* (Chicago: University of Chicago Press, 2006).

7. La Condamine to Voltaire, December 1735, in Voltaire, *Correspondance*, ed. Theodore Besterman (Paris: Gallimard, 1977–93), 1: D961.

8. Bernardo Gutierres Vocanegra, Don, "Inventario y reconocimiento del equipaje de los Académicos Franceses," 1 June 1736, in *Documentos para la historia de la Audiencia de Quito*, ed. José Rumazo (Madrid: Afrodisio Aguado, 1948–59), 6: 18.

9. Gutierres Vocanegra, "Inventario," 6: 18.

10. On the reading practices of European naturalists in South America, see Neil Safier, "'Every day that I travel . . . is a page that I turn': Reading and Observing in Eighteenth-Century Amazonia," *Huntington Library Quarterly* 70, no. 1 (2007): 103–28.

11. La Condamine, *Relation abrégée*, 42–43. On the heroic narrative of exploration as a genre, see Mary Terrall, "Heroic Narratives of Quest and Discovery," *Configurations* 6, no. 2 (1998): 223–42.

12. On the early history of Amazonian exploration and the gestation of the myths of the Amazons and El Dorado, see Jean-Pierre Sanchez, *Mythes et légendes de la conquête de l'Amérique,* 2 vols. (Rennes: Presses universitaires de Rennes, 1996).

13. Gaspar de Carvajal, *Relación del nuevo descubrimiento del Rio Grande de las Amazonas (1541–42)* (Quito: Gobierno del Ecuador, 1992), 127.

14. The account to which La Condamine had access was the translation by Marin le Roy, Sieur de Gomberville, of Father Acuña's account, published in 1682 as *Relation de la rivière des Amazones, traduite par seu Mr de Gomberville de l'Academie Françoise. Sur l'original espagnol du P. Christophle d'Acuña Jesuite. Avec une dissertation sur la rivière des Amazones pour servir de preface. . .* , 4 vols. (Paris: Claude Barbin, 1682). In addition to a "dissertation" and historical information, Gomberville's *Relation* included a "Lettre escrite de l'Isle de Cayenne," which was a journal of the travels of Jean Grillet and François Bechamel to Guyana in 1674. While the Spanish and Portuguese were the central protagonists in the European exploration of the Amazon and the primary producers of the narratives upon which most early European knowledge was based, the French Crown had also taken an interest in the lands of the Amazon. This interest peaked in the seventeenth century with the publication of Blaise François de Pagan's *Relation historique et geographique de la Grande Rivière des Amazones dans l'Amérique* (Paris: Cardin Bessongne, 1656), which included a new map of the region (by Nicolas Sanson) and a preface that proposed the French annexation of several important sites along the river to establish a new empire in America. The French lost any claim to the Amazon with the 1713 Treaty of Utrecht, which settled the border of French Guyana north of the Oyapoc River. For more on early European exploration of the Land of Canelos, see Neil Safier, "Fruitless Botany: Joseph de Jussieu's

South American Odyssey," in *Science and Empire in the Atlantic World*, ed. James Delbourgo and Nicholas Dew (New York: Routledge, 2007), 203–24.

15. La Condamine, *Journal du voyage*, 122.

16. La Condamine, *Journal du voyage*, 122.

17. La Condamine, *Journal du voyage*, 122–23.

18. La Condamine, *Journal du voyage*, 122–23.

19. Marqués de Valleumbroso to La Condamine, Cuzco, 12 March 1742, British Library (BL), Add. MSS 20793: 322–40. This letter, an annotated copy of the original between Valleumbroso and La Condamine, was sent by La Condamine to Joseph Antonio de Armona. La Condamine included several comments to Armona about Valleumbroso's letter, including a discussion of Gumilla's *Orinoco ilustrado* and the connection of the Orinoco and Amazon rivers that La Condamine himself confirmed. The journals of Pablo Maroni were later compiled and published as *Noticias autenticas del famoso rio Marañon [1738]* (Iquitos: IIAP-CETA, 1988); they include a nearly complete version of Samuel Fritz's journal as well.

20. Marqués de Valleumbroso to La Condamine, 322b.

21. Marqués de Valleumbroso to La Condamine, 327.

22. Marqués de Valleumbroso to La Condamine, 330.

23. La Condamine, *Relation abrégée*, 15.

24. Marqués de Valleumbroso to La Condamine, 325.

25. La Condamine, *Relation abrégée*, 14–15.

26. Marqués de Valleumbroso to La Condamine, 331.

27. Marqués de Valleumbroso to La Condamine, 333–34.

28. Ronald Meek, *Social Science and the Ignoble Savage* (Cambridge: Cambridge University Press, 1976).

29. La Condamine, *Relation abrégée*, citations on 111, 52, 74, 63.

30. Magnin later accompanied his French companion to the mission at La Laguna, at which point La Condamine was to continue his journey with Pedro Vicente Maldonado, who had descended the Bobonaza River from the central Andean cordillera along the Pastaza corridor through Baños and Canelos. On Magnin, see Jean Magnin, "Description de la Province et des missions de Maynas au Royaume de Quito," in Thomas Henkel, ed., *Chronique d'un chasseur des âmes: Un jésuite suisse en Amazonie au XVIIIe siècle* (Fribourg: Éditions de l'Hèbe, 1993); and Constantin Bayle, "Descubridores jesuitas del Amazonas," *Revista de Indias* 1 (1940): 121–85. This manuscript was first acknowledged in Marie-José Pillon, "Le voyage de La Condamine en Amazonie au XVIIIe siècle, étude critique de certains aspects" (Ph.D. diss, Paris-III Sorbonne la Nouvelle, 1986).

31. Magnin, "Description," 95.

32. La Condamine, *Relation abrégée*, 72.

33. Magnin, "Description," 151.

34. La Condamine, *Relation abrégée*, 51.

35. La Condamine, *Relation abrégée*, 52–53.

36. Magnin, "Description," 124; La Condamine, *Relation abrégée*, 53.

37. Magnin, "Description," 200.

38. La Condamine, *Relation abrégée*, 67.

39. Jacques-François Artur, "Histoire des colonies françoises de la Guianne," 2: 557–81, BN, MS N. Acq. Fr. 2572, reproduced as "Pièce Justificative No. 1, Relation du D.r Artur" in *Observations scientifiques de La Condamine pendant son séjour à Cayenne (1744)*, ed. Henri Froidevaux (Paris: Imprimerie Nationale, 1898), 23.

40. La Condamine, *Relation abrégée*, 58.

41. Artur, "Histoire," 23.

42. La Condamine, *Relation abrégée*, 84.

43. Gomberville's *Relation* (see n. 14) contained the map by Guillaume Sanson. La Condamine, *Relation abrégée*, 13.

44. On Fritz's map and diary, see Hernán Rodríguez Castelo, *El Padre Samuel Fritz: Diario* (Quito: Academia Ecuatoriana de la Lengua, 1997).

45. La Condamine, *Relation abrégée*, 14. Once again, this information was also present in Valleumbroso's account. Cf. Marqués de Valleumbroso to La Condamine, 334.

46. La Condamine, *Relation abrégée*, 64–65.

47. La Condamine, *Relation abrégée*, 25, 35, 43. A toise measures approximately six feet in present-day measure. For more on the importance of measurement and instrumentation to the scientific voyages of the eighteenth century, see Marie-Noëlle Bourguet and Christian Licoppe, "Voyage, Mesures, et Instruments: Une nouvelle expérience du monde au Siècle des lumières," *Annales: Histoire, Sciences Sociales* 52, no. 5 (1997): 1115–51; Marie-Noëlle Bourguet, Christian Licoppe, and H. Otto Sibum, eds., *Instruments, Travel and Science: Itineraries of Precision from the Seventeenth to the Twentieth Century* (London: Routledge, 2002). On the importance of new machines and instruments at the Academy, see Robin Briggs, "The Académie Royale des Sciences and the Pursuit of Utility," *Past and Present* 131 (1991): 38–88. On the importance of the scientific method at the Academy, see L. M. Marsak, *Bernard de Fontenelle: The Idea of Science in the French Enlightenment*, Transactions of the American Philosophical Society, 49, pt. 7 (Philadelphia: American Philosophical Society, 1959), esp. 40–45.

48. La Condamine, *Relation abrégée*, 116–17, emphasis added.

49. Christian Licoppe, *La formation de la pratique scientifique: Le discours de l'expérience en France et en Angleterre, 1630–1820* (Paris: La Découverte, 1996), 94. Focusing on the *récits expérimentaux* as recounted within the *Mémoires de l'Académie Royale des Sciences* and the *Philosophical Transactions of the Royal Society*, Licoppe attempts to understand the language of experimental practice and observation as a form of representation, searching for the conditions that permit the "enunciation and production" of experimental proof.

50. Licoppe describes this tradition as relying on "narrative structures that are particular to the experiential account," including an important emphasis on

the "visual testimony that provides proof of an empirical fact as much for the author as for the reader, who through the text is placed in the position of virtual witness." *La formation de la pratique scientifique*, 95.

51. La Condamine, *Relation abrégée*, 7–8, emphasis added.

52. La Condamine, *Relation abrégée*, 80.

53. La Condamine, *Relation abrégée*, 7–8.

54. Christopher Columbus, "First Voyage" (13 January 1493), as cited in Sanchez, *Mythes et légendes de la conquête de l'Amérique*, 1: 131.

55. For more detail concerning this and other myths, see Enrique de Gandía, *Historia crítica de los mitos de la conquista americana* (Buenos Aires: J. Roldán, 1929); and Wolfgang Haase and Meyer Reinhold, eds., *The Classical Tradition and the Americas* (Berlin: Walter de Gruyter, 1994). On the origins of the myth of El Dorado specifically, and Sir Walter Ralegh's 1595 journey to discover the "'city of gold,'" consult Juan Gil, *Mitos y utopias del descubrimiento*, vol. 3, *El Dorado* (Madrid: Alianza Editorial, 1989); and Charles Nicholl, *The Creature in the Map* (London: Jonathan Cape, 1995); see also the introduction and two analytical chapters by Neil Whitehead in Walter Ralegh, *The Discoverie of the Large, Rich, and Bewtiful Empyre of Guiana*, ed. Neil Whitehead (Norman: University of Oklahoma Press, 1997).

56. La Condamine, *Relation abrégée*, xvi.

57. La Condamine, *Relation abrégée*, 99.

58. La Condamine, *Relation abrégée*, 126, 127.

59. La Condamine, *Relation abrégée*, 129.

60. La Condamine, *Relation abrégée*, 129.

61. Claude-Marie Guyon, *Histoire des Amazones anciennes et modernes* (Paris: J. Villette, 1740), 204.

62. La Condamine, *Relation abrégée*, 16.

63. For an overview of indigenous life along the Amazon and its subsequent devastation following European contact and colonization, see John Hemming, *Amazon Frontier: The Defeat of the Brazilian Indians* (London: Macmillan, 1987).

64. See Neil Whitehead, introduction to Ralegh, *Discoverie*.

65. La Condamine, *Relation abrégée*, xiii.

66. La Condamine, *Relation abrégée*, 121–22, emphasis added.

67. La Condamine, *Relation abrégée*, 119–20. On the debate over Gumilla's account, see Cañizares-Esguerra, *How to Write the History*, 28–29, and chapter 3 below.

68. La Condamine, *Relation abrégée*, 122–24.

69. La Condamine's Amazonian map is not the only cartographic representation to use this graphic device to emphasize its original contributions. For a contemporary example of this overlaying device, see *Carte de France corrigée par ordre du Roy sur les observations de Mrs. de l'Académie des Sciences*, Bibliothèque Nationale de France, Paris, Ge. DD. 2987-777.

70. Bourguet and Licoppe, "Voyage," 1150.

71. Bourguet and Licoppe, "Voyage," 1150. The authors tell an interesting anecdote about Alexander von Humboldt's ire and indignation when, following

in the footsteps of La Condamine some fifty years later between the Amazon and Orinoco rivers, he discovered that there were still places in the world that had "escaped measure, and whose maps were false to the point of approximating the illustration of geographic myths." While it may be true that "a universalizing and cartographic appropriation of the world linked to an instrumental and quantifying approach had become legitimated to such a degree" by Humboldt's day as to justify his anger at the errors and falsifications of contemporary maps, it should also be clear that the instrumental process so often proclaimed and heralded by the intrepid eighteenth-century explorers incorporated enough speculation, and relied on a significant enough degree of noninstrumental data, to challenge the idea that the scientific method had succeeded in removing geographic myths from the cartographic record.

72. Denis Diderot and Jean Le Rond d'Alembert, eds., *Encyclopédie, ou Dictionnaire raisonné des sciences, des arts et des métiers, par une société de gens de lettres,* 17 vols. (Paris, 1751–72), 1: 318, s.v. "Encyclopédie."

73. Marqués de Valleumbroso to La Condamine, 335b.

74. La Condamine explained this point to his friend Joseph de Armona when he sent him an annotated copy of Valleumbroso's letter. Marqués de Valleumbroso to La Condamine, 335b.

75. Review of "Relation abrégée d'un voyage fait dans l'intérieur de l'Amérique méridionale," *Mémoires de Trévoux* (April 1746): 611–45.

CHAPTER THREE

1. Francisco Javier Clavijero, *Historia antigua de México* (Mexico City: Editorial Porrúa, 2003), book 10, diss. 6, sec. 6, 772.

2. "Extrait d'un mémoire de M. de la Condamine, en réponse à quelques endroits de l'Article XXXIII. de nos Mémoires, (I. vol. d'Avril 1746) où l'on rend compte du voyage fait par le même Académicien sur la Riviére des Amazones," article 116, *Mémoires pour l'Histoire des Sciences & des Beaux Arts* (October 1746): 2275.

3. "Lettre de M. La Condamine aux Auteurs de ce Journal," *Mémoires pour l'Histoire des Sciences & des Beaux Arts* (February 1748): 370–85.

4. Jardine and Grafton, "'Studied for Action': How Gabriel Harvey Read His Livy," *Past and Present* 129 (1990): 30–78.

5. *Suite de la Clef, ou Journal Historique sur Les Matieres du Tems* (June 1746): 410.

6. For detailed studies of the reception in the periodical press of the *Relation abrégée* and other texts written following the Franco-Hispanic expedition, see Chouillet, "Rôle de la presse périodique de langue," 171–89; Marcil, "La presse et le compte rendu de recits de voyage scientifique," 285–304, and idem, "Récits de voyage et presse périodique au XVIIIe siècle."

7. Pierre Massuet to La Condamine. This letter is cited in the auction catalog of Jean-Emmanuel Prunier, Louviers, 7 July 1991, "Dossier La Condamine," Archives de l'Académie des Sciences, Paris.

8. *Bibliothèque Raisonnée des Ouvrages des Savans de l'Europe* (January, February, and March 1746): 210–37. For information on the *Bibliothèque Raisonnée*, which was published in Amsterdam from 1728 to 1753, see Jean Sgard, ed., *Dictionnaire des Journaux* (Paris: Universitas, 1991), no. 169.

9. *Bibliothèque Raisonnée* (1746): 217.

10. "Séance publique de l'Académie des Sciences," *Mercure de France* (August 1745): 99–116. This language of a "long & dangerous voyage" would be recycled and repeated nearly verbatim in the *Encyclopédie*'s article on La Condamine's journey, showing the authority and resonance that periodical journals often had on printed texts that followed in their wake. The repetition of this phrase in the *Encyclopédie* is not surprising if we consider that the author of this review may have been Diderot himself. De Booy claims that Diderot is the author, and we know that Diderot occasionally wrote for the *Mercure* during this period. Elisabeth Badinter follows De Booy's opinion. See J. de Booy, "Denis Diderot, Ecrits de jeunesse," in *Studies on Voltaire and the Eighteenth Century*, esp. no. 119 (1975): 69–106; Badinter, *Les passions intellectuelles*, 1: 281, note 2.

11. On Rousseau's exceptional view of La Condamine's empiricism, see Rousseau, *Oeuvres politiques*, as quoted in Duchet, *Anthropologie*, 98. Duchet corroborates the view that La Condamine's account was privileged by contemporary writers in Europe: "For Prévost, Buffon, de Pauw, [La Condamine's] testimony is authoritative." She argues that the great merit of the *Relation abrégée* was La Condamine's ability to show that "one can instruct, and please, through reliable narrative where the need for truth prevails over the need for agreement" (91, 110). Some of the more notable texts that reproduced portions of La Condamine's text include Antoine-François Prévost's *Histoire générale des voyages,* 20 vols. (Paris: Didot, 1746–89); Guillaume-Thomas Raynal's *Histoire . . . des deux Indes* (Geneva: Jean-Léonard Pellet, 1783); and William Robertson, *Histoire de l'Amérique* (Maestricht: Jean-Edme Dufour & Philippe Roux, 1777). On the representation of the American "savage" more generally, see Chinard, *L'Amérique et le rêve exotique*, which discusses Louis Armand de Lom d'Arce Lahontan's *Nouveaux voyages de M. le baron de Lahontan* (The Hague: Frères Honoré, 1703); Joseph-François Lafitau's *Moeurs des sauvages amériquains* (Paris: Saugrain l'aîné, 1724); Voltaire's *Alzire* (Paris: Jean-Baptiste-Claude Bauche, 1736); Françoise d'Issembourg d'Happoncourt Graffigny's *Lettres d'une péruvienne* (Paris, 1748); and Montesquieu's *L'esprit des lois* (1748), among others.

12. It was the *Mercure de France* that commented on "the utility and number of discoveries that M. de la Condamine was to bring back with him." See "Séance publique de l'Académie des Sciences," *Mercure de France* (November 1744): 47–80.

13. La Condamine, "Relation abrégée," *Mémoires de l'Académie Royale des Sciences* (1745): 490.

14. La Condamine to de Pinto, 25 June 1747, British Library (BL), Egerton MSS 1745, fol. 188. On de Pinto's life, his ancestors, and his primary works, see Jacob Samuel Wijler, *Isaac de Pinto, sa vie et ses œuvres* (Apeldoorn: C. M. B. Dixon, 1923). For the most recent article treating the debate between Voltaire and de Pinto, and

more broadly, the Enlightenment tension between ideologies of universalism and cultural particularism, see Adam Sutcliffe, "Can a Jew be a Philosophe? Isaac de Pinto, Voltaire, and Jewish Participation in the European Enlightenment," *Jewish Social Studies* 6, no. 3 (2000): 31–51. On de Pinto more generally, see Ida J. A. Nijenhuis, "The Passions of an Enlightened Jew: Isaac de Pinto (1717–1787) as a 'Solliciteur du Bien Public,'" in *The Low Countries and Beyond*, ed. Robert S. Kirsner (Lanham: University Press of America, 1993), 47–54; idem, *Een Joodse Philosophe: Isaac de Pinto (1717–1787)* (Amsterdam: NEHA, 1992); Richard H. Popkin, "Hume and Isaac de Pinto," *Texas Studies in Literature and Language* 12 (1970): 417–30. The only time in La Condamine's printed works that he discusses the savants with whom he had contact while in Holland is his description of the experiments with poisoned arrows, which he conducted in Cayenne and then repeated in the presence of M. van Musschenbroek, van Swieten, and Albinus, all professors at Leiden University. See La Condamine, *Relation abrégée*, 208–10.

15. De Pinto to La Condamine, undated, BL, Egerton MSS 1745, fol. 181.

16. De Pinto to La Condamine, undated, fol. 184.

17. De Pinto to La Condamine, undated, fol. 185.

18. De Pinto to La Condamine, undated, fol. 182v. Here de Pinto is citing directly from La Condamine, *Relation abrégée d'un voyage fait dans l'intérieur de l'Amérique méridionale* (Paris: Veuve Pissot, 1745). On the question of the adaptability of human nature, cf. Sutcliffe, "Can a Jew be a Philosophe?" 44–45. Sutcliffe somewhat surprisingly fails to account for the important role climatic theory played in the elaboration of de Pinto's universalism, despite having seen Nijenhuis's interpretation where a similar argument to the one I have laid out is advanced. See Nijenhuis, "The Passions of an Enlightened Jew," 47–54. The theme of New World degradation would be picked up by Buffon, de Pauw, and others, and would eventually lead to a dispute engaging intellectuals on both sides of the Atlantic, including Thomas Jefferson and the Jesuit Francisco Xavier Clavijero. The classic account of this "Dispute of the New World" is Antonello Gerbi's *La disputa del Nuovo Mundo: Storia di una polemica, 1750–1900*, new ed. (Milan: Adelphi, 2000). For a more recent assessment taking into account a far broader range of source materials, see Cañizares-Esguerra, *How to Write the History of the New World*.

19. De Pinto to La Condamine, undated, fol. 184.

20. De Pinto claimed that he was "of the same sentiment as the famous Mr. Du Bos, that physical causes are more influential than moral causes in changing an entire nation." De Pinto to La Condamine, undated, fol. 182. Du Bos's climatic explanations for the variability in human creativity and talent drew on several parallel strains of classical and early modern thought, including ideas about "clusters" of genius drawn from classical antiquity and notions of distinctive national characters coming from Barclay, Chardin, and Sir William Temple. On eighteenth-century climate theory, see Clarence J. Glacken, *Traces on the Rhodian Shore: Nature and Culture in Western Thought from Ancient Times to the End of the Eighteenth Century* (Berkeley: University of California Press, 1967), 562; Gerbi,

La disputa del Nuovo Mundo, 55; Armin Hajman Koller, *The Theory of Environment: An Outline of the History of the Idea of Milieu, and Its Present Status* (Menasha: George Banta Publishing, 1918). On Du Bos, see Alfred Lombard, *L'abbé Du Bos: Un initiateur de la pensée moderne, 1670–1742* (1913; Geneva: Slatkine reprints, 1969). On the development of climatic theory between Du Bos and Montesquieu, see Roger Mercier, "La théorie des climats des 'Réflexions critiques' à 'L'esprit des lois,'" in *Revue d'Histoire Littéraire de la France* 53 (1953): part 1, 17–37 and part 2, 159–74.

21. Abbé du Bos, *Réflexions critiques sur la poësie & sur la peinture*, 3 vols. (1770; facsimile reprint, Geneva; Paris: Slatkine reprints, 1982), 2: 287.

22. De Pinto to La Condamine, undated, fol. 182. It should not be forgotten that they, too, were to a certain degree Creoles as well. Descendants of the one of the most tightly knit communities in seventeenth-century Amsterdam, that of the Sephardic Jews of Portuguese extraction, de Pinto and his fellow sephardim would have readily understood the challenges and particularities of living as members of a separate cultural and social grouping based upon claims to a particular blood line.

23. De Pinto to La Condamine, undated, fol. 184.

24. La Condamine later rejected this claim. "I do not believe that . . . the revolutions that will certainly arrive someday in America will be caused by the natives (*naturels*) of the country, according to the ideas that I developed [during my journey]." La Condamine to de Pinto, 17 September 1746, BL, Egerton MSS 1745, fol. 187.

25. De Pinto to La Condamine, undated, fol. 185.

26. The most recent biography of Maldonado is Carlos Ortiz Arellano, *Pedro Vicente Maldonado, forjador de la patria ecuatoriana, 1704–1748* (Quito: Casa de la Cultura Ecuatoriana, 2004). Other sources on his life include: Neptalí Zúñiga, *Pedro Vicente Maldonado, Un científico de América* (Madrid: Publicaciones Españolas, 1951); Alfredo Costales Cevallos, *El sabio Maldonado ante la posteridad* (Quito: Casa de la Cultura Ecuatoriana, 1948); Piedad and Alfredo Costales, *Los Maldonado en la Real Audiencia de Quito* (Quito: Banco Central del Ecuador, 1987); and Carlos Manuel Larrea, *La Real Audiencia de Quito y su Territorio* (Quito: Imprenta del Ministerio de Relaciones Exteriores, 1987).

27. See Pedro Vicente Maldonado, "Memorial a nombre de San Pedro de Riobamba" in *Documentos para la Historia de la Audiencia de Quito*, 8 vols., ed. José Rumazo (Madrid: Afrodisio Aguado, 1948–50), 1: 7–13. See also Maldonado, "Representación de Pedro Vicente Maldonado a S. M.," 1: 47–52. On 2 September 1747, the Fiscal del Consejo de Indias conferred this title upon the town of Maldonado's birth.

28. On the impact of Maldonado's project to construct this road from Quito to the Rio Esmeraldas, see Frank Salomon, *Los Yumbos, Niguas y Tsatchila o "Colorados" durante la Colonia Española: Etnohistoria del Noroccidente de Pichincha, Ecuador* (Quito: Ediciones Abya-Yala, 1997), 99–113.

29. Archives of the Royal Society (London), minutes dated 27 October 1748, 379. He was made a corresponding member of the Paris Academy of Sciences on

24 March 1747, but due to his premature death in London two weeks following his nomination on 23 October 1748, Maldonado was removed from consideration to become a fellow of the Royal Society.

30. La Condamine to de Pinto, 25 June 1747, fol. 188. On Maldonado's purchase of an atlas, see the letter from the Duke of Huescar to Don José de Carvajal, 12 June 1747, reproduced in Didier Ozanam, ed., *La diplomacia de Fernando VI* (Madrid: CSIC, 1975), 203.

31. De Pinto to La Condamine, undated, fol. 191. Answering this letter, La Condamine wrote that "I do not doubt that you were pleased to see a cultured man (*homme d'esprit*) born in a climate so different from our own and who until his arrival in Europe had only seen nature in its most naked form." La Condamine to de Pinto, undated, fol. 190. The idea of a climatic influence on the character of Maldonado has lost none of its vigor even to the present day. In one of the most recent articles on Maldonado's life, Celín Astudillo Espinosa wrote that "[in] few men did environment have such a decisive and evident influence over biology, personality, mental aptitude, scientific and geographical vocation, and the decision to take on not only grand intellectual projects but also technological and patriotic deeds, as it did in the illustrious eighteenth-century citizen of Riobamba, don Pedro Vicente Maldonado." Celín Astudillo Espinosa, "Don Pedro Vicente Maldonado," *Revista geografica* (Quito) 30 (June 1992): 13–31.

32. On the perception of Franklin as an exemplary British American, see Stacy Schiff, *A Great Improvisation: Franklin, France, and the Birth of America* (New York: Henry Holt, 2005).

33. De Pinto to La Condamine, undated, fol. 192.

34. Ibid., fol. 192.

35. Ibid., fol. 192.

36. Ibid., fol. 192.

37. On the Atlantic Creole, see Ira Berlin, "From Creole to African: Atlantic Creoles and the Origins of African-American Society in Mainland North America," *William and Mary Quarterly*, third series, 53, no. 2 (April 1996): 251–88.

38. La Condamine to de Pinto, 12 October 1747, fol. 196.

39. See Wijler, *Isaac de Pinto.*

40. De Pinto to La Condamine, undated, fol. 191.

41. La Condamine to de Pinto, 12 October 1747, fol. 196.

42. In addition to de Pinto, La Condamine sent a copy of the *Relation abrégée* to his friend Daniel Bernouilli in Basel, who found it an "interesting, instructive, and agreeable work" and complimented La Condamine on his "successful return from such a great, difficult, and dangerous voyage." Bernouilli to La Condamine, 10 July 1746, private collection, Paris.

43. La Condamine, *Extracto del diario de observaciones hechas en el viage de la provincia de Quito al Parà, por el rio de las Amazonas* (Amsterdam: Joan Catuffe, 1745), "Aviso del lector." All page numbers refer to the original text, not the modern page numbers in the facsimile reproduction. In a letter dated 14 June 1746,

Partyet writes that "I waited for Monsieur's arrival here to find out how we should transport the copies that you want me to send to Peru of the Spanish account of your Amazon River journey." Partyet to La Condamine, private collection, Paris. On Partyet's career in Cadiz, see Didier Ozanam, "Un consul de France à Cadix: Pierre-Nicolas Partyet (1716–1729)" in *L'ouvrier, l'Espagne, la Bourgogne et la vie provinciale. Parcours d'un historien (Mélanges offert à Pierre Ponsot)* (Madrid: Casa de Velasquez, 1994). Catuffe himself sent several copies of this edition to members of the Sephardic literary society frequented by de Pinto. While the *Extracto* and the *Relation abrégée* are similar, they are not identical, and there has been no serious study comparing the two. It is interesting, for instance, to note that the title of the *Extracto* includes references to stops La Condamine made in both Suriname and Amsterdam, which did not appear on the title page of the *Relation abrégée*. The majority of Spanish and Portuguese explorers in the Amazon region cite this Amsterdam edition, which seems to have enjoyed a wider circulation among Iberian readers. Juan de Velasco, for example, had a copy of the *Extracto* rather than the *Relation abrégée*. See Juan de Velasco, *Historia del Reino de Quito*, 3 vols. (1788–89; Quito: Casa de la Cultura Ecuatoriana, 1977). For a facsimile edition of the somewhat rare *Extracto*, see Antonio Lafuente and Eduardo Estrella, introduction to *Viaje a la América Meridional por el río de las Amazonas*, ed. Antonio Lafuente and Eduardo Estrella (Barcelona: Editorial Alta Fulla, 1986).

44. "Inventario de los Libros del Archivo del Pueblo de la Laguna pertenecientes a la Misión" (1768), MS 17614, fol. 5, Biblioteca Nacional, Madrid.

45. Dom Pedro de Almeida Portugal, first Count of Alorna, was Portuguese governor in India during the 1740s, while his son, the fourth Count of Assumar, was in Paris participating in the scientific life of the French capital. In a letter dated 13 April 1749, the young Count of Assumar sent his father a copy of La Condamine's account: "I send Your Excellency the Journey to the Amazon River by Mr. de la Condamine, one of the academicians that went to Peru to measure a degree of the meridian in order to resolve the question of the Earth's shape." João de Almeida Portugal, Count of Assumar, to his father, Dom Pedro de Almeida Portugal, Count of Alorna, 13 April 1749, reproduced in Nuno Gonçalo Monteiro, ed., *Meu pai e meu senhor muito do meu coração* (Lisbon: Instituto de Ciências Sociais da Universidade de Lisboa/Quetzal, 2000), 119.

46. For an overview of the Treaty of Tordesillas and territorial politics in the period leading up to the 1750 Treaty of Madrid, see *El Tratado de Tordesillas y su época* ([Valladolid]: Junta de Castilla y León, 1995); Mário Clemente Ferreira, *O Tratado de Madrid e o Brasil Meridional* (Lisbon: CNCDP, 2001); "A Demarcação do Território Brasileiro: O Tratado de Madrid e o Mapa das Cortes" in Antônio Gilberto Costa, *Cartografia da Conquista do Território das Minas* (Lisbon: Kapa Editorial, 2004); Mário Clemente Ferreira, "Cartografia e Diplomacia: O Mapa das Cortes e o Tratado de Madrid," unpublished paper; Inácio Guerreiro, ed., *Cartografia e diplomacia no Brasil do século XVIII* (Lisbon: CNCDP, 1997); Jorge Costa, ed., *A Terra de Vera Cruz: Viagens, descrições e mapas do século XVIII* (Porto: BPMP, 2000), 9–65; Guillermo Kratz, *El tratado hispano-portugués de límites de 1750 y sus*

consecuencias (Rome: IHSI, 1954). The map of the Amazon produced by La Condamine and published alongside his *Relation abrégée* served as one of the primary documents for both the 1750 map and the Madrid treaty. On the issue of boundaries in Portuguese America more broadly, see the special issue of the journal *Oceanos* entitled "A formação territorial do Brasil," no. 40 (October–December 1999).

47. Antonio de Ulloa and Jorge Juan, *Dissertación histórica y geographica sobre el meridiano de demarcación entre los Dominios de España, y Portugal, y los parages por donde passa en la America Meridional, conforme à los tratados, y derechos de cada estado, y las mas seguras, y modernas observaciones* (Madrid: Antonio Marin, 1749), 70.

48. Ulloa and Juan, *Dissertación histórica*, 89.

49. See Neil Safier, "At the Confines of the Colony," in *The Imperial Map: Cartography and the Mastery of Empire*, ed. James R. Akerman (Chicago: University of Chicago Press, 2008).

50. Jozé Monteiro de Noronha, "Roteiro da viagem da cidade do Pará até às Ultimas Colonias dos Dominios Portuguezes em os Rios Amazonas e Negro. Illustrado com algumas Noticias, que pódem interessar a curiosidade dos navegantes, e dar mais claro conhecimento das suas capitanias do Pará, e de S. José do Rio Negro," *Collecção de Noticias para a Historia e Geografia das Nações Ultramarinas, que vivem nos Dominios Portuguezes, ou lhes são vizinhas* (Lisbon: Academia das Sciencias, 1856), 4.

51. Noronha's description fits into the tradition of Edenic narratives described by Sérgio Buarque de Holanda. See his classic *Visão do paraíso: Os motivos edênicos no descobrimento e colonização do Brasil*, 2nd rev. ed. (São Paulo: Companhia Editora Nacional, 1969). On the trope of the panegyric in Portuguese America, see Júnia Ferreira Furtado, "As índias do conhecimento, ou a geografia imaginária da conquista do ouro," *Anais de História do Além Mar* 4 (2003): 155–212. For the economic aspects of Amazonian colonization, see Luiz Felipe de Alencastro, "Economic Network of Portugal's Atlantic World," in *Portuguese Oceanic Expansion, 1400–1800*, ed. Diogo Ramada Curto and Francisco Bethencourt (Cambridge: Cambridge University Press, 2006). A biography of Monteiro de Noronha can be found in "Biographia dos brasileiros distinctos por letras, armas, virtudes, etc.: Jozé Monteiro de Noronha," *Revista do Instituto Histórico e Geográphico Brasileiro* 2, no. 6 (1840): 252–58.

52. Monteiro de Noronha, "Roteiro," 10.

53. On the work of Humboldt, Lamarck, and de Candolle, the origins of biogeography, and its correlation to early mapping, see Malte C. Ebach and Daniel F. Goujet, "The First Biogeographical Map," *Journal of Biogeography* 33 (2006): 761–69.

54. Monteiro de Noronha, "Roteiro," 75.

55. Monteiro de Noronha, "Roteiro," 42.

56. For a more extended account of Sampaio's reading practices, see Safier, "'Every day that I travel.'"

57. Francisco Xavier Ribeiro de Sampaio, *Diario da viagem, que em visita, e correição das povoações da Capitania de S. Jozé do Rio Negro, fez o Ouvidor, e Intendente Geral da mesma Francisco Xavier Ribeiro de Sampaio no anno de 1774, e 1775* (Lisbon: Academia das Sciencias, 1825), 83.

58. Sampaio, *Diario da viagem*, 37.

59. Sampaio, *Diario da viagem*, 39.

60. Sampaio, *Diario da viagem*, 38. He even insisted that a certain letter written by the Portuguese governor João de Abreu de Castelo Branco would have been part of La Condamine's preexpedition reading: "[The Portuguese governor's] response made a big splash in [Quito], and earned him the title of 'Man of Arms, and Man of Letters' (*Homem de Espada, e Pluma*) in the Royal Academy of said city." Sampaio, *Diario da viagem*, 44.

61. Sampaio, *Diario da viagem*, 26.

62. Sampaio, *Diario da viagem*, 29.

63. Sampaio, *Diario da viagem*, 29.

64. *Journal Encyclopédique* (June 1770): 278–86; Pierre-Jean Grosley, *Londres* (Lausanne, 1774), ix. A year later, three years before his death, La Condamine also took the time to counter questions raised by Simon-Henri-Nicolas Linguet's *Canaux navigables*, referencing the "large number of readers" that had been exposed to his own text. Linguet's accusations wilted in the face of La Condamine's public claim: not a single reader in the twenty-four years that had passed since the initial publication of his work had ever sought to question the evidentiary basis of his assertions. To the aged traveler-turned-philosophe, the silence of this public tribunal was clear and absolute vindication of the legitimacy of his observations.

65. "Lettre de M. de la Condamine, aux Auteurs de ce Journal, contenant des observations sur la rivière des Amazones," *Journal Encyclopédique* (February 1770): 446–56, quotation on 450. An extract of La Condamine's journey was published in volume 13 of Prévost's *Histoire générale des voyages* (Paris, 1746–59), to which La Condamine made reference in this letter.

66. On geopolitical knowledge hierarchies in the early modern Atlantic, see Bauer, *Cultural Geography of Colonial American Literatures*; Parrish, *American Curiosity*; and Delbourgo, *A Most Amazing Scene of Wonders*.

67. Sutcliffe, "Can a Jew Be a Philosophe?"

CHAPTER FOUR

1. For a discussion of eighteenth-century cartouches, see Pedley, *Commerce of Cartography*, 56–63. Christian Jacob emphasizes the "deictic" or pointing function of the cartographic cartouche, which brings attention to particular aspects of the map's geographic features through ornament: "These ornamental cartouches, which have a deictic function, are not limited to the edges of the map. They can also occasionally be found in the interior of continents." Jacob, *L'empire des cartes*, 156. The cartouche for d'Anville's *L'Amérique méridionale* (1748), executed by Gravelot and engraved by Delahaye, shows four generic figures—three Indians and a woman holding a cross. Gravelot also produced the cartouche for the *Carte*

des parties principales du globe terrestre . . . par l'abbé Luneau de Boisjermain, where a scene showed angels unveiling a world map and four female figures representing the continents. In 1741, Gravelot illustrated Bernard Picart's *Histoire générale des cérémonies religieuses de tous les peuples* (Paris: Rollin fils, 1741), an eight-volume work containing various illustrations of native peoples. See Edmond de Goncourt and Jules de Goncourt, *L'art du dix-huitième siècle* (Paris : A. Quantin, 1880).

2. Jean-Baptiste Bourguignon d'Anville to La Condamine, 14 April 1750, private collection, Paris. For a study of d'Anville in the context of the Vaugondy family, see Mary Pedley, *Bel et Utile: Work of the Robert de Vaugondy Family of Mapmakers* (Tring: Map Collectors Publications, 1992). On d'Anville's Jesuit training and the compilatory practices of *géographes de cabinet* in France more broadly, see Pedley, *Commerce of Cartography*, 26–31; Nelson-Martin Dawson, *L'atelier Delisle: L'Amérique du Nord sur la table de dessin* (Sillery: Diffusion, Dimedia, [2000]).

3. While few studies have addressed the social practices undergirding editorial correction in a cartographic context, those that have include David Woodward, *Maps as Prints in the Italian Renaissance: Makers, Distributors and Consumers* (London: British Library, 1996); Dennis Reinhartz, *The Cartographer and the Literati: Herman Moll and His Intellectual Circle* (Lewiston: E. Mellen Press, 1997); Mary Pedley, "New Light on an Old Atlas: Documents concerning the Publication of the *Atlas Universel* (1757)," *Imago Mundi* 36 (1984): 48–63, idem, *Bel et utile*, idem, *Commerce of Cartography*.

4. For more on the importance of new ideas, especially scientific ones, for Creole culture in Latin America, see Diana Soto Arango, Miguel-Ángel Puig-Samper, and Luis Carlos Arboleda, *La ilustración en América colonial* (Madrid: CSIC/Ediciones Doce Calles, 1995); Diana Soto Arango, Miguel-Ángel Puig-Samper, Maria Dolores González-Ripoll, eds., *Científicos criollos e ilustración* (Madrid: Doce Calles, 1999); David Brading, *The First America: The Spanish Monarch, Creole Patriots, and the Liberal State, 1492–1867* (Cambridge: Cambridge University Press, 1993); Cañizares-Esguerra, *How to Write the History of the New World*; and Ruth Hill, *Sceptres and Sciences in the Spains* (Liverpool: Liverpool University Press, 2000). These studies have emphasized three broad features of Creole culture comprising a particular category of social, cultural, and political identification. First, scientific progress and the utility of knowledge were part and parcel of a larger program of social and political renovation. Second, Spanish American Creoles represented themselves as mediators between the pre-Columbian indigenous past and their predominantly European colonial present. Third, and finally, Creoles maintained an almost mythic attachment to the exuberance of American nature. On the Creole use of Amerindian traditions in the eighteenth century, see Cañizares-Esguerra, *How to Write the History of the New World*, esp. chapters 4 and 5. On Creole, mestizo, and indigenous territorial conceptions, see Barbara Mundy, *The Mapping of New Spain: Indigenous Cartography and the Maps of the Relaciones Geográficas* (Chicago: University of Chicago Press, 1996); David Woodward and G. Malcolm Lewis, eds., *The History of Cartography: Cartography in the traditional African, American, Arctic, and Pacific societies*, vol. 2, no. 3; Neil L. Whitehead,

Histories and Historicities in Amazonia (Lincoln: University of Nebraska Press, 2003); Dana Leibsohn, "Colony and Cartography: Shifting Signs on Indigenous Maps of New Spain," in *Reframing the Renaissance: Visual Culture in Europe and Latin America, 1450–1650*, ed. Claire J. Farago (New Haven: Yale University Press, 1995); Walter Mignolo, *The Darker Side of the Renaissance: Literacy, Territoriality, and Colonization* (Ann Arbor: University of Michigan Press, 1995); Serge Gruzinski, "Colonial Indian Maps in Sixteenth-Century Mexico: An Essay in Mixed Cartography," *Res* 13 (1987): 46–61; and Alessandra Russo, *El realismo circular: Tierras, espacios y paisajes de la cartografía novohispana, siglos XVI y XVII* (Mexico: Universidad Autónoma de México and Instituto de Investigaciones Estéticas, 2005).

5. On the sociology of the printed text, see Donald F. McKenzie, *Bibliography and the Sociology of Texts* (1986; Cambridge: Cambridge University Press, 1999). The examination of corrections in the European printing house is a relatively under-researched subfield of the history of the book. On early modern correction generally, see Adrian Johns, *The Nature of the Book: Print and Knowledge in the Making* (Chicago: University of Chicago Press, 1999), 90–91; see also Percy Simpson, *Proof-reading in the Sixteenth, Seventeenth and Eighteenth Centuries* (1935; Oxford: Oxford University Press, 1970); McKenzie, "Printers of the Mind," 1–76; and Grafton, "Correctores corruptores?" On cartographic correction, see Woodward, *Maps as Prints*; Pedley, *Bel et Utile*, 54–61; and Reinhartz, *Cartographer and the Literati*. On the "instability" of texts more generally, see E. A. J. Honigmann, *The Stability of Shakespeare's Texts* (London: Edward Arnold, 1965); Stephen Orgel, "What Is a Text?" in *Staging the Renaissance: Essays on Elizabethan and Jacobean Drama*, ed. David Scott Kastan and Peter Stallybrass (New York: Routledge, 1991). On the history of print in colonial Latin America, see the works of José Toribio Medina, including his *Historia de la imprenta en los antiguos dominios españoles de América y Oceania* (Santiago: Fondo Histórico y Bibliográfico José Toribio Medina, 1958); Julie Green Johnson and Susan L. Newbury, *The Book in the Americas: The Role of Books and Printing in the Development of Culture and Society in Colonial Latin America: Catalogue of an Exhibition* (Providence: John Carter Brown Library, 1988); José Luis Martínez, *El libro en hispanoamérica: Origen y desarrollo* (Madrid: Fundación Germán Sánchez Ruipérez: Ediciones Pirámide, 1986); *La imprenta en Iberoamérica, 1539–1833* (Madrid: [Biblioteca Nacional], 1983); and José Torre Revello, *El libro, la imprenta y el periodismo en América durante la dominación española* (Buenos Aires: Casa Jacobo Peuser, 1940).

6. Maldonado, "Memorial a nombre de San Pedro de Riobamba," 1: 7–13. See chapter 3 for additional biographical information on Maldonado.

7. La Condamine, *Journal du voyage*, 142, note.

8. D'Anville to La Condamine, 14 April 1750.

9. The series of proofs that form the primary corpus of archival materials for this chapter are held by the Département des Cartes et Plans of the Bibliothèque Nationale, Paris. More than two dozen cartographic proofs exist of the *Carta de la Provincia de Quito*, each of which portrays a different state of one of four final sheets of the Quito map.

10. Proof of sheet 2 of the *Carta de la Provincia de Quito*, BNF (Bibliothèque Nationale de France, Paris), Ge.D. 10878 (2). Compared to the stylized hand of d'Anville, this cursive form in typically Spanish handwriting is less formal and more harried.

11. The lake in question was likely the Lago La Cocha, east of the Colombian city of Pasto and not far from the present-day border with Ecuador.

12. Ulloa, *Relación histórica*, book 6, chapter 10, p. 606. See Julio F. Guillén, "Don Antonio de Ulloa y el descubrimiento del platino," *Anales de la Asociación Española para el Progreso de las Ciencias* 2 (1939): 413–16.

13. At the same time Maldonado's map was being produced, the *Philosophical Transactions of the Royal Society* published a series of articles on platinum. William Watson, in a letter, explained that some pieces called "Inca stones" (*Piedras de Inga*) had been brought to London "with several other Curiosities from America, by that excellent person, and my much-lamented friend, Don Pedro Maldonado." "Several Papers concerning a New Semi-Metal, Called Platina; Communicated to the Royal Society by Mr. Wm. Watson F.R.S.," *Philosophical Transactions of the Royal Society* 46 (1749–50): quotation on 588.

14. Proof of sheet 2 of the *Carta de la Provincia de Quito*. On the theory of Llanganate being a site of great mineralogical wealth and Inca treasure, see Byron Uzcátegui Andrade, *Los llanganates y la tumba de Atahualpa* (Quito: Ediciones Abya-Yala/Instituto Panamericano de Geografía e Histoira, 1992); and Mark Honigsbaum, *Valverde's Gold: A True Tale of Greed, Obsession and Grit* (New York: Macmillan, 2004).

15. John Lynch, *Bourbon Spain, 1700–1808* (Oxford: Blackwell, 1989), 157–95; Joaquín F. Quintanilla, *Naturalistas para una corte ilustrada* (Madrid: Ediciones Doce Calles, 1999).

16. Fernández de Pinedo, Alberto Gil Novales, and Albert Dérozier, *Centralismo, ilustracion y agonía del Antiguo Régimen* (Barcelona: Labor, 1984), as cited in Quintanilla, *Naturalistas para una corte ilustrada*, 16.

17. On Tomás López, see Gabriel Marcel, *El geógrafo Tomás López y sus obras; ensayo de biografía y de cartografía* (Madrid: Impr. del Patronato Huérfanos de Administración Militar, 1908); on Cruz Cano y Olmedilla and his 1775 map of South America, see Thomas R. Smith, "Cruz Cano's Map of South America, Madrid, 1775: Its Creation, Adversities, and Rehabilitation," *Imago Mundi* 20 (1966): 49–78.

18. La Condamine to unknown, 25 October 1759, BL, Add. MSS 20793: 303r–v.

19. In a letter to La Condamine, d'Anville exhorted La Condamine to keep the map's publication hidden from the other members of the Academy of Sciences: "I salute in the most humble way Monsieur de La Condamine, and before all else take advantage of the moment to pray of him not to say a single word to Messieurs Delisle and Buache, whom he introduced to us yesterday in the entryway of the Academy to see the map, etc." D'Anville to La Condamine, 14 April 1750. Both Joseph-Nicolas Delisle and Philippe Buache were members of the

Academy and involved with Delahaye; plagiarism of the Quito map would have been a serious consideration. In a public assembly the previous week, Delisle had presented his "Nouvelles Découvertes au Nord de la Mer du Sud"; it may have been Buache's map based on Delisle's *mémoires*, the *Carte des nouvelles découvertes au nord de la mer du sud*, that d'Anville had been summoned to see on Monday, April 13, 1750, a day otherwise devoid of formal Academic proceedings.

20. Burial records, Saint James Church, Westminster Archives, London. Many, including La Condamine himself, have confused Maldonado's date of death, but the burial records show clearly that he was buried in St. James's Church in Piccadilly (Westminster), along with Sarah Slaughter and a child, Phillis Hunt, on November 7, 1748.

21. For an analysis of d'Anville's role in the compilation of the *Carte de l'Amérique méridionale* (1748), see Júnia Ferreira Furtado and Neil Safier, "O sertão das Minas como espaço vivido: Luís da Cunha e Jean-Baptiste Bourguignon d'Anville na construção da cartografia européia sobre o Brasil," in *Brasil-Portugal: Sociedades, culturas e formas de governar no mundo português (séculos XVI–XVIII)*, ed. Eduardo França Paiva (São Paulo: Annablume, 2006).

22. D'Anville to La Condamine, 14 April 1750.

23. These included narratives of Antonio de Ulloa, Pierre Bouguer, Jean Magnin, La Condamine, Maldonado, and a colonial administrator named Miguel de Santisteban, whose work is discussed below.

24. On the administrative life of Miguel de Santisteban and his travels from Lima to Caracas, see David J. Robinson, *Mil leguas por América: De Lima a Caracas, 1740–41* (Bogota: Banco de la República, 1992). Santisteban wrote a manuscript travel account that he later gave to La Condamine, and La Condamine deposited this manuscript at the Bibliothèque du Roi along with his other papers. Miguel de Santisteban, "Journal du voyage de Don Miguel de Santi-Estevan ancien corregidor de Canciecas de Vilcabamba dans le haut Perou de Lima a Caracas en 1740 et 1741," MS Esp. 160 (microfilm no. 13033), Département des Manuscrits, Bibliothèque Nationale de France, Paris. In his *Journal du voyage*, La Condamine described having received this manuscript from Santisteban, who became a corresponding member of the Academy of Sciences thanks to La Condamine's efforts. See La Condamine, *Journal du voyage*, 142.

25. D'Anville to La Condamine, 14 April 1750.

26. McClellan, *Specialist Control*, 51–53. In addition to the unpublished materials of the *Comité de librairie* studied by McClellan, La Condamine and Bouguer published a series of texts defending their respective positions in the context of this infamous feud. These include Pierre Bouguer, *Lettre à Monsieur *** dans laquelle on discute divers points d'astronomie pratique, et où l'on fait quelques remarques sur le supplément au Journal historique du voyage à l'équateur de M. de la C.* (Paris: Guerin & Delatour, 1754), idem, *Justification des Mémoires de l'Académie Royale des Sciences de 1744* (Paris: Jombert, 1752); and La Condamine, *Supplément au journal historique du voyage à l'Equateur, et au livre de la mesure des trois premiers degrés du meridian, servant de réponse à quelques objections* (Paris: Durand/Pissot, 1752–54). Special thanks

to Jim McClellan for allowing me to see his unpublished paper on this dispute at an early stage of my research.

27. For a discussion of some of the strategies Guillaume Delahaye used in his correcting practice, and a particularly illuminating citation from his "Rapport sur les travaux du Citoyen Delahaye, graveur en géographie et en topographie," see Pedley, *Commerce of Cartography*, 49–51.

28. Proof of sheet 1 of the *Carta de la Provincia de Quito*, BNF, Ge.D. 10878 (1)A. It is interesting to note the geographic proximity of Delahaye's workshop near the church of St. Germain l'Auxerrois to d'Anville's atelier in the Galeries du Louvre. See the map by Didier Robert de Vaugondy, *Tablettes Parisiennes*, reproduced in Pedley, *Commerce of Cartography*, 98.

29. Pedley, *Commerce of Cartography*, 49. There may have also been an economic element—corrections cost time and money. For a case study of the economics of cartographic correction and the legal scuffles that could ensue, see Pedley, "New Light."

30. Proof of sheet 2 of the *Carta de la Provincia de Quito*. The original French may help clarify some of these shorthand commentaries: "Le trait de Cotopaxi C est trop dur il faut ladoucir et blanchir le haut ainsi tout blanc et le bas en crevasses dentelles . . . X adoucir ce milieu marque X faire un trait a gauche pour marquer la cote de la grosse montagne au dessous de lecriture quelandana faire quelques tailles pour lier a la grosse montagne les petites plus basses a droite . . . quelques traites a propos p. donner de la force a la chaine de montagnes de cet endroit et faire une suite."

31. As cited in Pedley, *Commerce of Cartography*, 50.

32. J. B. Harley, "Silences and Secrecy: The Hidden Agenda of Cartography in Early Modern Europe" in Harley, *The New Nature of Maps* (Baltimore: Johns Hopkins University Press, 2001), 86.

33. According to Salomon, Joseph de Resavala was governor of the region of the "colorados" of Santo Domingo, and had attempted earlier to open a route in the region but had been unsuccessful due to the muddy conditions, which did not allow mules to pass with cargo. Salomon, *Los Yumbos*, 100.

34. In one of the proofs, the original text is shown as "Por aqui vive la Nacion de los Malaguas que se *revelo* antiguamente" (emphasis added), which shows how changing a single letter can transform the meaning from *revealing itself* to *rebelling*.

35. Santisteban, "Journal du voyage," fol. 21.

36. The Yumbos were the indigenous group that occupied "the western mountain that drained to the Rio Guayllabamba, from the 'mouths of the mountain' of Nono and Calacalí to the vanished village of Bola Niguas." Frank Salomon, "Yumbo Ñan: La vialidad indígena en el noroccidente de Pichincha y el trasfondo aborigen del camino de Pedro Vicente Maldonado," *Cultura: Revista del Banco Central del Ecuador* 8, no. 24B (January–April 1986): 612. In the "auto" discussing the road from Quito to Esmeraldas, the Yumbos comment on Maldonado's treatment: "Don Pedro Maldonado, resident of this city, treats us harshly (*nos hase*

continuas violencias) by forcing us to work in the project of discovering a road without paying us a salary." Cited in Salomon, "Yumbo Ñan," 619. On this topic more generally, see Salomon, *Los Yumbos*.

37. On the idea of textual "instability" versus textual fixity, see Peter Stallybrass and Margreta DeGrazia, "The Materiality of the Shakespearean Text," *Shakespeare Quarterly* 44, no. 3 (Autumn 1993): 255–83.

38. La Condamine to Carvajal y Lancaster, 27 April 1750, Estado Legajo 2320, Archivo Histórico Nacional, Madrid.

39. Ibid.

40. This version of the map can be found at the Biblioteca Nacional de Madrid and the Museo Naval (Madrid), while the second version is held by both the Bibliothèque Nationale in Paris and the Royal Society in London.

41. La Condamine to unknown, 25 October 1759, BL, Add. MSS 20793: 302v.

42. D'Anville may have collaborated in this strategy. Writing to La Condamine, d'Anville expressed that he was complicit in ensuring that the circumstances were propitious for the proper elaboration of the Quito map: "I understand what M. de La Condamine asked of me with regard to the person of the ambassador and of M. de Luzán . . . I have made it my personal obligation to see that the plates are sent in twelve days." D'Anville to La Condamine, 14 April 1750.

43. Robert de Vaugondy, "Géographie," in *Encyclopédie, ou Dictionnaire raisonné des sciences, des arts et des métiers*, ed. Denis Diderot and Jean Le Rond d'Alembert (Paris: Briasson, 1757), 7: 611.

44. On Ulloa and Juan's role in the mapping of Spain and the rise and fall of the "Casa de la Geografía," see Quintanilla, *Naturalistas*, 21–36; and Horacio Capel, "Geografía y Cartografía," in *Carlos III y la ciencia de la Ilustración*, ed. Manuel A. Sellés, José Luis Peset, and Antonio Lafuente (Madrid: Alianza Editorial, 1988), 101–3.

45. Martin Folkes, "A Map of the province of Quito in South America transmitted to the President by M.r le Chevalier de la Condamine at Paris," 14 June 1750, Minutes of the Proceedings of the Royal Society, 142 (1750), fol. 2, Archives of the Royal Society, London.

46. This position supports J. B. Harley's work related to the impact of blank space and geographic silences on the imperial imaginary. Cf. Harley, "Silences and Secrecy," 105.

47. Folkes, "Map of the province of Quito," 2–3.

48. Zúñiga, *La expedición científica*, 106.

49. Alexander von Humboldt also heaped the highest praise on Maldonado's map, calling it "the most precise [map] we know of Europe's American possessions." Cited in Dario Lara, "L'amitié de deux hommes de sciences: Charles-Marie de la Condamine et Pedro Vicente Maldonado, et l'origine de l'amitié de deux peuples," in *Colloque International "La Condamine": Paris, 22/23 Novembre 1985* (Mexico: IPGH, 1987). In the nineteenth century, Manuel Villavicencio called it "the best map of the country, . . . drawn up by our learned compatriot Don Pedro Maldonado." Villavicencio, *Geografía de la República del Ecuador* (New York:

Impr. De R. Craighead, 1858). The most complete list of the reactions to the *Carta de la Provincia de Quito* is provided by Ortiz Arellano, *Pedro Vicente Maldonado*, 110–17.

50. Cited in Costales Cevallos, *El Sabio Maldonado ante la Posteridad*, 66. A native of Quito and editor of the periodical *Primicias de la cultura de Quito*, Santa Cruz y Espejo was sent into exile in Bogotá for speaking out against the colonial government, and upon his return to Quito, was jailed by Spanish authorities and died in prison.

51. Prévost, *Histoire générale des voyages*, 13: 635.

52. La Condamine, *Journal du voyage*, 110.

53. Review of *Journal du voyage*, *Journal des Sçavans* (1753): 50–51.

54. La Condamine, *Journal du voyage*, 141–42, footnote *. For a contemporary example of even more extended cartographic memoirs of this nature, see d'Anville's memoir published in the *Journal des savants*: "Lettre de Monsieur d'Anville, à Messieurs du Journal des Sçavans, sur une carte de l'Amérique méridionale qu'il vient de publier," *Journal des Sçavans* (March 1750): 522–63; "Seconde lettre de M. d'Anville à Messieurs du Journal des Sçavans, sur la carte qu'il a publiée de l'Amérique méridionale," *Journal des Sçavans* (April 1750): 625–73.

55. La Condamine to unknown, 25 October 1759, BL, Add. MSS 20793: 302v.

56. La Condamine to unknown, received 2 February 1750, private collection, Paris.

57. Entry for 2 September 1750, procès-verbaux de l'Académie Royale des Sciences, vol. 69.

58. On the practices of print culture and correction, see McKenzie, *Bibliography and the Sociology of Texts*; Chartier, *Inscrire et effacer*; Blair, *Information Overload in the Renaissance* (forthcoming); Johns, *Nature of the Book*; Peter Stallybrass, Roger Chartier, J. Franklin Mowry, and Heather Wolfe, "Hamlet's Tables and the Technologies of Writing in Renaissance England," *Shakespeare Quarterly* 55, no. 4 (2004): 379–419; Stallybrass and DeGrazia, "Materiality of the Shakespearean Text"; Grafton, "Correctores Corruptores?"; and Juliet Fleming, *Graffiti and the Writing Arts of Early Modern England* (Philadelphia: University of Pennsylvania Press, 2001).

59. Adorno, "Discursive Encounter of Spain and America," 210–28; Greenblatt, *Marvelous Possessions*; Pagden, *European Encounters with the New World*; Patricia Seed, *Ceremonies of Possession in Europe's Conquest of the New World, 1492–1640* (Cambridge: Cambridge University Press, 1995); Pratt, *Imperial Eyes*; and Stuart Schwartz, ed., *Implicit Understandings: Observing, Reporting, and Reflecting on the Encounters between Europeans and Other Peoples in the Early Modern Era* (Cambridge: Cambridge University Press, 1994).

60. Michel de Certeau, *L'invention du quotidien. I. Arts de faire,* introduction by Luce Giard (1980; reprint, Paris: Gallimard, 1990), 170.

61. "D. Joaquin de Lamo y Zúñiga," *Diccionario histórico-biográfico del Peru*, ed. Manuel de Mendiburu (Lima: Imprenta Gil, 1933), 6: 404–5.

62. Letter from Joaquin de Lamo y Zúñiga to Joseph de Jussieu, 13 April 1751, MS 179: 46, Bibliothèque du Musée National d'Histoire Naturelle (Paris).

63. Pedro Vicente Maldonado to La Condamine, Madrid, 28 March 1746, as cited in La Condamine, "Pièces justificatives," *Journal du voyage*, xx.

CHAPTER FIVE

1. La Condamine to unknown, 25 October 1759, BL, Add. MSS 20793: 301.

2. The Real Gabinete de Historia Natural had its roots in a 1752 proposal by Antonio de Ulloa to create both a geographical academy and a cabinet of natural history under Spanish auspices. On Dávila's collection and the history of the formation of the Real Gabinete de Historia Natural, see Pimentel, *Testigos del mundo*, 149–78; María de los Angeles Calatayud Arinero, *Pedro Franco Dávila y el Real Gabinete de Historia Natural* (Madrid: CSIC, 1988); idem, "Pedro Franco Dávila: Aspectos de una vida," in *Científicos criollos e Ilustración*, ed. Diana Soto Arango, Miguel Ángel Puig-Samper, and María Dolores González-Ripoll (Madrid: Doce Calles, 1999); Quintanilla, *Naturalistas*, 113–95.

3. The classic account of this "dispute" is Gerbi, *La disputa del Nuovo Mondo*; subsequent analyses of the debate can be found in Brading, *First America*; and Cañizares-Esguerra, *How to Write the History of the New World*.

4. The anonymous critique, "Juicio Imparcial sobre los Indios de Quito," and the counter-critique, "Defensa à favor de los Indios de Quito, contra la crítica que se publicó en Paris," are held by the Clements Library, University of Michigan, Peru MS 125. I would like to give special thanks to Barbara DeWolfe for suggesting I peruse the Clements's small but rich collection of Latin American manuscripts, where I first came across these two texts. On the inside front cover, thought to be in the hand of the American collector Obadiah Rich, is written "Crítica de un anonymo contra el Cap. 6.° Lib. 6.° de la Relacion de Juan y Ulloa, en que trata de los Indios de Quito: Y defensa de estos." This edition of the "Juicio Imparcial" contains 160 pages numbered in pencil, while the "Defensa" contains 184 pages. These manuscripts were originally purchased by Thomas Phillips from the collection of Edward King, Lord Kingsborough, sometime following the latter's death in 1837. Duplicate versions of these sources are also held by the Special Collections of the Harlan Hatcher Graduate Library at the University of Michigan, under the title of "Confutation of Ulloa" (F2221.U43 C75); neither copy is dated and it is unclear if one version preceded the other. I have been unable to locate additional copies of these manuscripts, in Spain or elsewhere.

5. "Defensa à favor de los Indios de Quito, contra la crítica que se publicó en Paris," Clements Library, fol. 7v. The fact that the title of the counter-critique states that the original manuscript may have been "published" in Paris is a clue to the possible provenance of the original critique's author. This clue, however, is far from certain since authors often used anonymity and a fictitious site of publication to foil potential enemies searching for an author's true identity.

6. On scribal publication and manuscript circulation in early modern Spain, see Fernando Bouza, *Corre manuscrito: Una historia cultural del Siglo de Oro* (Madrid: Marcial Pons, 2002); Antonio Castillo, *Entre la pluma y la pared: Una historia social de la escritura en los siglos de oro* (Madrid: AKAL, 2006).

7. The terms *anthropology* and *ethnography* are used advisedly as links to the "prehistory" of these disciplines in the eighteenth century. These are not actors' categories but rather protodisciplinary practices that coalesced to form the fields that we recognize today.

8. The relationship between early modern travel narratives and the origins of anthropology as a coherent discipline is a long-acknowledged, albeit still somewhat polemical, subject. On the use of anthropological observation and ethnographic description in a number of early modern contexts, see Schwartz, *Implicit Understandings*; Margaret T. Hodgen, *Early Anthropology in the Sixteenth and Seventeenth Centuries* (Philadelphia: University of Pennsylvania Press, 1998); Han F. Vermeulen and Arturo Alvarez Roldán, eds., *Fieldwork and Footnotes: Studies in the History of European Anthropology* (New York: Routledge, 1995); Michael Harbsmeier, "Towards a Prehistory of Ethnography: Early Modern German Travel Writing as Traditions of Knowledge," in Vermeulen and Alvarez Roldán, eds., *Fieldwork and Footnotes*.

9. The abundant documentary collection that exists in the Archivo General de Simancas (AGS) has allowed Jose P. Merino Navarro and Miguel M. Rodriguez San Vicente to trace the production history of the *Relación histórica* in its transformation from a series of observations made in situ to an elegant series of five volumes, although the rich archival record begs for further analysis and exploration. See their 1978 introduction to the facsimile edition of the *Relación histórica del viaje a la América meridional* (Madrid: Fundación Universitaria Española, 1978).

10. On the early eighteenth-century intellectual milieu in Spain, see Francisco Sánchez-Blanco, *La mentalidad ilustrada* (Madrid: Taurus, 1999); Jesús Perez Magallón, *Construyendo la modernidad: La cultura española en el tiempo de los novatores, 1675–1725* (Madrid: CSIC/Anejos de Revista de Literatura, 2002). On Ulloa in South America, see Brading, *First America*, esp. 422–46. The rhetorical purpose of the *Relación histórica* is in marked contrast to Ulloa's critical assessment of Spanish administration in the *Noticias secretas de América*. However, since the *Noticias secretas* was not discovered until 1826, Ulloa and Juan's secret account lies necessarily outside the purview of this book, as does Ulloa's *Noticias americanas* (Madrid: Manuel de Mena, 1772).

11. As cited in Luis J. Ramos Gomez, *Epoca, genesis y texto*, 1: 22–25.

12. Marqués de la Regalía to Marqués de la Ensenada, as cited in Merino Navarro and Rodriguez San Vicente, introduction, xxxii.

13. As cited in Merino Navarro and Rodriguez San Vicente, introduction, xxxii.

14. As cited in Merino Navarro and Rodriguez San Vicente, introduction, xxxiii.

15. Andrés Marcos Burriel, "Apuntamientos de algunas ideas para fomentar las letras," as cited in Bernabé Bartolomé Martínez, "Andrés Marcos Burriel: Un pionero de reformas en investigación y enseñanza," *Revista Complutense de Educación* 2, no. 3 (1991): 481.

16. Ibid., 482.

17. On the life and early education of Burriel, see Alfonso Echánove Tuero, *La preparación intelectual del P. Andrés Marcos Burriel, S.J.* (Madrid and Barcelona: Instituto Enrique Flórez/CSIC, 1971); Martínez, "Andrés Marcos Burriel"; José Simón Díaz, "Un erudito español: El P. Andrés Marcos Burriel," *Revista bibliográfica y documental* 3 (1949): 5–52.

18. Andrés Marcos Burriel to the Marqués de la Ensenada, 29 June 1747, as cited in Merino Navarro and Rodriguez San Vicente, introduction, xxxviii.

19. As cited in Merino Navarro and Rodriguez San Vicente, introduction, passim.

20. As cited in Merino Navarro and Rodriguez San Vicente, introduction, xxxix.

21. Juan Sempere y Guarinos, *Ensayo de una biblioteca española de los mejores escritores del reynado de Carlos III* (Madrid: Imprenta real, 1785–89), s.v. "Ulloa," 166.

22. "Distribuzion de ejemplares de los libros de observaziones y historia del viage a los Reinos del Perù en 5. tomos," AGS, Marina, MS 712: 174 (4 fols.).

23. *Mémoires pour l'Histoire des Sciences & des Beaux Arts* (March 1749): 453.

24. Don Josef Borrull to Gregorio Mayáns y Siscar, 28 September 1748, letter no. 726, Biblioteca-Archivo Hispano Mayansiana (Colegio de Corpus Christi, Valencia), 39.

25. *Mémoires pour l'Histoire des Sciences & des Beaux Arts* (March 1749): 453–54.

26. Louis Feuillée, preface, *Journal des observations physiques, mathematiques et botaniques, faites par ordre du Roi sur les côtes orientales de l'Amérique méridionale, & aux Indes Occidentales* (Paris: P. Giffart, 1714–25).

27. "Juicio Imparcial," sect. 5.

28. "Juicio Imparcial," sect. 2.

29. "Juicio Imparcial," sect. 69.

30. "Juicio Imparcial," sect. 2.

31. On Creole patriotism and "patriotic epistemologies," see Brading, *First America*; Cañizares-Esguerra, *How to Write the History of the New World*, chap. 4.

32. "Juicio Imparcial," sect. 1.

33. "Juicio Imparcial," sect. 6, emphasis added.

34. "Juicio Imparcial," sect. 8.

35. "Juicio Imparcial," sects. 64, 67.

36. "Juicio Imparcial," sect. 108.

37. "Juicio Imparcial," sect. 97.

38. "Juicio Imparcial," sect. 84. Contradicting Ulloa, the critic wrote that "it is absolutely false that they have constructed an entirely new convent. The convent

is old [and] beautiful, with an upper and lower cloister in which all they had to do was transform the infirmaries and hospital wards into living spaces and rooms for the Nuns."

39. "Juicio Imparcial," sect. 43.

40. The roiling political context at the time of Ulloa and Juan's appearance in Quito may have been another formative factor in motivating the anonymous author to compose his critique. Don Dionisio de Alsedo y Herrera, who served as president of the Audiencia of Quito from 1728 to 1736, was openly hostile to his successor, José de Araujo y Río, not only because of his status as a Creole but also because he was rumored to have engaged in illicit trade. Ulloa and Juan were also predisposed against Araujo y Río, as they were to most Creole administrators, and they became embroiled in a series of juridical conflicts over how much deference they were obliged to show him. In one particularly dramatic instance, Ulloa denounced Araujo y Río's treatment of the two naval officers by entering his house and offending the president of the Audiencia in his private chamber. Araujo y Río in turn attempted to incarcerate Ulloa over this imbroglio, at which point Ulloa took refuge in the Jesuit College in Quito along with Jorge Juan, who had likewise refused to submit to the authority of the Audiencia. By the end of their stay in Quito, Ulloa and Juan had accumulated a significant number of potential enemies, all of whom might have had ample reason to produce a critique of Ulloa's description of the region's Amerindian natives. On this conflict, see Ramos Gómez, *Epoca, genesis y texto*, 70–81; Herzog, *Upholding Justice*, 221–225; and Carlos Freile Granizo, "La misión científica franco española y la iglesia," in *Historia de la iglesia católica en el Ecuador*, ed. Jorge Salvador Lara (Quito: Abya-Yala, 2001), 3: 1680–87. Another intriguing, though admittedly unlikely, scenario is that the critique was penned by a local cacique, such as Francisco de Zamora. See Bernard Lavallé, *Al filo de la navaja: Luchas y derivas caciquiles en Latacunga, 1730–1790* (Quito: IFEA, 2002). I thank Sergio Miguel Huarcaya for bringing Zamora's case to my attention.

41. On the "Ratio studiorum," see John O'Malley, S.J., *The Ratio Studiorum: The Official Plan for Jesuit Education* (St. Louis: Institute for Jesuit Sources, 2005). On Burriel's formation as a Jesuit, see Echánove Tuero, *La preparación intelectual*.

42. On early eighteenth-century intellectual influences in Spain, see Sánchez-Blanco, *La mentalidad ilustrada*; and Perez Magallón, *Construyendo la modernidad*.

43. Benito Jerónimo Feijóo, "Mapa intellectual, y cotejo de naciones," in Feijóo, *Teatro crítico universal, discursos varios en todo género de materias . . .* (Madrid: D. Joachin Ibarra, 1779), 2: 314. The *Teatro crítico universal* was originally published in eight volumes from 1726 to 1740 and translated and reissued on several occasions over the course of the eighteenth century. According to Gregorio Marañón, the esteemed Spanish historian, "no other work of this quality, not meant to entertain but to instruct, ever reached the degree of popularity and renown as did the *Teatro crítico*." As cited in Gregorio Marañón Posadillo, "Consideraciones sobre Feijóo," *La Nueva España* (Oviedo), March 31, 1954.

44. Feijóo, "Mapa intellectual," 2: 301.

45. Feijóo, "Mapa intellectual," 2: 316.

46. On the idea of spatial taxonomy in the eighteenth century, see P. J. Marshall and Glyndwr Williams, *The Great Map of Mankind: Perceptions of New Worlds in the Age of Enlightenment* (Cambridge: Harvard University Press, 1982); and Meek, *Social Science and the Ignoble Savage.*

47. Feijóo, "Mapa intellectual," 2: 311.

48. Feijóo, "Mapa intellectual," 2: 320.

49. Ulloa, *Relación histórica*, sect. 931.

50. "Juicio Imparcial," sect. 6.

51. "Juicio Imparcial," sect. 4.

52. "Juicio Imparcial," sect. 5.

53. "Juicio Imparcial," sect. 5.

54. "Juicio Imparcial," sects. 4–5.

55. "Juicio Imparcial," sect. 11.

56. "Juicio Imparcial," sect. 7.

57. "Juicio Imparcial," sect. 8.

58. René-Antoine Ferchault de Réaumur, "Regles pour construire des thermometres dont les degrés soient comparables," *Mémoires de l'Académie Royale des Sciences* (1730 [1732]): 452–507. For more on the early history of the thermometer, see W. E. Knowles Middleton, *A History of the Thermometer and Its Use in Meteorology* (Baltimore: Johns Hopkins University Press, 1966). On the use of thermometers in the study of late-eighteenth-century botany, see Marie-Noëlle Bourguet, "Measurable Difference: Botany, Climate, and the Gardener's Thermometer in Eighteenth-Century France," in *Colonial Botany: Science, Commerce, and Politics in the Early Modern World*, ed. Londa Schiebinger and Claudia Swan (Philadelphia: University of Pennsylvania Press, 2005).

59. Juan and Ulloa, *Relación histórica*, bk. 5, chap. 1: 283–86, passim.

60. Ulloa, *Relación histórica*, bk. 5, chap. 6: 380.

61. Ulloa, *Relación histórica*, bk. 5, chap. 6: 381.

62. "Juicio Imparcial," sect. 33.

63. "Juicio Imparcial," sect. 35.

64. The rhetorical importance in eighteenth-century European scientific culture of relying on eyewitness observation corroborated by instrumental devices has been studied in Bourguet, Licoppe, and Sibum, *Instruments, Travel and Science.* On the history of thermometers in particular, see Bourguet, "Measurable Difference." On climatology and instruments, see Jan Golinski, *British Weather and the Climate of Enlightenment* (Chicago: University of Chicago Press, 2007).

65. Ilona Katzew, *Casta Painting: Images of Race in Eighteenth-Century Mexico* (New Haven: Yale University Press, 2005); and Susan Kellogg, "Depicting Mestizaje: Gendered Images of Ethnorace in Colonial Mexican Texts," *Journal of Women's History* 12, no. 3 (2000): 69–92.

66. Ulloa, *Relación histórica*, bk. 5, chap. 5: 364.

67. "Juicio Imparcial," sect. 85.

68. "Juicio Imparcial," sect. 113.

69. Two and a half decades later, in the chapter of his *Noticias americanas* entitled "De los indios de las dos Américas, y de sus costumbres, y usos," Ulloa begins by emphasizing the "extraordinary variety" and "diversity" of the Amerindians from various regions. This apparent nod toward a more nuanced approach to cultural description is the only evidence in Ulloa's writing that would imply a potential impact of the anonymous critique in print. See Ulloa, *Noticias americanas*, 305.

70. Juan and Ulloa, *A Voyage to South America*, iv.

71. Ibid.

72. On the early history of the Gabinete de Historia Natural, see María de los Angeles Calatayud Arinero, "Antecedentes y creación del Real Gabinete de Historia Natural de Madrid," *Arbor* 482 (February 1986): 9–33.

73. "Defensa à favor de los Indios de Quito," fol. 7v.

74. On manuscript culture in early modern Spain, see Bouza, *Corre manuscrito*; Gomez, *Entre la pluma y la pared*.

75. Benito Jerónimo Feijóo, *Glorias de España*, in *Biblioteca de autores españoles* (Madrid: Ediciones Atlas, 1952), 56: 209–10, as cited in Antonio Mestre, "La imagen de España en el siglo XVIII: apologistas, críticos y detractores," *Arbor* 449 (May 1983): 49–73.

CHAPTER SIX

1. Antonio de Ulloa, "Resumen histórico del origen, y succession de los Incas," in Jorge Juan and Antonio de Ulloa, *Relación histórica del viage a la América meridional* (Madrid: Antonio Marin, 1748), vi; Juan and Ulloa, *Relación histórica*, 541.

2. Inca Garcilaso de la Vega, *Comentarios reales de los Incas. Primera parte* (Lisbon: Crasbeeck, 1609), "Proemio al Lector."

3. Garcilaso was born Gómez Suárez de Figueroa to the Spanish military captain Sebastián Garcilaso de la Vega and the Inca princess Ñusta Isabel Chimpu Ocllo. For a biography of Garcilaso, see Aurelio Miró Quesada, *El Inca Garcilaso* (Lima: Pontificia Universidad Católica del Peru, 1994); and, in English, John Grier Varner, *El Inca: The Life and Times of Garcilaso de la Vega* (Austin: University of Texas Press, 1968).

4. Earlier translations of the *Comentarios reales* include *Le commentaire royal, ou L'histoire des Yncas, roys du Peru*, trans. Jean Baudoin (Paris: Augustin Courbé, 1633); *The Royal Commentaries of Peru, in Two Parts*, trans. Paul Rycaut (London: Miles Flesher, 1688); *Histoire des Yncas, rois du Perou*, trans. Jean Baudoin (Amsterdam: Gerard Kuyper, 1704); Histoire des Yncas, rois du Perou, trans. Jean Baudoin (Amsterdam: J. Desbordes, 1715); and *Histoire des Yncas, rois du Perou, depuis le premier Ynca Manco Capac, fils du soleil, jusqu'à Atahualpa dernier Ynca* (Amsterdam: J. F. Bernard, 1737). For a discussion of the structural and linguistic variations across these texts, and the representation of the Inca more broadly in the eighteenth century, see Fernanda Macchi, "Imágenes de los Incas en el siglo XVIII" (Ph.D. diss., Yale University, 2003).

5. Previous attempts to assess the representation of the Inca in eighteenth-century Europe have tended to paint in broad brushstrokes and across genres, rather than focusing on transformations through a single text. See Poole, "The Inca Operatic," in *Vision, Race, and Modernity: A Visual Economy of the Andean Image World* (Princeton: Princeton University Press, 1997); Cañizares-Esguerra, *How to Write the History of the New World*; and Macchi, "Imágenes." The period in which the *Histoire des Incas* (1744) was composed is one of the most poorly studied periods in the history of the Parisian Jardin du Roi, largely neglected by more classic accounts of French scientific culture in the early modern period. On the history of the Jardin du Roi, see Spary, *Utopia's Garden*; and Charles Coulston Gillispie, *Science and Polity in France at the End of the Old Regime* (Princeton: Princeton University Press, 1980).

6. On the collecting practices of early natural historians, see Findlen, *Possessing Nature*; and Krzysztof Pomian, *Collectionneurs, amateurs et curieux: Paris, Venise, XVIe–XVIIIe siècle* (Paris: Gallimard, 1987).

7. Voltaire to Jean-Baptiste Nicolas Formont, 17 April 1735, in *Correspondance*, ed. Theodore Besterman (Paris: Gallimard, 1977), 1: 587.

8. Garcilaso, *Comentarios reales*, "Proemio."

9. Garcilaso, *Comentarios reales*, "Proemio." On the treatises of Father José de Acosta, Pedro de Cieza de León, Francisco López de Gómara, and others, see Luís Millones Figueroa's *Pedro de Cieza de León y su Crónica de Indias: La entrada de los Incas en la historia universal* (Lima: Fondo Editorial de la Pontificia Universidade, 2001). On the many interpretations of Garcilaso's intended meaning, see David Brading, "The Incas and the Renaissance: The Royal Commentaries of Inca Garcilaso de la Vega," *Journal of Latin American Studies* 18 (1986): 1–23; Margarita Zamora, *Language, Authority, and Indigenous History in the* Comentarios reales de los Incas (Cambridge: Cambridge University Press, 1988); José Antonio Mazzotti, *Coros mestizos del Inca Garcilaso: Resonancias andinas* (Mexico: Fondo de Cultura Economica, 1996); Macchi, "Imágenes"; Christian Fernández, *Inca Garcilaso: Imaginación, memória e identidad* (Lima: Fondo Editorial/ UNMSM, 2004); and Marie Elise Escalante Adaniya, *Un estudio sobre la nominación en las crónicas de Garcilaso de la Vega y Guamán Poma* (Lima: Fondo Editoral/UNMSM, 2004).

10. On the slight variations between Garcilaso's original edition and this new French translation, see Macchi, "Imágenes," 103–11.

11. Garcilaso de la Vega, preface, *Histoire des Yncas, rois du Perou* (1704).

12. Ibid.

13. On Barcia's project, see Jonathan Earl Carlyon, *Andrés González de Barcia and the Creation of the Colonial Spanish American Library* (Toronto: University of Toronto Press, 2005).

14. Voltaire, to Algarotti, titled "Sur M. de La Condamine, qui était occupé de la mesure d'un degré du méridien au Pérou, lorsque M. de Voltaire faisait Alzire," in *The Complete Works of Voltaire*, ed. W. H. Barber and Ulla Kölvig (Oxford: Voltaire Foundation), 14: 50.

15. On the dramatic representations of the Inca, see Poole, *Vision, Race, and Modernity*, chap. 2. On the significant (and sometimes subtle) differences in the various editions of Garcilaso's translations circulating in Europe, see Macchi, "Imágenes," 86–164.

16. Judging from the date of her first reading, Graffigny appears to have used both Baudoin's 1737 Amsterdam edition and the 1744 *Histoire des Incas*. Curiously, however, in a letter dated May 6, 1743, she refers to the text she is reading as the *Histoire des Incas*, using the spelling associated with the 1744 text. On the genesis and composition of the *Lettres d'une péruvienne*, see English Showalter, "Les *Lettres d'une Péruvienne*: Composition, Publication, Suites," *Archives et Bibliothèques de Belgique* 54 (1983): 14–28; and Showalter, *Françoise de Graffigny: Her Life and Works*, Studies on Voltaire and the Eighteenth Century, 11 (Oxford: Voltaire Foundation, 2004). The letters in which Graffigny refers to the *Histoire des Incas* can be found in Graffigny, *Correspondance de Madame de Graffigny*, ed. J. A. Dainard, English Showalter et al. (Oxford: Voltaire Foundation, 1985–), 4: 265, 283. On the epistolary form as political critique in Montesquieu, see Dena Goodman, *Criticism in Action: Enlightenment Experiments in Political Writing* (Ithaca: Cornell University Press, 1989), 22–27.

17. Guy de la Brosse, *L'Ovvertvre dv Iardin Royal de Paris, Povr la Demonstration des Plantes Medecinales* (Paris: Jacques Dugast, 1640), 16.

18. For a succinct overview of the early interest of the French Crown in the botanical bounty of extra-European territories, see Antoine Schnapper, *Le géant, la licorne, la tulipe: Collections françaises au XVIIe siècle* (Paris: Flammarion, 1988), esp. chap. 4. On the economic, aesthetic, and natural historical motivations for pursuing New World exotica, see Pamela H. Smith and Paula Findlen, eds., *Merchants and Marvels: Commerce, Science, and Art in Early Modern Europe* (New York: Routledge, 2002). Biographical information on Dalibard is scanty, especially for the period of his career prior to his experiments with electricity. For the few details that have come to light, see Robert-Gustave-Marie Triger, "Le Collège de Crannes et Thomas-François Dalibard, naturaliste et physicien," *Bulletin de la Société d'agriculture, sciences et arts de la Sarthe* 30 (1885): 189–204. See also "Thomas François Dalibard," *Dictionary of Scientific Biography*, ed. Charles Coulston Gillispie (New York: Scribner's, 1971), 3: 535; J. F. and L. G. Michaud, eds., *Biographie universelle, ancienne et moderne* (Paris: Desplaces, 1854), 10: 43; and I. Bernard Cohen, *Franklin and Newton* (Cambridge: Harvard University Press, 1966).

19. Footnotes did not appear at all in Baudoin's 1633 edition or in the Amsterdam edition of 1737, but their use increased over the course of the eighteenth century. For a famous eighteenth-century example, see Jean-Jacques Rousseau, *Discours sur l'origine et les fondements de l'inégalité parmi les hommes* (Amsterdam: Marc Michel Rey, 1755), note 8.

20. Historians have mysteriously given exclusive credit to Dalibard for carrying out the translation and editorial production. The inclusion of Dalibard's

name in the preface, and Dalibard's name alone, may have amounted to an eighteenth-century code for the attribution of authorship, or in this case "editorship," of anonymously produced texts. But it is more likely, judging from the inclusion of the anonymous astronomer as a collaborator in the translation project, that the reediting of this text was a collective endeavor, achieved with the assistance of others who collated, translated, and ultimately printed this revised edition.

21. Garcilaso de la Vega, translator's preface, *Histoire des Incas,* ix. Unless otherwise noted, references to the *Histoire des Incas* are to the 1744 edition.

22. Garcilaso de la Vega, translator's preface, *Histoire des Incas,* ix. Among other things, the editor discussed the shared doctrines of a single God, the deity's eternal and immutable presence, the immortality of the soul, and the notion of resurrection. On the paternal metaphor in eighteenth-century France, see Lynn Hunt, *The Family Romance of the French Revolution* (Berkeley and Los Angeles: University of California Press, 1992); and David A. Bell, *The Cult of the Nation in France: Inventing Nationalism,* 1680–1800 (Cambridge: Harvard University Press, 2001), 19.

23. Garcilaso de la Vega, translator's preface, *Histoire des Incas,* xiv–xv.

24. René Louis de Voyer de Paulmy d'Argenson, *Journal et mémoires,* ed. E.-J.-B. Rathery (Paris: J. Renouard, 1859), 3: 159 (19 May 1739).

25. Garcilaso de la Vega, translator's preface, *Histoire des Incas,* xx.

26. Cited in Steven L. Kaplan, "The Famine Plot Persuasion in Eighteenth-Century France," *Transactions of the American Philosophical Society* 72, no. 3 (1982): 33–34. The link between bread and cultural politics during the early modern period, as well as the wide-ranging impact of acute subsistence crises that occurred during the years leading up to the Revolution, are recognized as staple features of early modern French society and have also been studied by Kaplan in *Bread, Politics, and Political Economy in the Reign of Louis XV,* 2 vols. (The Hague: Martinus Nijhoff, 1976) and idem, *The Bakers of Paris and the Bread Question, 1700–1775* (Durham: Duke University Press, 1996). As Kaplan argues persuasively, a "famine plot persuasion" and the social convulsions that accompanied it took "a deep hold of the French consciousness (and unconsciousness) in the old regime and [became] a durable part of the collective memory and mentality." Kaplan, "Famine Plot Persuasion," 62.

27. Marie-Noëlle Bourguet has shown the explicit connection between overseas travel and subsistence problems in the French metropole. See Bourguet, "La collecte du monde: Voyage et histoire naturelle (fin XVIIème siècle–début XIXème siècle)," in *Le Muséum au premier siècle de son histoire,* ed. Claude Blanckaert et al. (Paris: Muséum national d'Histoire naturelle, 1997), 184. For a discussion of the importance of alimentary products to the French and Spanish overseas expeditions, see also Jean-Pierre Clément, "Réflexions sur la politique scientifique française vis-à-vis de l'Amérique espagnole au siècle des Lumières," in *Nouveau Monde et renouveau de l'histoire naturelle* (Paris: Presses de l'Université de la Sorbonne nouvelle-Paris III, 1994), 3: 131–59.

28. On early Iberian botanical expeditions, see Antonio Barrera-Osorio, *Experiencing Nature: The Spanish American Empire and the Early Scientific Revolution* (Austin: University of Texas Press, 2006); Daniela Bleichmar, "Books, Bodies, and Fields: Sixteenth-Century Transatlantic Encounters with New World *Materia Medica*," in *Colonial Botany: Science, Commerce, and Politics in the Early Modern World*, ed. Londa Schiebinger and Claudia Swan (Philadelphia: University of Pennsylvania Press, 2005); and C. R. Boxer, *Two Pioneers of Tropical Medicine: Garcia d'Orta and Nicolás Monardes* (London: Hispanic and Luso-Brazilian Councils, 1963).

29. On these early expeditions, see Paul Fournier, *Voyages et découvertes scientifiques des missionnaires naturalistes français à travers le monde pendant cinq siècles, XVe à XXe siècles* (Paris: P. Lechevalier & Fils, 1932); and McClellan, *Colonialism and Science*. McClellan has argued convincingly that as plantation systems expanded during the eighteenth century, European scientific communities, including the French, British, and Dutch, became increasingly interested in cultivating non-European species for incorporation into large-scale production systems. McClellan calls this system "applied" or economic botany and describes how nations often sought to create applied botanical gardens to wrest control of particular species from other nations that held effective monopolies over certain products (47–62). On Pierre Poivre's role in implementing French economic and "environmental" policy in the Mauritius Islands, see Grove, *Green Imperialism*, esp. 168–263. It was during Fagon's tenure as royal physician that the intendancy of the Jardin du Roi came under the physician's jurisdiction. In 1732 the intendancy was permanently separated from the office of the royal physician.

30. Feuillée, *Journal des observations physiques, mathematiques et botaniques*, 3: 4.

31. Feuillée, *Histoire des plantes medicinales*, in *Journal des observations physiques*, 2: 707.

32. Feuillée, *Histoire des plantes medicinales*, 2: 7.

33. Benjamin Franklin, *Expériences et observations sur l'électricité faites à Philadelphie en Amérique*, trans. Thomas-François Dalibard (Paris: Durand, 1752). For more on Dalibard's role in the electricity experiments in France, see Cohen, *Franklin and Newton;* and Michel Lopez, "La caractérisation de l'électricité dans la foudre du XVIIIe siècle par Thomas-François Dalibard, un physicien français méconnu," *Actes du XIe colloque Forum Industrie de l'Université du Maine* (Le Mans, 1999). For Dalibard's other work around Paris, see Thomas-François Dalibard, *Florae Parisiensis Prodromus, ou Catalogue des plantes qui naissent dans les environs de Paris, rapportées sous les dénominations modernes & anciennes, & arrangées suivant la méthode séxuelle de M. Linnaeus* (Paris: Durand, Pissot, 1749).

34. As cited in Yves Laissus, "Jardin du Roi," in *Enseignement et diffusion des sciences au XVIIIe siècle*, ed. René Taton (Paris: Hermann, 1964), 293.

35. Bernard de Fontenelle, "Éloge de M. du Fay," *Mémoires de l'Académie Royale des Sciences* (1739 [1741]): 77.

36. Laissus, "Jardin du Roi," 294.

37. Fontenelle, "Éloge de M. du Fay," 78.

38. Fontenelle, "Éloge de M. du Fay," 78–79.

39. Garcilaso de la Vega, *Histoire des Incas*, 2: 201, note e; 2: 198.

40. Garcilaso de la Vega, *Histoire des Incas*, 2: 195, note a.

41. Garcilaso de la Vega, *Histoire des Incas*, quotations on 2: 200, note c; 2: 204, note f; 2:204, note g; 2: 207, note i; 2: 212.

42. Garcilaso de la Vega, *Histoire des Incas*, 2: 217.

43. Garcilaso de la Vega, *Histoire des Incas*, 2: 229.

44. The same was true for the banana tree, which first flowered in Paris in 1741 according to the description in the same chapter: "[I]t is the first [tree] to which this has happened in France. The fruit even became fairly large, and almost reached maturity, even though it was closed up in the Hothouse." Garcilaso de la Vega, *Histoire des Incas*, 2: 242.

45. Fontenelle, "Éloge de M. du Fay," 79.

46. Garcilaso de la Vega, translator's preface, *Histoire des Incas*, xxi. The concern for "[t]he present state of this vast country" is yet another indication of the interest in contemporary affairs.

47. Garcilaso de la Vega, preface to the second volume, *Histoire des Incas*, 2: vii.

48. Garcilaso de la Vega, preface to the second volume, *Histoire des Incas*, 2: viii.

49. The *Dictionnaire de l'Académie françoise* (Paris: J. B. Coignard, 1694) defines "Asterisme" in the following manner: "Asterism. Astronomical term, constellation, a collection of many stars. Asterisms are shown on the celestial globe." *Dictionnaire de l'Académie françoise* (Paris: J. B. Coignard, 1694). D'Alembert, in his article "Astérisme" for the *Encyclopédie*, tersely defines it in a similar way: "In Astronomy, asterism has the same meaning as constellation." Jean Le Rond d'Alembert, "Astérisme," in *Encyclopédie, ou Dictionnaire raisonné des sciences, des arts et des métiers*, ed. Denis Diderot and Jean Le Rond d'Alembert (Paris: Briasson, 1751–72), 1: 776.

50. Barcia used brackets for his 1729 edition of Gregorio Garcia's *Origen de los Indios del Nuevo Mundo* in a similar fashion, to dramatize his library rather than a museum: "The page," according to Carlyon, became "a stage." See Carylon, *Andrés González de Barcia*, 140–54. Likewise, the 1704 Amsterdam edition of the *Histoire des Yncas* occasionally employed asterisks as footnote markers within the text. See Garcilaso, *Histoire des Yncas, rois du Perou* (1704), 60.

51. Martin-Dominique Fertel, *La science pratique de l'imprimerie. Contenant des instructions tres-faciles pour se perfectionner dans cet art* (Saint Omer: Martin Dominique Fertel, 1723).

52. Fertel, preface, *La science pratique de l'imprimerie*.

53. Garcilaso de la Vega, *Histoire des Incas*, 2: 33.

54. There are several possible candidates for the authorship of this revised section on Inca astronomy, including Etienne-Simon de Gamaches (1672–1756), who published his *Astronomie physique, ou Principes généraux de la nature appliqués au mécanisme astronomique, et comparés aux principes de la philosophie de M. Newton* (Paris: Jombert, 1740) around this time; Maupertuis, who in 1732 had published

his *Discours sur les différentes figures des astres* (Paris: Imprimerie Royale, 1732); and Pierre Bouguer, who had published a set of texts entitled *Entretiens sur la cause de l'inclinaison des orbites des planetes* in 1734, although his absence from Paris during the period 1735–43 makes him the least likely of the three.

55. This section corresponds roughly to chaps. 9–15 of book 8 in Garcilaso's original text. A number of literary scholars have addressed the issue of organization in Garcilaso's *Comentarios reales*. They include Zamora, *Language, Authority, and Indigenous History*; Mazzotti, *Coros mestizos*; Macchi, "Imágenes"; and Fernández, *Inca Garcilaso*.

56. Garcilaso de la Vega, *Histoire des Incas*, 2: 207, footnote a.

57. One such example includes a discussion of how the "Indiennes" dyed their hair dark by literally boiling it in a cauldron filled with herbs, part of the section that describes the *Maugei* tree. Footnote (a) on p. 210 of the *Histoire des Incas* notes that "[t]his custom is nevertheless not universal, since the women from Collas cover their head because of the cold temperatures in their land." In the original Spanish text, Garcilaso had provided the following description: "The Indian women of Peru wear their hair long and loose . . . with the exception of the [women from] Collas, who because of the cold temperatures in their land, wear a head covering." Garcilaso, *Comentarios reales*, book 8, chap. 13.

58. Feuillée, *Journal des observations*; Amédée François Frézier, *Voyage de la mer du sud* (Paris: Jean-Geoffroy Nyon, Etienne Ganeau, Jacque Quillau, 1717); Marggraf and Willem Piso, *Historia naturalis Brasiliae* (1648); and Carolus Clusius, *Histoire des plantes en laqvelle est contenve la description entiere des herbes cest a dire, leurs especes, forme, noms, temprerament, vertus & operations: Non seulement de celles qui croissent en ce pais, mais aussi des autres estrangeres qui viennent en vsage de medecine* (Antwerp: Jen Loë, 1557). In fact, Feuillée's *Journal des observations* consists of three separate volumes. The first two volumes were published in 1714, the final volume in 1725. The *Histoire des plantes medicinales* is part of the second volume.

59. Garcilaso de la Vega, *Histoire des Incas*, 2: 216.

60. La Condamine, "Sur l'Arbre du Quinquina," *Mémoires de l'Académie Royale des Sciences* (1738): 226–43.

61. Garcilaso de la Vega, *Histoire des Incas*, 2: 270.

62. Garcilaso de la Vega, *Histoire des Incas*, 2: 274.

63. Brosse, *L'Ovvertvre dv Iardin Royal de Paris*, 7.

64. Garcilaso de la Vega, *Histoire des Incas*, 2: 249–50.

65. La Condamine, "Mémoire sur quelques anciens monumens du Perou, du tems des Incas," 456.

CHAPTER SEVEN

1. Denis Diderot, *Supplément au voyage de Bougainville*, ed. Michel Delon (Paris: Gallimard, 2002), 35. On the *Supplément* as political critique, see Dena Goodman, *Criticism in Action: Enlightenment Experiments in Political Writing* (Ithaca: Cornell University Press, 1989), chaps. 6–8.

2. Diderot, as cited in the preface to *Supplément*, 13; Denis Diderot, *Oeuvres complètes* (Paris: Club français du livre, 1969–73), 15: 497. As quoted in Gerhard Stenger, *Le magazine littéraire* (France) 391 (2000): 30–32.

3. Jaucourt, "Voyageur," *Encyclopédie*, 17: 477.

4. *Mercure de France*, August, 1745, 101.

5. Diderot to La Condamine, July 1751, as cited in Anne-Marie Chouillet, "Trois lettres inédites de Diderot," *Recherches sur Diderot et sur l'Encyclopédie (RDE)* 11 (October 1991).

6. La Condamine, *Relation abrégée*, 166.

7. La Condamine, *Relation abrégée*, 166–67.

8. The order of this article seems to have been drawn from Mathurin-Jacques Brisson, *Le regne animal divisé en IX classes; ou, Méthode contenant la division generale des animaux en IX classes, & la division particuliere des deux premieres classes, sçavoir de celle des quadrupedes & celle des cetacées . . . Aux quelles on a joint une courte description de chaque espèce . . .* (Paris: Jean-Baptiste Buache, 1756).

9. For a history of the *Encyclopédie*, see John Lough, *The Encyclopédie* (Geneva: Slatkine reprints, 1989); Jean Haechler, *L'Encyclopédie. Les combats et les hommes* (Paris: Les Belles Lettres, 1998); François Moureau, *Le roman vrai de l'Encyclopédie* (Paris: Découvertes Gallimard, 1990).

10. Diderot, "Prospectus," *Discours préliminaire des éditeurs de 1751 et articles de l'Encyclopédie*, ed. Martine Groult (Paris: Champion, 1999).

11. This idea of "intellectual uniformitarianism" is drawn from William Clark, Jan Golinski, and Simon Schaffer, introduction, *The Sciences in Enlightened Europe* (Chicago: University of Chicago Press, 1999). On travel and truth, see Dorinda Outram, "On Being Perseus: New Knowledge, Dislocation, and Enlightenment Exploration," in *Geography and Enlightenment*, ed. David N. Livingstone and Charles W. J. Withers (Chicago: University of Chicago Press, 1999). This chapter does not include details on the expedition's representation in the realm of geodesy, the specific purpose for which the expedition was undertaken, an area in which its contributions were more widely acknowledged, and about which the members' participation has been more widely discussed.

12. On the place of the exotic in early modern Europe, see Smith and Findlen, *Merchants and Marvels*; Henry Vyverberg, *Human Nature, Cultural Diversity, and the French Enlightenment* (Oxford: Oxford University Press, 1989); Peter Mason, *Infelicities: Representations of the Exotic* (Baltimore: Johns Hopkins University Press, 1998); and G. S. Rousseau and Roy Porter, eds., *Exoticism in the Enlightenment* (Manchester: Manchester University Press, 1990). The classic account is Chinard, *L'Amérique et le rêve exotique*. As G. S. Rousseau and Roy Porter have written, the exotic in early modern Europe was defined primarily as strangeness and otherness, not only through geographical distance but also because of distinctive customs and behaviors of non-European peoples, beyond the boundaries of the mundane and the quotidian.

13. On science and the public sphere, see especially Michael R. Lynn, "Enlightenment in the Public Sphere: The Musée de Monsieur and Scientific

Culture," *Eighteenth-Century Studies* 32, no. 4 (1999): 463–76; Lynn, *Popular Science and Public Opinion in Eighteenth-Century France* (Manchester: Manchester University Press, 2006); Robbins, *Elephant Slaves and Pampered Parrots*; Thomas H. Broman, "The Habermasian Public Sphere and Eighteenth-Century Historiography: A New Look at 'Science in the Enlightenment,'" *History of Science* 36 (1998): 123–49. On the public sphere, see Reinhard Kosseleck, *Critique and Crisis: Enlightenment and the Pathogenesis of Modern Society* (Cambridge: MIT Press, 1988); Jürgen Habermas, *The Structural Transformation of the Public Sphere: An Inquiry into a Category of Bourgeois Society*, trans. Thomas Burger (Cambridge: MIT Press, 1989); Dena Goodman, *The Republic of Letters* (Ithaca: Cornell University, 1994); and Margaret Jacob, *Living the Enlightenment: Freemasonry and Politics in Eighteenth-Century Europe* (Oxford and New York: Oxford University Press, 1991).

14. Denis Diderot and Louis-Jean-Marie Daubenton, "Cabinet d'Histoire naturelle," in *Encyclopédie, ou Dictionnaire raisonné des sciences, des arts et des métiers*, ed. Denis Diderot and Jean Le Rond d'Alembert (Paris: Briasson, 1751–72), 2: 490. For publication dates of the individual volumes, see Frank A. Kafker and Serena L. Kafker, *The Encyclopedists as Individuals: A Biographical Dictionary of the Authors of the Encyclopédie*, Studies on Voltaire and the Eighteenth Century, 257 (Oxford: Voltaire Foundation, 1988), xxv. Of course, the *Encyclopédie* was produced over the course of two decades, which meant that more published material was available for the latter volumes than for the earlier ones. This may, in some cases, account for disparities between multiple articles on the same subject, such as the "Inca" and "Ynca" articles, published respectively in the eighth volume (1765) and the seventeenth volume (also 1765).

15. On some of the more audacious scientific spectacles of late-eighteenth-century France, including de Rozier's final balloon flight, and the attention lavished upon them by eighteenth-century periodicals, see Robert Darnton, *Mesmerism and the End of the Enlightenment in France* (Cambridge: Harvard University Press, 1968), esp. 40–45.

16. Goodman, *Republic of Letters*, 23–33. Paula Findlen has shown that physical spaces such as museums, laboratories, and gardens played a fundamental role in "the transposition of knowledge from the discursive to the visual arena" during the sixteenth and seventeenth centuries, but this transposition largely returned to the discursive frame in the eighteenth century. Cf. Findlen, *Possessing Nature*, 199.

17. Cf. Keith Michael Baker, "Épistémologie et politique: Pourquoi l'*Encyclopédie* est-elle un dictionnaire?" in *L'Encyclopédie: Du réseau au livre et du livre au réseau*, ed. Robert Morrissey et Philippe Roger (Paris: Honoré Champion, 2001), 52, 53.

18. Daubenton, "Description du Cabinet du Roy," in Buffon, *Histoire naturelle* (1749), 3: 2.

19. Daubenton, "Description du Cabinet du Roy," 3: 7.

20. Daubenton, "Cabinet d'Histoire naturelle," in *Encyclopédie, ou Dictionnaire raisonné*, 2: 489.

21. Diderot, "Prospectus," 20. On the early history of the *Encyclopédie*, see *L'Encyclopédie*, ed. Jeanne Charpentier and Michel Charpentier (Paris: Éditions Bordas, 1967), 3–10; Lough, *Encyclopédie*; Haechler, *L'Encyclopédie*.

22. Diderot, "Prospectus," passim.

23. Cited in Anne-Marie Chouillet, "Trois lettres inédites," 13.

24. "Amérique," in *Supplément à l'Encyclopédie* (Amsterdam: M. M. Rey, 1776–77), 1: 343–62. See also Richard Switzer, "America in the *Encyclopédie*," *SVEC* 58 (1967): 1481–99.

25. For specific statistical comparisons, see James Llana, "Natural History and the *Encyclopédie*," *Journal of the History of Biology* 33 (2000): 1–25. For biographical information on the *Encyclopédistes*, see Kafker and Kafker, *Encyclopedists as Individuals*.

26. Letter from d'Alembert to Le Breton, n.d., as cited in Badinter, *Les passions intellectuelles*, 2: 23, note 1.

27. Diderot wrote to La Condamine: "I accept with great pleasure the offer you make to help us perfect our dictionary." Chouillet points out that Kafker was unaware who recruited La Condamine for the *Encyclopédie*, and writes that "[t]he response is in this letter: La Condamine 'recruited himself.'" Chouillet, "Trois lettres inédites." Apparently, this strategy of self-recruitment was practiced widely enough that, in 1757, the *Mercure Danois* made explicit accusations against those who sought personal profit by associating themselves with the *Encyclopédie* as authorial pinch hitters. "There are those among the *Encyclopédistes* who, seized with a burning desire to give themselves immortality at a low price, apply for the chance to insert in the grand dictionary some small article with their name below written in large characters . . . The number of these pretenders to the honor of being *Encyclopédistes* is larger than one could imagine, and from time to time you will seem them appear [in print] . . . at the expense of the perfection of the work in question." *Mercure danois* (October 1757), cited in François Moureau, *Le roman vrai de l'Encyclopédie* (Paris: Gallimard, 1990), 150–53.

28. Robert de Vaugondy, "Géographie," in *Encyclopédie, ou Dictionnaire raisonné*, 7: 611.

29. Jaucourt, "Quito," in *Encyclopédie, ou Dictionnaire raisonné*, 13: 726.

30. D'Alembert, "Discours préliminaire," in *Encyclopédie, ou Dictionnaire raisonné*, 1: xi.

31. Anonymous, "Inca ou Ynca," in *Encyclopédie, ou Dictionnaire raisonné*, 8: 641.

32. Abbé Mallet, "Amautas," in *Encyclopédie, ou Dictionnaire raisonné*, 1: 317.

33. Jaucourt, "Ynca," in *Encyclopédie, ou Dictionnaire raisonné*, 17: 671.

34. Translator's preface, *Histoire des Incas, rois du Pérou* (Paris: Prault fils, 1744), ix.

35. The dream of a French Amazon had been briefly evoked in the preface to the Comte de Pagan's *Relation historique et géographique du grand pays et rivière des Amazones en Amérique* (1655). Pagan's text was followed in 1682 by a French translation of Father Acuña's *Nvevo descvbrimiento del gran rio de las Amazonas* by

Gomberville. See Dominique Linchet, "La relation de la riviere des Amazones de M. de Gomberville: la traduction au service du colonialisme, du patriotisme et de la propagande sous Louis XIV" (Ph.D. diss., University of North Carolina, 1995).

36. Pierre Petit, *Traité historique sur les Amazones; Où l'on trouve tout ce que les auteurs, tant anciens que modernes, ont écrit pour ou contre ces heroines; Et où l'on apporte quantité de medailles & d'autres monumens anciens, pour prouver qu'elles ont existé* (Leiden, 1718), 74.

37. Louïs le Maingre de Bouciquault, *Les Amazones revoltées, roman moderne, en forme de parodie sur l'histoire universelle, & la fable. Avec des notes politiques sur les travaux d'Hercule, La Chevalerie-Militaire, et la decouverte du nouveau monde, &c. &c. &c.* (Rotterdam, 1737); and abbé Guyon, *Histoire des Amazones anciennes et modernes* (Paris, 1740), 191, 215. On the literary treatment of the Amazon myth, in both classical and early modern contexts, see Kathryn Schwarz, *Tough Love: Amazon Encounters in the English Renaissance* (Durham, N.C.: Duke University Press, 2000); Josine H. Blok, "The Early Amazons: Modern and Ancient Perspectives on a Persistent Myth," in *Religions in the Graeco-Roman World*, ed. R. Van den Broek (Leiden: E. J. Brill, 1995); and Karen L. Raber, "Warrior Women in the Plays of Cavendish and Killigrew," *SEL: Studies in English Literature* 40, no. 3 (Summer 2000): 413–33.

38. On the early history of the quinquina tree, see Jaime Jaramillo-Arango, "Estudio crítico acerca de los hechos básicos de la historia de la quina," *Revista de la Academía Colombiana de Ciencias Exactas Físicas y Naturales* (Bogotá) 8 (1951): 245–73; and Robert Finn Erickson, "The French Academy of Sciences Expedition to Spanish America, 1735–1744" (Ph.D. diss., University of Illinois, 1955), esp. 245–82. Londa Schiebinger and Claudia Swan include a brief but suggestive account of La Condamine's "discovery" of quinquina, or cinchona, in their introduction to *Colonial Botany: Science, Commerce, and Politics in the Early Modern World* (Philadelphia: University of Pennsylvania Press, 2005), 1–2. See also Schiebinger's contribution, "Prospecting for Drugs," in *Colonial Botany*, 124. Schiebinger makes a similar analysis in *Plants and Empire* (Cambridge: Harvard University Press, 2004).

39. "Mémoire sur une résine élastique, nouvellement découverte à Cayenne par M. Fresneau, et sur l'usage de divers suc laiteux d'arbres de la Guiane ou France Équinoctiale," *Mémoires de l'Académie Royale des Sciences* (1751 [1755]): 319–33. For details on La Condamine's role in bringing news of rubber to Europe, see François de Chasseloup Laubat, *François Fresneau, seigneur de la Gataudière, père du caoutchouc* (Paris: Plon, 1942), esp. 76–113; Auguste Chevalier, "Le deuxième centenaire de la découverte du Caoutchouc faite par Charles-Marie de La Condamine," *Revue de Botanique Appliquée et d'Agriculture Tropicale* 179 (1936): 519–29; *Revue Générale du Caoutchouc*, Organe Officiel du Syndicat du Caoutchouc. Revue Mensuelle—Treizième année, 125 (October 1936).

40. Jaucourt, "Résine," in *Encyclopédie, ou Dictionnaire raisonné*, 14: 173; Jaucourt, "Cahuchu," in *Encyclopédie, ou Dictionnaire raisonné*, 17: 760.

41. Erickson offers a cogent summary of the research conducted on the quinquina tree and other natural specimens by members of the Franco-Hispanic expedition. See "French Academy of Sciences Expedition to Spanish America, 1735–1744," esp. 245–82.

42. Eduardo Estrella, "Ciencia ilustrada y saber popular en el conocimiento de la quina," in *Saberes andinos. Ciencia y tecnología en Bolivia, Ecuador y Peru*, ed. Marcos Cueto (Lima: Instituto de Estudios Peruanos, 1995), 44 and passim.

43. La Condamine, "Sur l'Arbre du Quinquina," *Mémoires de l'Académie Royale des Sciences* (1738 [1740]): 226–44; Joseph de Jussieu, *Description de l'arbre à Quinquina* (Paris, 1936). Jussieu's manuscript papers on the quinquina tree can be found at the Bibliothèque Centrale du Muséum d'Histoire Naturelle (Paris), MS 1626.

44. The article "Quinquina" draws almost exclusively on La Condamine's *mémoire*. Jaucourt, "Quinquina," in *Encyclopédie, ou Dictionnaire raisonné*, 13: 717.

45. La Condamine, "Sur l'arbre du Quinquina," passim; Jaucourt, "Quinquina," passim.

46. La Condamine, "Sur l'arbre du Quinquina," passim, emphasis added; Jaucourt, "Quinquina," passim.

47. La Condamine, *Relation abrégée*, 153.

48. Ibid., 153–54.

49. I have used the Project for American and French Research on the Treasury of the French Language (ARTFL) database of Diderot and d'Alembert's *Encyclopédie* to measure the full impact of the Quito expedition across the range of the *Encyclopédie* (http://www.lib.uchicago.edu/efts/ARTFL). As Richard Switzer has said in reference to articles that relate to America in the *Encyclopédie*, "Frequently these mentions are lurking in unsuspected places." Richard Switzer, "America in the *Encyclopédie*," *SVEC* 58 (1967): 1481–99, quotation on 1496.

50. D'Alembert, "Amazones, Rivière des," in *Encyclopédie, ou Dictionnaire raisonné*, 1: 318.

51. Jaucourt, "Perroquet vert varié," in *Encyclopédie, ou Dictionnaire raisonné*, 12: 399.

52. The French translation, *Les saisons*, was published in 1759.

53. Jaucourt, "Zone Torride," in *Encyclopédie, ou Dictionnaire raisonné*, 17: 725.

54. Diderot, "Aguas," in *Encyclopédie, ou Dictionnaire raisonné*, 1: 191.

55. Vosgien, *Dictionnaire géographique-portatif, ou Description de tous les royaumes, provinces, villes, patriarchats, evéchés* (The Hague, 1748).

56. Bourguet, "La collecte du monde," 187.

57. Bronislaw Baczko, "Voyager avec l'*Encyclopédie*," in *L'Encyclopédie. Du réseau au livre et du livre au réseau*, ed. Robert Morrissey and Philippe Roger (Paris: Honoré Champion, 2001), 122–23.

58. J. A. G. Roberts, "L'image de la Chine, dans l'Encyclopédie," *Recherches sur Diderot et sur l'*Encyclopédie 22 (April 1997): 87–107.

59. Vyverberg, *Human Nature*, 121.

60. Daniel Mornet, *Les sciences de la nature en France, au XVIIIe siècle* (1911; Geneva: Slatkine Reprints, 2001), 173.

61. [Henri Gabriel Duchesne and P. J. Macquer], *Le petit cabinet d'histoire naturelle ou Manuel du naturaliste. Ouvrage utile aux voyageurs et à ceux qui visitent les cabinets d'histoire naturelle et de curiosités* (Paris, 1797).

62. Diderot refers to both of these empires at various times. See, for example, his "Prospectus," where he discusses "l'empire des sciences & des arts." *Discours préliminaire des éditeurs de 1751 et articles de l'Encyclopédie*, ed. Martine Groult (Paris: Champion, 1999), 25.

63. Henri Froidevaux, *Observations scientifiques de La Condamine pendant son séjour à Cayenne (1744)* (Paris: Imprimerie Nationale, 1898). On colonial power dynamics and their impact on the production of knowledge, see Ann Laura Stoler and Frederick Cooper, "Between Metropole and Colony: Rethinking a Research Agenda," in *Tensions of Empire: Colonial Cultures in a Bourgeois World,* ed. Frederick Cooper and Ann Laura Stoler (Berkeley: University of California Press, 1997), 1–56.

64. Diderot, "Encyclopédie," in *Encyclopédie, ou Dictionnaire raisonné,* 5: 644.

65. Jaucourt, "Singe," in *Encyclopédie, ou Dictionnaire raisonné,* 15: 209.

CONCLUSION

1. Juan de Velasco, *Historia del Reino de Quito,* 1: 143.

2. La Condamine, *Journal du voyage,* 173.

3. Jacques-François Artur, *Histoire des colonies françoises de la Guianne,* 2: 557–81, as cited in Froidevaux, *Observations scientifiques de La Condamine pendant son séjour à Cayenne (1744),* 19–20.

4. On the cross-border circulation of fugitive slaves between French Guyana and Grão-Pará in the eighteenth century, see Rosa Elizabeth Acevedo Marin and Flávio Gomes, "Reconfigurações coloniais: Tráfico de indígenas, fugitivos e fronteiras no Grão-Pará e Guiana francesa (sécs. XVII e XVIII)," *Revista de História* (São Paulo) 149 (2003): 69–107. On fugitives and *quilombos* in colonial Brazil and French Guyana more generally, see Flávio dos Santos Gomes, *A hidra e os pântanos: Mocambos, quilombos, e comunidades de fugitivos no Brasil (séculos XVII–XIX)* (São Paulo: Editora UNESP, 2005); and *Deux siècles d'esclavage en Guyane française, 1652–1848,* ed. Anne-Marie Bruleaux, Régime Calmon, and Serge Mam-Lam-Fouck (Cayenne: CEGER; Paris: L'Harmattan, 1986).

5. After returning from South America, Alexander von Humboldt described this reaction in his own *Relation historique*: "Since my return from the Orinoco and River of the Amazons, people in Paris have often asked me if I shared [La Condamine's] opinion or if I believed, like many of his contemporaries, that he undertook the defense of the *Cougnantainsecouima* [women warriors] . . . merely to take advantage of the generous reception of a public session of the academy, and their eagerness for [hearing] new things." Humboldt, *Relation historique du voyage aux régions équinoxiales du Nouveau continent,* book 8, chap. 23, 2: 484–85.

6. La Condamine, *Relation abrégée*, 110–11.

7. La Condamine, *Relation abrégée*, 110–11.

8. Humboldt, "Von Quito nach Lima," in *Reise auf dem Río Magdalena, durch die Anden und Mexico*, ed. Margot Faak (Berlin: Akademie-Verlag, 1990), 2: 122.

9. Velasco, *Historia del Reino de Quito*, 3: 240.

Bibliography

Archives

Académie des Inscriptions et Belles-Lettres, Paris (AIBL)
Archivo General de Simancas, Spain (AGS)
Archivo Histórico Nacional, Madrid (AHN)
Arquivo Histórico Ultramarino, Lisbon (AHU)
Bibliothèque Centrale du Muséum National d'Histoire Naturelle, Paris (BMNHN)
Bibliothèque de l'Institut de France, Paris
Bibliothèque de l'Observatoire de Paris
Biblioteca Nacional de España, Madrid (BNE)
Bibliothèque Nationale de France, Paris (BNF)
British Library, London (BL)
Clements Library, University of Michigan
Museo Naval, Madrid (MN)
National Archives, London
Private collection, Paris
Royal Society, London
Special Collections Library, University of Michigan
Westminster Archives, London

Primary Sources

Abbeville, Claude d'. *Histoire de la mission des pères capucins en l'isle de Maragnon et terres circonvoisines.* Paris: François Huby, 1614.
Acuña, Cristobal de. *Nuevo descubrimiento del gran rio de las Amazonas por el padre Christoval de Acuña . . . al qual fué, y se hizo por orden de Su Magestad, el año de 1639.* Madrid: Imprenta del Reyno, 1641.

Alembert, Jean Le Rond d'. "Discours préliminaire." In *Encyclopédie, ou Dictionnaire raisonné des sciences, des arts et des métiers*. Edited by Denis Diderot and Jean Le Rond d'Alembert, 1: xi. Paris: Briasson, 1751.

————. "Figure de la Terre." In *Encyclopédie, ou Dictionnaire raisonné des sciences, des arts et des métiers*. Edited by Denis Diderot and Jean Le Rond d'Alembert, 6: 754. Paris: Briasson, 1756.

————. "Méthode Générale pour déterminer les orbites & les mouvemens de toutes les Planètes, en ayant égard à leur action mutuelle." *Mémoires de l'Académie Royale des Sciences* (1745 [1749]): 365–90.

Anville, Jean-Baptiste Bourguignon d'. *L'Amérique Méridionale*. Map. Paris, 1748.

————. "Lettre de Monsieur d'Anville, à Messieurs du Journal des Sçavans, sur une Carte de l'Amérique Méridionale qu'il vient de publier." *Journal des Sçavans* (March 1750): 522–63.

————. "Seconde lettre de M. d'Anville à Messieurs du Journal des Sçavans, sur la Carte qu'il a publiée de l'Amérique Méridionale." *Journal des Sçavans* (April 1750): 625–73.

Argenson, René Louis de Voyer de Paulmy d'. *Journal et mémoires*. Edited by Rathéry. 8 vols. Paris: J. Renouard, 1859.

Artur, Jacques-François. "Histoire des colonies françoises de la Guianne." In *Observations scientifiques de La Condamine pendant son séjour à Cayenne (1744)*. Edited by Henri Froidevaux. Paris: Imprimerie Nationale, 1898.

Barbier, Edmond-Jean-François. *Chronique de la régence et du règne de Louis XV; ou, Journal de Barbier*. Paris: Charpentier, 1857.

Barbosa de Sá, José. *Dialogos geograficos coronologicos polliticos e naturais*. Lisbon, 1769.

Barrère, Pierre. *Essai sur l'histoire naturelle de la France Equinoxiale. Ou dénombrement des plantes, des animaux & des minéraux, qui se trouvent dans l'isle de Cayenne, les isles de Remire, sur les côtes de la mer, & dans le continent de la Guyane*. Paris: la Veuve Piget, 1743.

Baudelot de Dairval, Charles César. *Mémoire sur quelques observations qu'on peut faire pour ne pas voyager inutilement*. Brussels: Jean Leonard, 1688.

Berchtold, Leopold, comte de. *Essai pour diriger et étendre les recherches des voyageurs qui se proposent l'utilité de leur patrie*. Paris: Du Pont, 1797.

Berthier, Guillaume François. "Parallele de la branche philosophique du système de l'Encyclopédie, avec la partie philosophique du *Livre de la Dignité & de l'accroissement des Sciences*, Ouvrage du Chancelier Bacon." *Journal de Trévoux* 708 (March 1751): art. 37. Reproduced in *Discours préliminaire des éditeurs de 1751 et articles de l'Encyclopédie*. Edited by Martine Groult. Paris: Champion, 1999.

Bibliothèque raisonnée des ouvrages des savans de l'Europe. Amsterdam, 1728–53.

Bolivar, Simón. "Mi delirio sobre el Chimborazo." In *Papeles de Bolívar*. Edited by Vicente Lecuna. 2 vols. Madrid: Editorial America, 1920.

Bomare, Jacques Christophe Valmont de. *Dictionnaire raisonné, universel d'histoire naturelle. Contenant l'histoire des animaux, des végétaux et des minéraux, et celle des corps célestes, des météores, & des autres principaux phénomenes de la nature; avec l'histoire et la description des drogues simples tirées des trois regnes . . . plus, une table concordante des noms latins . . . .* 9 vols. Paris: Brunet, 1775.

Bouciquault, Louïs le Maingre de. *Les Amazones revoltées, roman moderne, en forme de parodie sur l'histoire universelle, & la fable. Avec des notes politiques sur les travaux d'Hercule, La Chevalerie-Militaire, et la decouverte du nouveau monde, &c. &c. &c.* Rotterdam, 1737.

Bouguer, Pierre. *La figure de la terre. Déterminée par les observations de Messieurs Bouguer, & de la Condamine, de l'Académie royale des Sçiences, envoyés par ordre du Roy au Pérou, pour observer aux environs de l'Equateur. Avec une relation abregée de ce voyage, qui contient la description du pays dans lequel les operations ont été faites.* Paris: Charles-Antoine Jombert, 1749.

———. *Justification des Memoires de l'Academie royale des sçiences de 1744 et du livre de la figure de la terre.* Paris: C. A. Jombert, 1752.

———. *Lettre à Monsieur *** dans laquelle on discute divers points d'astronomie pratique, et où l'on fait quelques remarques sur le supplément au Journal historique du voyage à l'équateur de M. de la C.* Paris: Guerin & Delatour, 1754.

———. *Nouveau traité de navigation, contenant la théorie et la pratique du pilotage.* Paris: H. L. Guerin, 1753.

———. "Relation abrégée du voyage fait au Pérou." *Mémoires de l'Académie Royale des Sciences* (1744 [1748]): 249–97.

Bouis, Baron de. *Parterre géographique et historique, ou nouvelle méthode d'enseigner la géographie et l'histoire.* Paris, 1753.

Bowen, Emanuel. *Atlas Minimus Illustratus: Containing Fifty-two Pocket Maps of the World. Drawn and Engraved by J. Gibson; Revised, Corrected, and Improved.* London: Carnan and Newbery, 1774.

Breves instrucções aos correspondentes da Academia das Sciencias de Lisboa sobre as remessas dos productos, e noticias pertencentes à Historia da Natureza, para formar hum Museo Nacional. Lisbon: Regia officina typografica, 1781.

Brisson, Mathurin-Jacques. *Le regne animal divisé en IX classes; ou, Méthode contenant la division generale des animaux en IX classes, & la division particuliere des deux premieres classes, sçavoir de celle des quadrupedes & celle des cetacées . . . Aux quelles on a joint une courte description de chaque espèce . . .* Paris: Jean-Baptiste Buache, 1756.

Brosse, Guy de la. *L'Ovvertvre dv Iardin Royal de Paris, Povr la demonstration des plantes medecinales.* Paris: Jacques Dugast, 1640.

Buffon, George-Louis Leclerc, comte de, Louis-Jean-Marie Daubenton, Philibert Guéneau deMontbeillard, and Gabriel-Léopold-Charles-Aimé Bexon. *Histoire naturelle, générale et particulière, avec la description du cabinet du Roi.* 36 vols. Paris: Imprimerie Royale, 1749–89.

———. "Réflexions sur la loi de l'attraction." *Mémoires de l'Académie Royale des Sciences* (1745–49): 493–500.

———. "Réponse à M. de la Condamine." In *Histoire naturelle, générale et particulière: Supplément,* 4: 24–26. Paris: L'Imprimerie Royale, 1774–76.

Caldas, Francisco José de. *Expedición botánica de José Celestino Mutis al Nuevo Reino de Granada y memórias inéditas de Francisco José de Caldas.* Edited by Diego Mendoza. Madrid: Librería General de Victoriano Suárez, 1909.

Carte de France corrigée par ordre du Roy sur les observations de Mrs. de l'Académie des Sciences. Paris, 1682.

Carvajal, Gaspar de. *Relación del nuevo descubrimiento del Rio Grande de las Amazonas.* 1540–41. Quito: Gobierno del Ecuador, 1992.

Chevalier, François. "Sur une manière de lever la carte d'un pays." *Histoire de l'Académie Royale des Sciences* (1707–30): 113–16.

Clavijero, Francisco Javier. *Historia antigua de México.* Mexico City: Editorial Porrúa, 2003.

Clusius, Carolus. *Histoire des plantes en laqvelle est contenve la description entiere des herbes cest a dire, leurs especes, forme, noms, temperament, vertus & operations: Non seulement de celles qui croissent en ce pais, mais aussi des autres estrangeres qui viennent en vsage de medecine.* Antwerp: Jen Loë, 1557.

Collecção de noticias para a historia e geografia das nações ultramarinas, que vivem nos dominios Portuguezes, ou lhes são vizinhas. 7 vols. Lisbon: Typografia da Academia Real das Sciencias, 1812–56.

Condorcet, Marie-Jean-Antoine-Nicolas Caritat, marquis de. "Éloge de M. de la Condamine." In *Éloge des Académiciens de l'Académie royale des sciences, morts depuis l'an 1666 jusqu'en 1790,* vol. 1. Paris: Frédéric Vieweg, et Fuchs, 1799.

———. *Esquisse d'un tableau historique des progress de l'esprit humain.* Paris: Agasse, year III [1794].

Coronelli, Vincenzo Maria. *America meridionale.* Map. Venice, c. 1691.

Da Silva Araujo e Amazonas, Lourenço. *Diccionario topographico, historico descriptivo da Comarca do Alto-Amazonas.* Recife, 1852.

Dalibard, Thomas-François. *Florae Parisiensis Prodromus, ou Catalogue des plantes qui naissent dans les environs de Paris, rapportées sous les dénominations modernes & anciennes, & arrangées suivant la méthode séxuelle de M. Linnaeus.* Paris: Durand, 1749.

Daubenton, Louis-Jean-Marie. "Description du Cabinet du Roy." In Buffon, et al., *Histoire naturelle, générale et particulière, avec la description du Cabinet du Roi,* 3: 2. Paris: Imprimerie Royale, 1749.

Delille, Jacques. "Discours de Réception à L'Académie Française." In *Oeuvres de Jacques Delille.* Paris: L. G. Michaud, 1825.

Delisle, Guillaume. *Carte de la Terre Ferme, du Perou, du Bresil, et du pays des Amazones.* Map. Paris, 1703.

Descartes, René. *Discours de la méthode.* Notre Dame: University of Notre Dame Press, 1994.

Dictionnaire de l'Académie Françoise. Paris: J. B. Coignard, 1694.

Diderot, Denis. "Aguas." In *Encyclopédie, ou Dictionnaire raisonné des sciences, des arts et des métiers.* Edited by Denis Diderot and Jean Le Rond d'Alembert, 1: 191. Paris: Briasson, 1751.

———. *De l'interpretation de la Nature.* 1754. In *Oeuvres Philosophiques.* Edited by P. Verniere. Paris: Garnier, 1956.

———. "Encyclopédie." In *Encyclopédie, ou Dictionnaire raisonné des sciences, des arts et des métiers.* Edited by Denis Diderot and Jean Le Rond d'Alembert, 5: 646. Paris: Briasson, 1755.

———. *Oeuvres Complètes.* Edited by Roger Lewinter. Paris: Club Français du Livre, 1971.

———. "Prospectus." In *Discours préliminaire des éditeurs de 1751 et articles de l'Encyclopédie.* Edited by Martine Groult. Paris: Champion, 1999.

———. *Supplément au Voyage de Bougainville.* Edited by Michel Delon. Paris: Gallimard, 2002.

———. *Sur la liberté de la presse.* Edited by Jacques Proust. Paris: Éditions sociales, 1964.

Diderot, Denis, and Jean Le Rond d'Alembert, eds. *Encyclopédie, ou Dictionnaire raisonné des sciences, des arts et des métiers, par une société de gens de lettres.* 17 vols. Paris: Briasson, 1751–72. Facsimile reprint, Oxford: Pergamon, 1969. Also consulted at http://www.lib.uchicago.edu/efts/ARTFL/projects/encyc/.

Diderot, Denis, and Louis-Jean-Marie Daubenton. "Cabinet d'Histoire naturelle." In *Encyclopédie, ou Dictionnaire raisonné des sciences, des arts et des métiers.* Edited by Denis Diderot and Jean Le Rond d'Alembert, 2: 490. Paris: Briasson, 1752.

Documentos para la historia de la Audiencia de Quito. Edited by José Rumazo. 8 vols. Madrid: Afrodisio Aguado, 1948–50.

[Duchesne, H. G., and P. J. Macquer.] *Le petit cabinet d'histoire naturelle ou Manuel du naturaliste. Ouvrage utile aux voyageurs et à ceux qui visitent les cabinets d'histoire naturelle et de curiosités.* 4 vols. 2nd edition. Paris: Rémont, 1797.

Duhamel du Monceau, Henri-Louis. *Avis pour le transport par mer des arbres, des plantes vivaces, des semences et de diverses autres curiositésd'histoire naturelle.* Paris: L'imprimerie Royale, 1753.

Dulard, Paul-Alexandre. *La grandeur de dieu dans les merveilles de la nature.* Paris: Belin, 1804.

Du Bos, Abbé. *Réflexions critiques sur la poësie & sur la peinture.* 1770. 3 vols. Facsimile reprint, Geneva; Paris: Slatkine reprints, 1982.

Épître à M. Bouguer sur ses démêlés avec M. de la Condamine. In La Condamine, *Supplément au journal historique du voyage à l'Equateur.* 2 vols. Paris: Durand, 1752–54.

Évreux, Yves d'. *Suitte de l'histoire des choses plus mémorables advenues en Maragnon ès années 1613 et 1614, second traité.* Paris: François Huby, 1615.

Feijóo, Benito Jerónimo. *Glorias de España.* In *Biblioteca de autores españoles,* 56. Madrid: Ediciones Atlas, 1952.

———. "Mapa intellectual, y cotejo de naciones." In Feijóo, *Teatro crítico universal, ó discursos varios en todo género de materias, para desengaño de errores communes: Escrito por el muy ilustre señor D. Fr. Benito Jerónimo Feijoo y Montenegro, Maestro*

General del Orden de San Benito, del Consejo de S.M. &c., vol. 2, discourse 15: 299–321. Madrid: D. Joachin Ibarra, 1779.

Fernandez de Medrano, Sebastian. *Breve tratado de geographia divido [sic] en tres partes, Que la una contiene la descripcion del Rio y Imperio de las Amazonas Americanas, con su carta geographica . . .* Brussels: Lamberto Marchant, 1700.

Fertel, Martin Dominique. *La science pratique de l'imprimerie. Contenant des instructions tres-faciles pour se perfectionner dans cet art.* Saint Omer: Martin Dominique Fertel, 1723.

Feuillée, Louis. *Journal des observations physiques, mathematiques et botaniques, faites par ordre du Roi sur les Côtes Orientales de l'Amérique Méridionale, & aux Indes Occidentales, et dans un autre voïage fait par le même ordre à la Nouvelle Espagne, & aux Isles de l'Amérique.* 3 vols. Paris: P. Giffart, 1714–25.

Franklin, Benjamin. *Expériences et observations sur l'électricité faites à Philadelphie en Amérique.* Translated by Thomas-François Dalibard. Paris: Durand, 1752.

Frézier, Amédée-François. *Relation du voyage de la Mer du Sud aux côtes du Chily et Pérou fait pendant les années 1712, 1713 et 1714.* Paris: Jean-Geoffroy Nyon, Etienne Ganeau, Jacque Quillau, 1717.

———. *A Voyage to the South-Sea, along the Coasts of Chili and Peru, in the Years 1712, 1713, and 1714.* London: Jonah Bowyer, 1717.

Gamaches, Etienne-Simon de. *Astronomie physique, ou, Principes généraux de la nature appliqués au mécanisme astronomique, et comparés aux principes de la philosophie de M. Newton.* Paris: Jombert, 1740.

Garcilaso de la Vega, Inca. *Comentarios reales de los Incas. Primera parte. Que tratan del origen de los Yncas, reyes que fueron del Peru, de su idolatria, leyes, y gouierno en paz y en guerra: De sus vidas y conquistas, y de todo lo que fue aquel Imperio y su Republica, antes que los Españoles passaran a el.* Lisbon: Crasbeeck, 1609.

———. *Le commentaire royal, ou L'Histoire des Yncas, roys du Peru; contenant leur origine, depuis le premier Ynca Manco Capac, leur establissement, leur idolatrie, leurs sacrifices, leurs vies, leurs loix, leur gouuernement en paix & en guerre, leurs conquestes; les merueilles du temple du soleil; ses incroyables richesses, & tout l'estat de ce grand empire, auant que les Espagnols s'en fissent maistres, au temps de Guascar, & d'Atahuallpa.* Translated by Jean Boudoin. Paris: Augustin Courbé, 1633.

———. *Histoire des Incas, rois du Pérou. Nouvellement traduite de l'Espagnol de Garcillasso de la Vega. Et mise dans un meilleur ordre; avec des notes & des additions sur l'histoire naturelle de ce Pays.* Paris: Prault fils, 1744.

———. *Histoire des Yncas, rois du Perou.* Translated by Jean Baudoin. Amsterdam: Gerard Kuyper, 1704.

———. *Histoire des Yncas, rois du Perou.* Translated by Jean Baudoin. Amsterdam: J. Desbordes, 1715.

———. *Histoire des Yncas, rois du Pérou, depuis le premier Ynca Manco Capac, fils du soleil, jusqu'à Atahualpa dernier Ynca: Où l'on voit leur établissement, leur religion, leurs loix, leurs conquêtes . . .* Translated by Jean Baudoin. 2 vols. in 1. Amsterdam: J.-F. Bernard, 1737.

———. *The Royal Commentaries of Peru, in Two Parts*. Translated by Paul Rycaut. London: Miles Flesher, 1688.

Gastelier, Jacques-Elie. *Lettres sur les affaires du temps (1738–1741)*. Geneva: Slatkine, 1993.

Gazette d'Amsterdam. Second series. Amsterdam, 1691–1796.

Gomberville, Marin Le Roy de. *Relation de la rivière des Amazones, traduite par seu Mr de Gomberville de l'Academie Françoise. Sur l'original espagnol du P. Christophle d'Acuña Jesuite. Avec une dissertation sur la rivière des Amazones pour servir de preface . . .* 4 vols. Paris: Claude Barbin, 1682.

Gonçalvez de Fonseca, José. "Navegação feita da Cidade do Gram Pará até a bocca do Rio da Madeira pela escolta que por este rio subio às Minas do Mato Grosso, por ordem mui recommendada de Sua Magestada Fidelissima no anno de 1749, escripta por José Gonsalves da Fonseca no mesmo anno." In *Collecção de noticias para a historia e geografia das nações ultramarinas*, vol. 4. Lisbon: Academia das Ciências de Lisboa, 1812–56.

Graffigny, Françoise d'Issembourg d'Happoncourt, madame de. *Correspondance de Madame de Graffigny*. Edited by J. A. Dainard, English Showalter, et al. Oxford: Voltaire Foundation, 1985–.

———. *Lettres d'une péruvienne*. Paris, 1748.

Grosley, Pierre-Jean. *Londres*. Lausanne, 1774.

Gumilla, José. *El Orinoco ilustrado, historia natural, civil, y geographica, de este gran rio, y de sus caudalosas vertientes: Govierno, usos, y costumbres de los Indios sus habitadores, con nuevas, y utiles noticias de animales, arboles, frutos, aceytes, resinas, yervas, y raìces medicinales : Y sobre todo, se hallaràn conversiones muy singulares à nuestra Santa Fè, y casos de mucha edificacion*. Madrid: Manuel Fernandez, 1741.

Guyon, Claude-Marie. *Histoire des Amazones anciennes et modernes*. Paris: Jean Villette, 1740.

[Hennebert, Jean–Baptiste-François, and Gaspard Guillard de Beaurieu]. *Cours d'histoire naturelle, ou Tableau de la Nature, considérée dans l'homme, les quadrupèdes, les oiseaux, les poissons & les insectes. Ouvrage propre à inspirer aux gens du monde le desir de connoître les merveilles de la nature*. Paris: Lacombe, 1770.

Herbert, Cl.-J. *Essai sur la police générale des grains, sur leurs prix, et sur les effets de l'agriculture (1755)*. Edited by Edgard Depitre. Paris: Librairie Paul Geuthner, 1910.

Histoire de l'Academie royale des sciences, avec les Mémoires de mathématique & de physique, pour la même année. Tirés des registres de cette Académie. Paris: Imprimerie Royale. 1666–1790.

Humboldt, Alexander von. "Besuch der Pyramiden von Yaruquí." In *Reise auf dem Río Magdalena, durch die Anden und Mexico*. Edited by Margot Faak. Berlin: Akademie-Verlag, 1990.

———. *Reise auf dem Río Magdalena, durch die Anden und Mexico*. Edited by Margot Faak. 3 vols. Berlin: Akademie-Verlag, 1986–2003.

———. *Relation historique du voyage aux régions équinoxiales du nouveau continent, fait en 1799, 1800, 1802, 1803, et 1804.* 3 vols. Stuttgart: F. A. Brockhaus, 1970.

———. *Vue des cordillères, et monumens des peuples indigènes de l'Amérique.* Paris: F. Schoell, 1810.

"Inca ou Ynca." In *Encyclopédie, ou Dictionnaire raisonné des sciences, des arts et des métiers.* Edited by Denis Diderot and Jean Le Rond d'Alembert, 8: 641. Paris: Briasson, 1765.

Jaucourt, Louis, comte de. "Cahuchu." In *Encyclopédie, ou Dictionnaire raisonné des sciences, des arts et des métiers.* Edited by Denis Diderot and Jean Le Rond d'Alembert, 17: 760. Paris: Briasson, 1765.

———. "Perroquet vert varié." In *Encyclopédie, ou Dictionnaire raisonné des sciences, des arts et des métiers.* Edited by Denis Diderot and Jean Le Rond d'Alembert, 12: 399. Paris: Briasson, 1765.

———. "Quinquina." In *Encyclopédie, ou dictionnaire raisonné des sciences, des arts et des métiers.* Edited by Denis Diderot and Jean Le Rond d'Alembert, 13: 717. Paris: Briasson, 1765.

———. "Quito." In *Encyclopédie, ou Dictionnaire raisonné des sciences, des arts et des métiers.* Edited by Denis Diderot and Jean Le Rond d'Alembert, 13: 726. Paris: Briasson, 1765.

———. "Résine." *Encyclopédie, ou Dictionnaire raisonné des sciences, des arts et des métiers.* Edited by Denis Diderot and Jean Le Rond d'Alembert, 14: 173. Paris: Briasson, 1765.

———. "Singe." In *Encyclopédie, ou Dictionnaire raisonné des sciences, des arts et des métiers.* Edited by Denis Diderot and Jean Le Rond d'Alembert, 15: 209. Paris: Briasson, 1765.

———. "Voyage." In *Encyclopédie, ou Dictionnaire raisonné des sciences, des arts et des métiers.* Edited by Denis Diderot and Jean Le Rond d'Alembert, 17: 476. Paris: Briasson, 1765.

———. "Voyageur." In *Encyclopédie, ou Dictionnaire raisonné des sciences, des arts et des métiers.* Edited by Denis Diderot and Jean Le Rond d'Alembert, 17: 477. Paris: Briasson, 1765.

———. "Ynca." In *Encyclopédie, ou Dictionnaire raisonné des sciences, des arts et des métiers.* Edited by Denis Diderot and Jean Le Rond d'Alembert, 17: 671. Paris: Briasson, 1765.

———. "Zone Torride." In *Encyclopédie, ou Dictionnaire raisonné des sciences, des arts et des métiers.* Edited by Denis Diderot and Jean Le Rond d'Alembert, 17: 725. Paris: Briasson, 1765.

Journal des Sçavans. First series. Amsterdam, 1665–1781.

Journal Encyclopédique. Liège, 1756–75. Facsimile reprint, Geneva: Slatkine, 1967.

Juan, Jorge. "Petizion de los dos officials de Marina Españoles nombrados para asistir a las observaciones de los Franceses." In *Documentos para la historia de la Audiencia de Quito.* Edited by José Rumazo. Vol 5. Madrid: Afrodisio Aguado, 1948–50.

Juan, Jorge, and Antonio de Ulloa. *Dissertación histórica, y geographica sobre el meridiano de demarcación entre los Dominios de España, y Portugal, y los parages por donde passa en la America Meridional, conforme à los tratados, y derechos de cada estado, y las mas seguras, y modernas observaciones.* Madrid: Antonio Marin, 1749.

———. *Noticias secretas de América.* Madrid: Ediciones Istmo, 1988.

———. *Relacion histórica del viage a la América meridional hecho de órden de S. Mag. para medir algunos grados de meridiano terrestre, y venir por ellos en conocimiento de la verdadera figura, y magnitud de la tierra, con otras varias observaciones astronomicas, y phisicas . . .* Madrid: Antonio Marin, 1748. Facsimile edition with an introduction by José P. Merino Navarro and Miguel M. Rodriguez San Vicente. Madrid: Fundación Universitaria Española, 1978.

Juan, Jorge and Antonio de Ulloa. *A Voyage to South America. Describing at large, the Spanish Cities, Towns, Provinces, &c. on that Extensive Continent.* Second Edition. Revised and Corrected. London: L. Davis and C. Reymers, 1760.

Jussieu, Joseph de. *Description de l'arbre à Quinquina.* Edited by François Felix Pancier. Paris, 1936.

La Condamine, Charles-Marie de. *Carte du cours du Maragnon, ou de la grande riviére des Amazones. Dans sa partie navigable depuis Jaen de Bracamoros jusqu'à son Embouchure et qui comprend la Province de Quito, et la côte de la Guiane depuis le Cap de Nord jusqu'à Essequebè. Levée en 1743 et 1744 et assujettie aux observations astronomiques par M. de la Condamine de l'Ac. Rle. des Sc. Augmenté du cours de la Riviére Noire et d'autres détails tirés de divers mémoires et routiers manuscrits de voyageurs modernes.* Map designed by Jean-Baptiste Bourguignon d'Anville; engr. by Guillaume Delahaye. Paris, 1745.

———. *Carte du Detroit appellé Pongo de Mansériché dans le Maragnon ou la Rivière des Amazones entre Saut-Iago [sic] et Borja, où le lit du Fleuve se retrécit de 250 Toises a 25 Toises.* Map. Paris, 1745.

———. *Extracto del diario de observaciones hechas en el viage de la Provincia de Quito al Parà, por el Rio de las Amazonas; Y del Parà a Cayana, Surinam y Amsterdam. Destinado para ser leydo en la Assemblea publica de la Academia Real de las Ciencias de Paris. Por Monsr. de la Condamine.* Amsterdam: Joan Catuffe, 1745. Reprint, with introduction by Antonio Lafuente and Eduardo Estrella. Barcelona: Editorial Alta Fulla, 1986.

———. *Histoire des pyramides de Quito. Élevées par les Académiciens envoyés sous l'Equateur par ordre du Roi.* [Paris], 1751.

———. *Journal du voyage fait par ordre du Roi, a l'Équateur, servant d'introduction historique, a la Mesure des trois premiers degrés du méridien. Par M. de la Condamine.* Vol. 1, *Introduction historique: Ou journal des travaux des Académiciens envoyés par ordre du Roi sous l'Équateur.* Vol. 2, *Mésure des trois premiers degrés du méridien dans l'hémisphere austral.* Paris: Imprimerie royale, 1751.

———. "Lettre à Madame sur l'émeute populaire excitée en la ville de Cuenca au Pérou, le 29 d'août 1739, contre les académiciens des sciences envoyés pour la mesure de la Terre" in *Relation abrégée.* Paris: Veuve Pissot, 1745.

———. "Mémoire sur quelques anciens monumens du Perou, du tems des Incas." *Histoire de l'Academie Royal des Sciences et Belles Lettres, année 1746*, 435–56. Berlin: Ambroise Haude, 1748.

———. "Mémoire sur l'Inoculation de la petite vérole." *Mémoires de l'Académie Royale des Sciences* (1754): 615–70.

———. "Nouveau projet d'une mesure invariable, propre à servir de mesure commune à toutes les Nations." *Mémoires de l'Académie Royale des Sciences* (1747 [1751]): 489–514.

———. "Peticion de Don Carlos de la Condamine." In *Documentos para la Historia de la Audiencia de Quito*, edited by José Rumazo. Madrid: Afrodisio Aguado, 1948–50.

———. "Observations mathématiques et physiques faites dans un voyage de Levant en 1731 & 1732." *Mémoires de l'Académie Royale des Sciences* (1732 [1735]): 295–322.

———. *Relation abrégée d'un voyage fait dans l'intérieur de l'Amérique méridionale. Depuis la côte de la mer du Sud, jusqu'aux côtes du Brésil & de la Guiane, en descendant la rivière des Amazones; lûe à l'assemblée publique de l'Académie des Sciences, le 28. avril 1745.* Paris: Veuve Pissot, 1745.

———. "Relation abrégée d'un voyage fait dans l'intérieur de l'Amérique meridionale." *Mémoires de l'Académie Royale des Sciences* (1745 [1749]): 391–492.

———. "Sur l'Arbre du Quinquina." *Mémoires de l'Académie Royale des Sciences* (1738 [1740]): 226–44.

———. "Sur la manière de déterminer astronomiquement la différence en longitude de deux lieux peu éloignés l'un de l'autre." *Mémoires de l'Académie Royale des Sciences* (1735 [1738]): 1–11.

———. "Second mémoire sur l'inoculation de la petite vérole, contenant la suite de l'histoire de cette méthode & de ses progrès, de 1754 à 1758." *Mémoires de l'Académie Royale des Sciences* (1758 [1763]): 439–82.

———. *Supplément au Journal Historique du voyage à l'Equateur, et au livre de la mesure des trois premiers degrés du Meridien: Servant de réponse à quelques objections.* Paris: Durand, 1752–54.

———. *Voyage sur l'Amazone.* Edited by Hélène Minguet. Paris: Maspero, 1981.

Lafitau, Joseph-François. *Moeurs des sauvages amériquains, comparées aux moeurs des premiers temps.* Paris: Saugrain l'ainé, 1724.

Lahontan, Louis Armand de Lom d'Arce, baron de. *Nouveaux voyages de M. le baron de Lahontan dans l'Amérique Septentrionale.* The Hague: Frères Honoré, 1703.

Lery, Jean de. *Histoire d'un voyage faict en la terre du Brésil.* 1578. Edited with an introduction by Frank Lestringant. Paris: Librairie Générale Française, 1994.

Lettres édifiantes et curieuses écrites des missions étrangères. 1707–76. 34 vols. Paris: N. Le Clerc.

Linguet, Simon-Nicolas-Henri. *Canaux navigables, ou Développement des avantages qui résulteraient de l'execution des plusieurs projets en ce genre pour la Picardie, l'Artois, la Bourgogne, la Champagne, la Bretagne, & toute la France en géneral.* Amsterdam: L. Cellot, 1769.

———. *Mémoire sur les propriétés et privilèges exclusifs de la librairie: Présenté en 1774.* Paris, 1774.

Lucas, Paul. *Troisième voyage du sieur Paul Lucas, fait en MDCCXIV, &c. par ordre de Louis XIV, dans La Turquie, L'Asie, La Sourie, La Palestine, La Haute et La Basse Egypte, &c.* 3 vols. Rouen: R. Machuel, 1719.

Maffei, Scipione. "Per I Signori Academici delle Science spediti al Peru. Sonneto, in forma d'Inscrizione, da porsi, nel sito dove le due linee che saranno da essi ritracciate, sotto l'Equatore s'intersecheranno." In La Condamine, *Histoire des pyramides de Quito.* [Paris], 1751.

Magnin, Jean. "Description de la Province et des missions de Maynas au Royaume de Quito." In Thomas Henkel, ed., *Chronique d'un chasseur des âmes: Un jésuite suisse en Amazonie au XVIIIe siècle.* Fribourg: Éditions de l'Hèbe, 1993.

Maillet, Benoît de. *Description d'Egypte, contenant plusieurs remarques curieuses sur la géographie ancienne et moderne de ce païs, sur ces monumens anciens, sur les moeurs . . .* Paris: L. Genneau and J. Rollin fils, 1735.

Maldonado, Pedro Vicente. "Memorial a nombre de San Pedro de Riobamba." In *Documentos para la Historia de la Audiencia de Quito*, 8 vols. Edited by José Rumazo, 1: 7–13. Madrid: Afrodisio Aguado, 1948–50.

———. *Carta de la Provincia de Quito y de sus adjacentes.* One map in four sheets; 120×81 cm. Engraved by Guillaume de la Haye. Scale [ca 1:900.000]. Spanish leagues 17 1/2 per degree. [Paris], 1750.

———. "Representación de Pedro Vicente Maldonado a S. M." In *Documentos para la Historia de la Audiencia de Quito*, 8 vols. Edited by José Rumazo, 1: 47–52. Madrid: Afrodisio Aguado, 1948–50.

Mallet, Abbé. "Amautas." In *Encyclopédie, ou Dictionnaire raisonné des sciences, des arts et des métiers.* Edited by Denis Diderot and Jean Le Rond d'Alembert, 1: 317. Paris: Briasson, 1751.

Marggraf, Georg and Willem Piso. *Historia naturalis Brasiliae.* Amsterdam, 1648.

Maroni, Pablo. *Noticias autenticas del famoso rio Marañon.* 1738. Reprint, Iquitos, Peru: IIAP-CETA, 1988.

Maupertuis, Pierre-Louis Moreau de. *Discours sur les différentes figures des astres; D'ou l'on tire des conjectures sur les étoiles qui paroissent changer de grandeur; & sur l'Anneau de Saturne. AVEC une exposition abbrégée des systemes de M. Descartes & de M. Newton.* Paris: L'Imprimerie Royale, 1732.

———. *Lettre sur le progrès des sciences.* Paris: [s.l.: s.n.], 1752.

Mémoires pour l'histoire des sciences & des beaux arts. Trévoux, Lyon, and Paris, 1701–67.

Mercure Danois. Copenhagen, 1753–59.

Mercure de France. Paris, 1724–78.

Mocquet, Jean. *Voyages en Afrique, Asie, Indes orientales et occidentales.* Paris: J. de Heuqueville, 1617.

Monteiro de Noronha, Jozé. "Roteiro da cidade do Pará até às ultimas colonias dos Dominios Portuguezes em os Rios Amazonas e Negro. Illustrado com algumas noticias, que pódem interessar a curiosidade dos navegantes, e dar mais claro conhecimento das suas capitanias do Pará, e de S. José do Rio Negro." In *Collecção de noticias para a historia e geografia das nações ultramarinas, que vivem nos Dominios Portuguezes, ou lhes são vizinhas,* vol. 6. Lisbon: Academia das Sciencias, 1856.

Montesquieu, Charles de Secondat, Baron de. *De l'esprit des lois, ou du rapport que les lois doivent avoir avec la constitution de chaque gouvernement, les moeurs, le climat religion, la commerce, &c.* Geneva: Barrillot & Fils, 1748.

Montfaucon, Bernard de. *Antiquity Explained and Represented in Sculptures: London, 1721–1722.* 5 vols. Translated by David Humphreys. New York: Garland, 1976.

Newton, Isaac. *The Principia: Mathematical Principles of Natural Philosophy.* Translated by I. Bernard Cohen and Anne Whitman, assisted by Julia Budenz. Berkeley: University of California Press, 1999.

Pagan, Blaise François, comte de. *Relation historique et geographique de la Grande Rivière des Amazones dans l'Amérique. Extraite de divers autheurs, et reduitte en meilleure forme. Avec la carte de la mesme Rivière, et de ses provinces.* Paris: Cardin Bessongne, 1656.

Petit, Pierre. *Traité historique sur les Amazones; Où l'on trouve tout ce que les auteurs, tant anciens que modernes, ont écrit pour ou contre ces heroines;Et où l'on apporte quantité de medailles & d'autres monumens anciens, pour prouver qu'elles ont existé.* Leide: J. A. Langerak, 1718.

Picart, Bernard. *Histoire générale des cérémonies religieuses de tous les peuples.* 8 vols. Paris: Rollin fils, 1741.

Pluche, Noel-Antoine. *Le spectacle de la nature, ou entretien sur les particularités de l'histoire naturelle, qui ont paru les plus propres à rendre les jeunes gens curieux, et à leur former l'esprit.* Second edition. 8 vols. Paris: Veuve Estienne, 1732–51.

Postel, Guillaume. *Les tres-merveilleuses victoires des femmes du nouveau monde, et comment elles doivent à tout le monde par raison commander, & même à ceulx qui auront la monarchie du monde vieil.* Paris: Jehan Ruelle, 1553.

Le Pour et Contre. Paris, 1733–40. Facsimile reprint, Geneva: Slatkine, 1967.

Prévost, Antoine-François, abbé. *Histoire générale des voyages, ou nouvelle collection de toutes les relations de voyages par mer et par terre qui ont été publiées jusqu'à présent dans les différentes langues de toutes les nations connues.* 20 vols. Paris: Didot, 1746–89.

Radonvilliers, l'Abbé de. "Réponse au discours de réception à L'Académie française de J. Delille." In *Oeuvres de J. Delille.* Paris: L. G. Michaud, 1825.

Ramsay, A.-M. *Les voyages de Cyrus.* Paris: G. F. Quillau, 1727.

Raynal, Guillaume-Thomas. *Histoire philosophique et politique des établissements et du commerce des Européens dans les deux Indes.*

Réaumur, René-Antoine Ferchault de. "Moyens d'empêcher l'evaporation des liqueurs spiritueuses dans lesquelles on veut conserver des productions de la Nature de differens genres." *Mémoires de l'Académie Royale des Sciences* (1746): 483–538.

———. "Regles pour construire des thermometres dont les degrés soient compara-bles." *Mémoires de l'Académie Royale des Sciences* (1730): 452–507.

"Representación de Pedro Vicente Maldonado a S. M." In *Documentos para la Historia de la Audiencia de Quito*. 8 vols. Edited by José Rumazo, 1: 47–52. Madrid: Afrodisio Aguado, 1948–50.

Robertson, William. *Histoire de l'Amérique*. Maestricht: Jean-Edme Dufour & Philippe Roux, 1777. Originally published as *History of America*. London: W. Strahan, 1777.

Rodriguez, Manuel. *El Marañon, y Amazonas. Historia de los descubrimientos, entradas, y reduccion de naciones. Trabajos malogrados de algunos conquistadores, y dichosos de otros, assi temporales, como espirituales, en las dilatadas montañas y mayores rios de la America*. Madrid: Antonio Gonçalez de Reyes, 1684.

Romagnesi, Jean-Antoine. *Les sauvages: Parodie de la tragedie d'Alzire*. London, 1736.

Rousseau, Jean-Jacques. *Discours sur l'origine et les fondements de l'inégalité parmi les hommes*. Amsterdam: Marc Michel Rey, 1755.

———. *Oeuvres politiques*. Paris: Bordas, 1989.

Sá, José Antonio de. *Compendio de observaçoens que fórmão o plano da viagem politica, e filosofica, que se deve fazer dentro da Patria*. Lisbon, 1783.

Sampaio, Francisco Xavier Ribeiro de. *Diario da viagem que, em visita e correição das povoações da capitania de S. Joze do Rio Negro, fez o ouvidor, e intendente geral da mesma Francisco Xavier Ribeiro de Sampaio, no anno de 1774 e 1775*. Lisbon: Academia das Sciencias, 1825.

Sempere y Guarinos, Juan. *Ensayo de una biblioteca española de los mejores escritores del reynado de Carlos III*. Madrid: Imprenta Real, 1785–89.

Sicard, Claude. *Lettres et relations inédites*. Edited by Maurice Martin. Cairo: Institut Français d'Archéologie Orientale du Caire, 1982.

———. *Parallèle géographique de l'ancienne Égypte et de l'Égypte moderne*. Edited by Serge Sauneron and Maurice Martin. Cairo: Institut Français d'Archéologie Orientale du Caire, 1982.

Suite de la clef, ou Journal historique sur les matieres du tems. Paris, 1717–76.

Terrasson, Jean, abbé. *Sethos: Histoire, ou, Vie tirée des monuments anecdotes de l'ancienne Egypte*. Paris: H. L. Guerin, 1731.

Thevet, André. *Les singularités de la France antarctique*. 1557. Published as *Le Brésil d'André Thévêt*. Edited with an introduction by Frank Lestringant. Paris: Chandeigne, 1997.

Thomson, James. *Les saisons*. Paris: Chaubert, 1759. Originally published as *The Seasons*. Edinburgh: Donaldson, 1724.

"Traduction du discours prononcé par S.E. le Président de la République de l'Équateur, à l'occasion de la restauration des Pyramides de Caraburu et d'Oyambaro, le 25 novembre 1836." Reproduced in "Histoire des pyramides de Quito. Documents Inédits." Edited by G. Perrier. *Journal de la Société des Américanistes (Paris)* 1943–46 (1947).

Turgot, Anne-Robert-Jacques-Marie. *Mémoire instructif sur la manière de rassembler, de préparer, de conserver et d'envoyer diverses curiosités d'histoire naturelle.* Paris: Bruyset, 1758.

Ulloa, Antonio de. *Noticias americanas: Entretenimientos phisicos-historicos sobre la América Meridional y la Septentrianal [sic] oriental.* Madrid: Manuel de Mena, 1772.

———. "Resumen histórico del origen, y succession de los Incas." In Jorge Juan and Antonio de Ulloa, *Relación histórica del viage a la América meridional.* Madrid: Antonio Marin, 1748.

Varenius, Bernard. *Géographie générale, composée en Latin par Bernard Varenius; Revue par Isaac Newton, augmentée par Jacques Jurin, traduite en Anglois d'après les editions latines données par ces auteurs, avec des additions sur les nouvelles découvertes; & présentement traduite de l'Anglois en François avec des figures en taille-douce.* Paris: Vincent, 1755.

Vau de Claye, Jacques de. *Carte d'une partie de l'Océan Atlantique.* Map. 1579.

Vaugondy, Robert de. "Géographie." In *Encyclopédie, ou Dictionnaire raisonné des sciences, des arts et des métiers.* Edited by Denis Diderot and Jean Le Rond d'Alembert, 7: 611. Paris: Briasson, 1757.

Velasco, Juan de. *Historia del Reino de Quito.* 1788–89. 3 vols. Quito: Casa de la Cultura Ecuatoriana, 1977.

Volney, Constantin-François Chasseboeuf de. *Les ruines, ou méditation sur les révolutions des empires.* Paris: Desenne, Volland, Plassan, 1791.

Voltaire, François-Marie Arouet de. *Alzire, ou les Americains.* Paris: Jean-Baptiste-Claude Bauche, 1736.

———. *The Complete Works of Voltaire.* Edited by Theodore Besterman, et. al. Geneva, Banbury, and Oxford, 1968–.

———. *Correspondance.* Edited by Theodore Besterman. 12 vols. Paris: Gallimard, 1977–93.

———. *Correspondence.* Edited by Theodore Besterman. 135 vols. Geneva: Institut et musée Voltaire, 1953–77.

———. *Dictionnaire philosophique.* 1764. Edited by René Pomeau. Paris: Flammarion, 1964.

———. *Élements de la philosophie de Newton.* 1738. In *The Complete Works of Voltaire.* Edited by Robert L. Walters and W.H. Barber, vol. 15. Oxford: Voltaire Foundation, 1992.

———. *Lettres philosophiques: Ou Lettres anglaises.* 1734. Paris: Garnier, 1964.

———. "Sur M. de La Condamine, qui était occupé de la mesure d'un degré du méridien au Pérou, lorsque M. de Voltaire faisait Alzire." In *The Complete Works of Voltaire.* Edited by W. H. Barber and Ulla Kölvina, 14: 540. Oxford: Voltaire Foundation, 1989.

Vosgien, [Jean-Baptiste Ladvocat]. *Dictionnaire géographique-portatif, ou Description de tous les royaumes, provinces, villes, patriarchats, evéchés.* The Hague: C. P. Gosse & J. Neaulme, 1748.

Voyages and Discoveries in South-America, the First Up the River of the Amazons, to Quito, in Peru, and Back Again to Brazil, Performed at the Command of the King of

Spain, by Christopher d'Acugna. The Second up the River of Plata, and Thence by Land to the Mines of Potozi, by M. Acarete. The Third from Cayenne, into Guiana, in Search of the Lake of Parima, Reputed the Richest Place in the World, by M. Grillet and Bechamel. Done into English from the Originals, Being the Only Accounts of Those Parts Hitherto Extant. London: S. Buckley, 1698

Warburton, William. *The Divine Legation of Moses Demonstrated: On the Principles of a Religious Deist, from the Omission of the Doctrine of a Future State of Reward and Punishment in the Jewish Dispensation.* London: Fletcher Gyles, 1738–42.

———. *Essai sur les hieroglyphes des Egyptiens, où l'on voit l'origine & le progrès du langage & de l'ecriture, l'antiquité des sciences en Egypte & l'origine du culte des animaux. Traduit de l'Anglois de M. Warburthon, avec des observations sur l'antiquité des hieroglyphes scientifiques & des remarques sur la chronologie & sur la première ecriture des Chinois.* Paris: Hippolyte Louis Guerin, 1744.

Watson, William. "Several Papers concerning a New Semi-Metal, Called Platina; Communicated to the Royal Society by Mr. Wm. Watson F.R.S." *Philosophical Transactions of the Royal Society* 46 (1749–50): 584–96.

Woodward, John. *Brief instructions for making observations in all parts of the world : As also, for collecting, preserving, and sending over natural things.* London: R. Wilkin, 1696.

Secondary Sources

Abercrombie, Thomas A. *Pathways of Memory and Power: Ethnography and History among an Andean People.* Madison: University of Wisconsin Press, 1998.

Acevedo Marin, Rosa Elizabeth, and Flávio Gomes. "Reconfigurações Coloniais: Tráfico de Indígenas, Fugitivos e Fronteiras no Grão-Pará e Guiana Francesa (sécs. XVII e XVIII)." *Revista de História* (São Paulo) 149 (2003): 69–107.

Adorno, Rolena. "The Discursive Encounter of Spain and America: The Authority of Eyewitness Testimony in the Writing of History." *William and Mary Quarterly*, third series, 49, no. 2 (April 1992): 210–28.

———. *From Oral to Written Expression: Native Andean Chronicles of the Early Colonial Period.* Syracuse, N.Y.: Maxwell School of Citizenship and Public Affairs, 1982.

Alencastro, Luiz Felipe de. "Economic Network of Portugal's Atlantic World." In *Portuguese Oceanic Expansion, 1400–1800.* Edited by Diogo Ramada Curto and Francisco Bethencourt. Cambridge: Cambridge University Press, 2006.

———. *O trato dos viventes: Formação do Brasil no Atlântico sul. Séculos XVI e XVII.* São Paulo: Companhia das Letras, 2000.

Alpers, Svetlana. *The Art of Describing: Dutch Art in the Seventeenth Century.* Chicago: University of Chicago Press, 1983.

Anderson, Wilda. *Between the Library and the Laboratory: The Language of Chemistry in Eighteenth-Century France.* Baltimore: Johns Hopkins University Press, 1984.

Andrien, Kenneth. *The Kingdom of Quito, 1690–1830*. Cambridge: Cambridge University Press, 1995.

Astudillo Espinosa, Celín. "Don Pedro Vicente Maldonado." *Revista geografica* (Quito) 30 (June 1992): 13–31.

Augé, Marc. "Voyage et ethnographie. La vie comme récit." *L'homme* 151 (1999): 11–20.

Auslander, Leora. *Taste and Power: Furnishing Modern France*. Berkeley: University of California Press, 1996.

Baczko, Bronislaw. "Voyager avec l'*Encyclopédie*." In *L'Encyclopédie: Du réseau au livre et du livre au réseau*. Edited by Robert Morrissey and Philippe Roger, 113–26. Paris: Honoré Champion, 2001.

Badinter, Elisabeth. *Les passions intellectuelles*. Vol. 1, *Désirs de gloire (1735–1751)*. Vol. 2, *Exigence de dignité (1751–1762)*. Paris: Fayard, 1999–2002.

Bailyn, Bernard. *Atlantic History: Concept and Contours*. Cambridge: Harvard University Press, 2005.

Baker, Keith Michael. "Épistémologie et politique: Pourquoi l'*Encyclopédie* est-elle un dictionnaire?" In *L'Encyclopédie: Du réseau au livre et du livre au réseau*. Edited by Robert Morrissey and Philippe Roger. Paris: Honoré Champion, 2001.

———. "Politics and Public Opinion under the Old Regime: Some Reflections." In *Press and Politics in Pre-Revolutionary France*. Edited by Jack R. Censer and Jeremy D. Popkin, 208–14. Berkeley: University of California Press, 1987.

Barrera-Osorio, Antonio. *Experiencing Nature: The Spanish American Empire and the Early Scientific Revolution*. Austin: University of Texas Press, 2006.

Basalla, George. "The Spread of Western Science." *Science*, new series, 156, no. 3775 (1967): 611–22.

Bauer, Ralph. *The Cultural Geography of Colonial American Literatures: Empire, Travel, Modernity*. Cambridge: Cambridge University Press, 2003.

Bayle, Constantin. "Descubridores jesuitas del Amazonas." In *Revista de Indias* 1 (1940): 121–85.

Bell, David A. *The Cult of the Nation in France: Inventing Nationalism*. Cambridge: Harvard University Press, 2001.

———. *Lawyers and Citizens: The Making of a Political Elite in Old Regime France*. New York: Oxford University Press, 1994.

———. "The 'Public Sphere,' the State, and the World of Law in Eighteenth-Century France." *French Historical Studies* 17, no. 4 (Autumn 1992): 912–34.

Bénassy, Marie-Cécile, Jean-Pierre Clément, Francisco Pelayo, and Miguel Ángel Puig-Samper, eds. *Nouveau monde et renouveau de l'histoire naturelle*. 3 vols. Paris: Presses de la Sorbonne Nouvelle, 1986–93.

Benavides Solís, Jorge. *La arquitectura y el urbanismo de Cochasquí*. [Quito]: Departamento de Cultura de la Universidad Central, 1986.

Benedict, Barbara M. *Curiosity: A Cultural History of Early Modern Inquiry*. Chicago: University of Chicago Press, 2001.

Berlin, Ira. "From Creole to African: Atlantic Creoles and the Origins of African-American Society in Mainland North America." *William and Mary Quarterly*, 3rd series, 53, no. 2 (April 1996): 251–88.

Besse, Jean-Marc. *Face au monde. Atlas, jardins, géoramas*. Paris: Desclée de Brouwer, 2003.

Bhabha, Homi. "Of Mimicry and Man." In *The Location of Culture*. London: Routledge, 1994.

Biagioli, Mario. *Galileo, Courtier: The Practice of Science in the Culture of Absolutism*. Chicago: University of Chicago Press, 1993.

Biagioli, Mario, and Peter Galison, eds. *Scientific Authorship: Credit and Intellectual Property in Science*. New York: Routledge, 2003.

Bicalho, Maria Fernanda. *A Cidade e o Império. O Rio de Janeiro no século XVIII*. Rio de Janeiro: Editora Civilização Brasileira, 2003.

Blair, Ann. "Humanist Methods in Natural Philosophy: The Commonplace Book." *Journal of the History of Ideas* 53, no. 4 (Oct.–Dec. 1992): 541–51.

———. "Reading Strategies for Coping with Information Overload, ca. 1550–1700." *Journal of the History of Ideas* 64, no. 1 (January 2003): 11–28.

———. *The Theater of Nature: Jean Bodin and Renaissance Science*. Princeton: Princeton University Press, 1997.

Blanckaert, Claude, Claudine Cohen, Pietro Corsi, and Jean-Louis Fischer, eds. *Le Muséum au premier siècle de son histoire*. Paris: Muséum national d'Histoire naturelle, 1997.

Bleichmar, Daniela. "Books, Bodies, and Fields: Sixteenth-Century Transatlantic Encounters with New World *Materia Medica*." In *Colonial Botany: Science, Commerce, and Politics in the Early Modern World*. Edited by Londa Schiebinger and Claudia Swan. Philadelphia: University of Pennsylvania Press, 2005.

Blok, Josine H. "The Early Amazons: Modern and Ancient Perspectives on a Persistent Myth." In *Religions in the Graeco-Roman World*. Edited by R. Van den Broek et al. Leiden: E. J. Brill, 1995.

Bollème, G., et al., eds. *Livre et société dans la France du XVIIIe siècle*. Paris: Mouton and Co., 1965.

Bono, James J. *The Word of God and the Languages of Man: Interpreting Nature in Early Modern Science and Medicine*. Vol. 1, *Ficino to Descartes*. Madison: University of Wisconsin Press, 1995.

Boone, Elizabeth Hill, and Walter D. Mignolo, eds. *Writing without Words: Alternative Literacies in Mesoamerica and the Andes*. Durham, N.C.: Duke University Press, 1994.

Booy, J. de, ed. *Denis Diderot, Ecrits inconnus de jeunesse*. Studies on Voltaire and the Eighteenth Century, vol. 119 (1975):69–106; 195–209 and vol. 178 (1979): (447–65).

Bourguet, Marie-Noëlle. "La collecte du monde: Voyage et histoire naturelle (fin XVIIème siècle–début XIXème siècle." In *Le Muséum au premier siècle de son histoire*. Edited by Claude Blanckaert et al., 163–96. Paris: Muséum national d'Histoire naturelle, 1997.

————. "Measurable Difference: Botany, Climate, and the Gardener's Thermometer in Eighteenth-Century France." In *Colonial Botany: Science, Commerce, and Politics in the Early Modern World*. Edited by Londa Schiebinger and Claudia Swan. Philadelphia: University of Pennsylvania Press, 2005.

————. "Voyage, mer et science au XVIIe siècle." *Bulletin de la Société d'Histoire Moderne et Contemporaine*, nos. 1–2 (1997): 39–56.

Bourguet, Marie-Noëlle, and Christian Licoppe. "Voyage, instruments et mesures: Une nouvelle expérience du monde au Siècle des lumières." *Annales. Histoire, Sciences sociales* 52, no. 5 (September–October 1997): 1115–51.

Bourguet, Marie-Noëlle, Christian Licoppe, and H. Otto Sibum. Introduction to *Instruments, Travel and Science: Itineraries of Precision from the Seventeenth to the Twentieth Century*. Edited by Marie-Noëlle Bourguet, Christian Licoppe, and H. Otto Sibum, 1–19. London: Routledge, 2002.

Bourguet, Marie-Noëlle, Christian Licoppe, and H. Otto Sibum, eds., *Instruments, Travel and Science: Itineraries of Precision from the Seventeenth to the Twentieth Century*. London: Routledge, 2002.

Bouza, Fernando. *Corre manuscrito: Una historia cultural del Siglo de Oro*. Madrid: Marcial Pons, 2002.

————. *Imagen y propaganda. Capítulos de historia cultural del reinado de Felipe II*. Madrid: Ediciones AKAL, 1998.

Brading, David. *The First America: The Spanish Monarchy, Creole Patriots, and the Liberal State, 1492–1867*. Cambridge: Cambridge University Press, 1993.

————. "The Incas and the Renaissance: The Royal Commentaries of Inca Garcilaso de la Vega." *Journal of Latin American Studies* 18, no. 1 (May 1986): 1–23.

Brenot, Anne-Marie. "Les voyageurs français dans la vice-royauté du Pérou au XVIIIe siècle." *Revue d'histoire moderne et contemporaine* 35 (1988): 240–61.

Brewer, Daniel. *The Discourse of Enlightenment in Eighteenth-Century France: Diderot and the Art of Philosophizing*. Cambridge: Cambridge University Press, 1993.

Brewer, John, and Susan Staves, eds. *Early Modern Conceptions of Property*. New York: Routledge, 1995.

Briggs, Robin. "The Académie Royale des Sciences and the Pursuit of Utility." *Past and Present* 131 (1991): 38–88.

Broc, Numa. *La géographie des philosophes: Géographes et voyageurs français au XVIIIe siècle*. Paris: Ophrys, 1975.

————. "Voyages et géographie au XVIIIe siècle." In *Revue d'histoire des sciences et de leurs applications* 22 (1969): 137–54.

Broman, Thomas. "The Habermasian Public Sphere and 'Science *in* the Enlightenment.'" *History of Science* 36 (1998): 123–49.

Brown, Gregory S. "A Field of Honor: The Cultural Politics of Playwriting in Eighteenth-Century France." Ph.D. diss., Columbia University, 1997.

————. "The Self-Fashionings of Olympe de Gouges, 1784–1789." *Eighteenth-Century Studies* 34, no. 3 (2001): 383–401.

Brown, Harcourt. "From London to Lapland and Berlin." In *Science and the Human Comedy: Natural Philosophy in French Literature from Rabelais to Maupertuis*, 167–206. Toronto: University of Toronto Press, 1976.

Bruleaux, Anne-Marie, Régime Calmon, and Serge Mam-Lam-Fouck, eds. *Deux siècles d'esclavage en Guyane française, 1652–1848*. Cayenne: CEGER; Paris: L'Harmattan, 1986.

Buarque de Holanda, Sérgio. *Visão do paraíso: Os motivos edênicos no descobrimento e colonização do Brasil*. São Paulo: Companhia Editora Nacional, 1969.

Burke, Peter. *A Social History of Knowledge: From Gutenberg to Diderot*. Cambridge: Cambridge University Press, 2000.

Burnett, D. Graham. *Masters of All They Surveyed: Exploration, Geography, and a British El Dorado*. Chicago: University of Chicago Press, 2000.

Calatayud Arinero, María de los Angeles. "Antecedentes y creación del Real Gabinete de Historia Natural de Madrid." *Arbor* 482 (February 1986): 9–33.

———. "Pedro Franco Dávila: Aspectos de una vida." In *Científicos criollos e Ilustración*. Edited by Diana Soto Arango, Miguel Ángel Puig-Samper, and María Dolores González-Ripoll. Madrid: Doce Calles, 1999.

———. *Pedro Franco Dávila y el Real Gabinete de Historia Natural*. Madrid: CSIC, 1988.

Calhoun, Craig. *Habermas and the Public Sphere*. Cambridge: MIT Press, 1992.

Campbell, Mary Baines. *Wonder and Science: Imagining Worlds in Early Modern Europe*. Ithaca: Cornell University Press, 1999.

Cañizares-Esguerra, Jorge. "How Derivative Was Humboldt? Microcosmic Nature Narratives in Early Modern Spanish America and the (Other) Origins of Humboldt's Ecological Ideas." In *Colonial Botany: Science, Commerce, and Politics in the Early Modern World*. Edited by Londa Schiebinger and Claudia Swan. Philadelphia: University of Pennsylvania Press, 2005.

———. *How to Write the History of the New World: Historiographies, Epistemologies, and Identities in the Eighteenth-Century Atlantic World*. Stanford: Stanford University Press, 2001.

———. *Puritan Conquistadors: Iberianizing the Atlantic, 1550–1700*. Stanford: Stanford University Press, 2006.

Canny, Nicholas, and Anthony Pagden, eds. *Colonial Identity in the Atlantic World. 1500–1800*. Princeton: Princeton University Press, 1986.

Capel, Horacio. "Geografía y Cartografía." In *Carlos III y la ciencia de la Ilustración*. Edited by Manuel A. Sellés, José Luis Peset, and Antonio Lafuente. Madrid: Alianza Editorial, 1988.

Carlyon, Jonathan Earl. *Andrés González de Barcia and the Creation of the Colonial Spanish American Library*. Toronto: University of Toronto Press, 2005.

Carney, Judith. *Black Rice: The African Origins of Rice Cultivation in the Americas*. Cambridge: Harvard University Press, 2001.

Casciato, Maristella, Maria Grazia Ianniello, and Maria Vitale. *Enciclopedismo in Roma barocca: Athanasius Kircher e il Museo del Collegio Romano tra Wunderkammere e museo scientifico*. Venice: Marsilio, 1986.

Castillo, Antonio. *Entre la pluma y la pared: Una historia social de la escritura en los siglos de oro*. Madrid: AKAL, 2006.

Céard, Jean. *La nature et les prodiges: L'insolite au 16e siècle, en France*. Geneva: Droz, 1977.

Certeau, Michel de. *L'invention du quotidien. I. Arts de faire*. Introduction by Luce Giard. 1980. Paris: Gallimard, 1990. Translated by Steven F. Rendall as *The Practice of Everyday Life* (Berkeley: University of California Press, 2002).

———. "L'oralité, ou l'espace de l'autre: Léry." In *L'écriture de l'histoire*, 215–48. Paris: Éditions Gallimard, 1975.

Charlton, D. G. *New Images of the Natural in France: A Study of European Cultural History, 1750–1800*. Cambridge: Cambridge University Press, 1984.

Charpentier, Jeanne, and Michel Charpentier, eds. *L'Encyclopédie*. Paris: Éditions Bordas, 1967.

Chartier, Roger. *The Cultural Origins of the French Revolution*. Durham, N.C.: Duke University Press, 1991.

———. *The Cultural Uses of Print in Early Modern France*. Trans. Lydia G. Cochrane. Princeton: Princeton University Press, 1987.

———. *Culture écrite et société. L'ordre des livres (XIVe–XVIIIe siècle)*. Paris: Albin Michel, 1996.

———. "Do Books Cause Revolutions?" In *Cultural Origins*, 67–91.

———. "Figures de l'auteur." In *Culture écrite et société*, 45–80.

———. "L'homme de lettres." In *L'homme des lumières*. Edited by Michel Vovelle, 159–209. Paris: Éditions du Seuil, 1996.

———. *Inscrire et effacer: Culture écrite et littérature, XIe–XVIIIe siècle*. Paris: Seuil, 2005. Translated by Arthur Goldhammer as *Inscription and Erasure: Literature and Written Culture from the Eleventh to the Eighteenth Century*. Philadelphia: University of Pennsylvania Press, 2007.

———. Introduction to Chrétien Guillaume de Lamoignon de Malesherbes, *Mémoires sur la librairie et sur la liberté de la presse*. Paris: Imprimerie nationale, 1994.

———. *Lectures et lecteurs dans la France d'Ancien régime*. Paris: Seuil, 1987.

———. *Publishing Drama in Early Modern Europe*. London: British Library, 1999.

———. "Texts, Printing, Readings." In *The New Cultural History*. Edited by Lynn Hunt, 154–75. Berkeley: University of California Press, 1989.

———. "Les représentations de l'écrit." In *Culture écrite et société. L'ordre des livres (XIVe–XVIIe siècle)*, 17–44. Paris: Albin Michel, 1996.

Chartier, Roger, and Pietro Corsi, eds. *Sciences et langues en Europe*. Paris: Centre Koyré/EHESS, 1996.

Chevalier, Auguste. "Le deuxième centenaire de la découverte du Caoutchouc faite par Charles-Marie de La Condamine." *Revue de Botanique Appliquée et d'Agriculture Tropicale* 179 (1936): 519–29.

Chiappelli, Fredi, ed. *First Images of America*. Berkeley: University of California Press, 1976.

Chinard, Gilbert. *L'Amérique et le rêve exotique dans la littérature française au XVIIe et XVIIIe siècle.* Paris: Droz, 1934.

Chouillet, Anne-Marie. "Rôle de la presse périodique de langue française dans la diffusion des informations concernant les missions en Laponie ou sous l'Équateur." In *La figure de la terre, du XVIIIe siècle à l'ère spatiale.* Edited by Henri Lacome and Pierre Costabel, 171–89. Paris: Gauthier-Villars, 1988.

Chouillet, Anne-Marie. "Trois lettres inédites de Diderot." *Recherches sur Diderot et sur l'Encyclopédie (RDE)* 11 (October 1991): 8–17.

Christóvão, Fernando, ed. *Condicionantes culturais da literatura de viagens. Estudos e bibliografias.* Lisbon: Edições Cosmos, 1999.

Clark, William, Jan Golinski, and Simon Schaffer. Introduction to *The Sciences in Enlightened Europe.* Edited by William Clark, Jan Golinski, and Simon Schaffer. Chicago: University of Chicago Press, 1999.

Clark, William, Jan Golinski, and Simon Schaffer, eds. *The Sciences in Enlightened Europe.* Chicago: University of Chicago Press, 1999.

Clément, Gilles. *Le jardin planétaire.* Paris: Albin Michel, 1999.

Clément, Jean-Pierre. "Réflexions sur la politique scientifique française vis-à-vis de l'Amérique espagnole au siècle des Lumières." In *Nouveau monde et renouveau de l'histoire naturelle.* Edited by Marie-Cécile Bénassy, Jean-Pierre Clément, Francisco Pelayo, and Miguel Ángel Puig-Samper, 3: 131–59. Paris: Presses de la Sorbonne Nouvelle, 1994.

Cohen, I. Bernard. *The Birth of a New Physics.* New York: Norton, 1985.

———. *Franklin and Newton.* Cambridge: Harvard University Press, 1966.

Colloque International "La Condamine": Paris, 22/23 Novembre 1985. Mexico: IPGH, 1987.

Conley, Tom. *The Self-Made Map: Cartographic Writing in Early Modern France.* Minneapolis: Minnesota University Press, 1996.

Conlon, Pierre. "La Condamine the Inquisitive." *Studies on Voltaire and the Eighteenth Century* 55 (1967): 361–93.

Cooter, Roger, and Stephen Pumphrey. "Separate Spheres and Public Places: Reflections on the History of Science Popularization and Science in Popular Culture." *History of Science* 32 (1994): 237–67.

Corbin, Alain. *Le territoire du vide. L'Occident et le désir du rivage, 1750–1840.* Paris: Flammarion, 1988.

Costa, Jorge, ed. *A Terra de Vera Cruz: Viagens, descrições e mapas do século XVIII.* Porto: BPMP, 2000.

Costales Cevallos, Alfredo. *El sabio Maldonado ante la posteridad.* Quito: Casa de la Cultura Ecuatoriana, 1948.

Costales, Piedad, and Alfredo. *Los Maldonado en la Real Audiencia de Quito.* Quito: Banco Central del Ecuador, 1987.

Crosby, Alfred W. *Ecological Imperialism: The Biological Expansion of Europe, 900–1900.* Cambridge: Cambridge University Press, 1986.

Cueto, Marcos, ed. *Saberes andinos. Ciencia y tecnología en Bolivia, Ecuador y Perú.* Lima: Instituto de Estudios Peruanos, 1995.

Curl, James Stevens. *Egyptomania: The Egyptian Revival; A Recurring Theme in the History of Taste*. Manchester: Manchester University Press, 1994.

Darnton, Robert. *The Business of Enlightenment: A Publishing History of the Encyclopédie, 1775–1800*. Cambridge: Belknap Press, 1979.

———. *The Forbidden Best-Sellers of Eighteenth-Century France*. New York: Norton, 1995.

———. *The Great Cat Massacre and Other Essays in French Cultural History*. New York: Vintage Books, 1985.

———. "The High Enlightenment and the Low-Life of Literature." In *The Literary Underground of the Old Regime*. Cambridge: Harvard University Press, 1982.

———. *Mesmerism and the End of the Enlightenment in France*. Cambridge: Harvard University Press, 1968.

Darnton, Robert, and Daniel Roche, eds. *Revolution in Print: The Press in France, 1775–1800*. Berkeley: University of California Press, 1989.

Daston, Lorraine. "Baconian Facts, Academic Civility, and the Prehistory of Objectivity." *Annals of Scholarship* 8, nos. 3–4 (1991): 337–63.

———. "The Factual Sensibility." *Isis* 79 (1988): 452–70.

———. "Marvelous Facts and Miraculous Evidence in Early Modern Europe." *Critical Inquiry* 18 (Autumn 1991): 93–124.

———. "The Nature of Nature in Early Modern Europe." *Configurations* 6, no. 2 (1998): 149–72.

Daston, Lorraine, and Katherine Park. *Wonders and the Order of Nature, 1150–1750*. New York: Zone Books, 1998.

Davis, Natalie Zemon. *Fiction in the Archives: Pardon Tales and Their Tellers in Sixteenth-Century France*. Stanford: Stanford University Press, 1987.

———. *Trickster Travels: A Sixteenth-Century Muslim between Worlds*. New York: Hill and Wang, 2006.

Dawson, Nelson-Martin. *L'atelier Delisle: L'Amérique du Nord sur la table de dessin*. Sillery: Diffusion, Dimedia, [2000].

Dear, Peter. "Narratives, Anecdotes, and Experiments: Turning Experience into Science in the Seventeenth Century." In *The Literary Structure of Scientific Argument*. Edited by Peter Dear. Philadelphia: University of Pennsylvania Press, 1991.

Dear, Peter, ed. *The Literary Structure of Scientific Argument*. Philadelphia: University of Pennsylvania Press, 1991.

Del Pino, Fermín. "Por una antropología de la ciencia. Las expediciones ilustradas españolas como 'potlach' reales." *Revista de Indias* 47, no. 180 (May–August 1987): 533–46.

Delbourgo, James. "Leviathan and the Atlantic." *History of Science* 43 (2005): 101–7.

———. *A Most Amazing Scene of Wonders: Electricity and Enlightenment in Early America*. Cambridge: Harvard University Press, 2006.

"A Demarcação do Território Brasileiro: O Tratado de Madrid e o Mapa das Cortes" in *Cartografia da Conquista do Território das Minas*. Edited by Antônio Gilberto Costa. Lisbon: Kapa Editorial, 2004.

Dening, Greg. *Mr. Bligh's Bad Language*. Cambridge: Cambridge University Press, 1992.

———. "The Theatricality of Observing and Being Observed: Eighteenth-Century Europe "Discovers' the ?-Century 'Pacific.'" In *Implicit Understandings: Observing, Reporting, and Reflecting on the Encounters between Europeans and Other Peoples in the Early Modern Era*. Edited by Stuart B. Schwartz, 451–83. Cambridge: Cambridge University Press, 1994.

Díaz, José Simón. "Un erudito español: El P. Andrés Marcos Burriel." *Revista bibliográfica y documental* 3 (1949): 5–52.

"La diffusion des sciences au XVIIIe siècle." Special issue, *Revue d'histoire des sciences* 44, nos. 3–4 (July–December 1991).

Donvez, Jacques. *De quoi vivait Voltaire?* [Paris]: Deux Rives, [1949].

Drayton, Richard. *Nature's Government: Science, Imperial Britain, and the "Improvement" of the World*. New Haven: Yale University Press, 2000.

Drouin, Jean-Marc. *L'écologie et son histoire: Réinventer la nature*. Paris: Flammarion, 1993.

———. "La moisson des voyages scientifiques: Les singularités, l'inventaire, la loi et l'histoire." In *Anais do VI Seminário Nacional de Historia da Ciência e da Tecnologia*. Edited by Isidoro Alves and Elena Moraes Garcia. Rio de Janeiro: Sociedade Brasileira para a História da Ciência, 1997.

Duchet, Michèle. *Anthropologie et histoire au siècle des lumières: Buffon, Voltaire, Rousseau, Helvétius, Diderot*. Paris: Maspero, 1971.

Duviols, Jean-Paul. *L'Amérique espagnole vue et rêvée: Les livres de voyages de Christophe Colomb à Bougainville*. [Paris]: Éditions Promodis, 1985.

Eamon, William. *Science and the Secrets of Nature: Books of Secrets in Medieval and Early Modern Culture*. Princeton: Princeton University Press, 1994.

Ebach, Malte C., and Daniel F. Goujet. "The First Biogeographical Map." *Journal of Biogeography* 33 (2006): 761–69.

Echánove Tuero, Alfonso. *La preparación intelectual del P. Andrés Marcos Burriel, S.J.* Madrid-Barcelona: Instituto Enrique Flórez/CSIC, 1971.

Edney, Matthew. *Mapping an Empire: The Geographical Construction of British India, 1765–1843*. Chicago: University of Chicago Press, 1997.

Ehrard, Jean. *L'idée de nature en France dans la première moitié du XVIIIe siècle*. Paris: Albin Michel, 1994.

Eisenstein, Elizabeth L. *The Printing Press as an Agent of Change*. 2 vols. Cambridge: Cambridge University Press, 1979.

———. Review of Adrian Johns, *The Nature of the Book: Print and Knowledge in the Making. American Historical Review* 91, no. 2 (June 2000): 316–17.

———. "An Unacknowledged Revolution Revisited." *American Historical Review* 107, no. 1 (February 2002): 87–105.

Elliott, John H. *The Old World and the New: 1492–1650*. Cambridge: Cambridge University Press, 1970. Revised edition in 1992.

Engstrand, Iris. *Spanish Scientists in the New World: The Eighteenth-Century Expeditions*. Seattle: University of Washington Press, 1981.

Epstein, Richard. "Possession as the Root of Title." *Georgia Law Review* (1979): 1221–43.

Erickson, Robert Finn. "The French Academy of Sciences Expedition to Spanish America, 1735–1744." Ph.D. diss., University of Illinois, 1955.

Escalante Adaniya, Marie Elise. *Un estudio sobre la nominación en las crónicas de Garcilaso de la Vega y Guamán Poma*. Lima: Fondo Editoral/UNMSM, 2004.

Fay, Bernard. *La Franc-Maçonnerie et la révolution intellectuelle du XVIIIe siècle*. Paris: Éditions de Cluny, 1935.

Fernández, Christian. *Inca Garcilaso: Imaginación, memória e identidad*. Lima: Fondo Editorial/ UNMSM, 2004.

Ferrand de Almeida, André. "Samuel Fritz and the Mapping of the Amazon." *Imago Mundi* 55 (2003): 103–12.

Ferreira, Mário Clemente. "Cartografia e Diplomacia: O Mapa das Cortes e o Tratado de Madrid." Unpublished paper.

———. *O Tratado de Madrid e o Brasil Meridional*. Lisbon: CNCDP, 2001.

Findlen, Paula. *Possessing Nature: Museums, Collecting, and Scientific Culture in Early Modern Italy*. Berkeley: University of California Press, 1994.

———. "Possessing the Past: The Material World of the Italian Renaissance." *American Historical Review* 103 (1998): 83–114.

———. "Science as a Career in Enlightenment Italy: The Strategies of Laura Bassi." *Isis* 84 (1993): 441–69.

Fleming, Juliet. *Graffiti and the Writing Arts of Early Modern England*. Philadelphia: University of Pennsylvania Press, 2001.

"A formação territorial do Brasil." Special issue, *Oceanos* 40 (October–December 1999).

Foucault, Michel. *The Order of Things: An Archaeology of the Human Sciences*. 1970. Reprint, New York: Vintage, 1994.

———. "What Is an Author?" In *The Foucault Reader*. Edited by Paul Rabinow, 101–20. New York: Pantheon Books, 1984.

Fournier, Paul. *Voyages et découvertes scientifiques des missionnaires naturalistes français à travers le monde pendant cinq siècles, XVe à XXe siècles*. Paris: P. Lechevalier & fils, 1932.

Fox-Genovese, Elizabeth. *The Origins of Physiocracy: Economic Revolution and Social Order in Eighteenth-Century France*. Ithaca: Cornell University Press, 1976.

Frasca-Spada, Marina, and Nick Jardine, eds. *Books and the Sciences in History*. Cambridge: Cambridge University Press, 2000.

Freile Granizo, Carlos. "La misión científica franco-española y la iglesia." In *Historia de la Iglesia Católica en el Ecuador*. Edited by Jorge Salvador Lara, 3: 1680–87. Quito: Abya-Yala, 2001.

Friedland, Paul. *Political Actors: Representative Bodies and Theatricality in the Age of the French Revolution*. Ithaca: Cornell University Press, 2002.

Frontières entre le Brésil et la Guyane Française. Mémoire présenté par les États-Unis du Brésil au gouvernement de la Confédération Suisse, arbitre choisi selon les stipulations du traité conclu à Rio-de-Janeiro, le 10 avril 1897 entre le Brésil et la France . . . Paris: A. Lahure, 1899–1900.

Furtado, Júnia Ferreira. "As índias do conhecimento, ou a geografia imaginária da conquista do ouro." *Anais de História do Além Mar* 4 (2003): 155–212.

Furtado, Júnia Ferreira, and Neil Safier. "O sertão das Minas como espaço vivido: Luís da Cunha e Jean-Baptiste Bourguignon d'Anville na construção da cartografia européia sobre o Brasil." In *Brasil-Portugal: Sociedades, culturas e formas de governar no mundo português (séculos XVI–XVIII)*. Edited by Eduardo França Paiva. São Paulo: Annablume, 2006.

Games, Alison. "Atlantic History: Definitions, Challenges, and Opportunities." *American Historical Review* 111, no. 3 (June 2006): 741–57.

Gandía, Enrique de. *Historia crítica de los mitos de la conquista americana*. Buenos Aires: J. Roldán, 1929.

Gans, Mozes Heiman. *Memorbook: History of Dutch Jewry from the Renaissance to 1940*. Baarn: Bosch & Keuning, 1977.

García Cárcel, Ricardo. *Felipe V y los españoles: Una visión periférica del problema de España*. Madrid: Plaza Janés, 2002.

Gascoigne, John. *Science in the Service of Empire: Joseph Banks, the British State, and the Uses of Science in the Age of Revolution*. Cambridge: Cambridge University Press, 1998.

Gerbi, Antonello. *La disputa del Nuovo Mundo: Storia di una polemica, 1750–1900*. New edition. Milan: Adelphi, 2000.

Gil, Juan. *Mitos y utopias del descubrimiento*. Vol. 3, *El Dorado*. Madrid: Alianza Editorial, 1989.

Gillispie, Charles Coulston. *Science and Polity in France at the End of the Old Regime*. Princeton: Princeton University Press, 1980.

Glacken, Clarence J. *Traces on the Rhodian Shore: Nature and Culture in Western Thought from Ancient Times to the End of the Eighteenth Century*. Berkeley: University of California Press, 1967.

Goldberg, Jonathan. *Writing Matter: From the Hands of the English Renaissance*. Stanford: Stanford University Press, 1990.

Goldgar, Anne. *Impolite Learning: Conduct and Community in the Republic of Letters, 1680–1750*. New Haven: Yale University Press, 1995.

Golinski, Jan. "Barometers of Change: Meteorological Instruments as Machines of Enlightenment." In *The Sciences in Enlightened Europe*. Edited by William Clark, Jan Golinski, and Simon Schaffer, 69–93. Chicago: University of Chicago Press, 1999.

———. *British Weather and the Climate of Enlightenment*. Chicago: University of Chicago Press, 2007.

———. *Making Natural Knowledge: Constructivism and the History of Science*. Cambridge: Cambridge University Press, 1998.

Gomes, Flávio dos Santos. *A hidra e os pântanos: Mocambos, quilombos, e comunidades de fugitivos no Brasil (séculos XVII–XIX)*. São Paulo: Editora UNESP, 2005.

Gomez, Nelson. *La misión geodésica y la cultura de Quito*. Quito: Ediguias Ltda., 1987.

Goncourt, Edmond de, and Jules de Goncourt. *L'art du dix-huitième siècle*. Paris: A. Quantin, 1880.

Gondim, Neide. *A invenção da Amazônia*. Sao Paulo: Marco Zero, 1994.

Goodman, Dena. *Criticism in Action: Enlightenment Experiments in Political Writing*. Ithaca: Cornell University Press, 1989.

———. "Introduction: The Public and the Nation." *Eighteenth-Century Studies* 29, no. 1 (Fall 1995): 1–4.

———. *The Republic of Letters*. Ithaca, N.Y.: Cornell University Press, 1994.

Gordon, Daniel. "Philosophy, Sociology, and Gender in the Enlightenment Conception of Public Opinion." *French Historical Studies* 17, no. 4 (Autumn 1992): 882–911.

Grafton, Anthony. "The Ancient City Restored: Archaeology, Ecclesiastical History, and Egyptology." In *Rome Reborn: The Vatican Library and Renaissance Culture*. Edited by Anthony Grafton. New Haven: Yale University Press, 1993.

———. "Correctores corruptores? Notes on the Social History of Editing." In *Editing Texts/Texte Edieren*. Edited by Glenn W. Most. Göttingen: Vandenhoeck and Ruprecht, 1998.

———. *Defenders of the Text: The Traditions of Scholarship in an Age of Science, 1450–1800*. Cambridge: Belknap Press, 1994.

———. *The Footnote: A Curious History*. Cambridge: Harvard University Press, 1997.

———. *New Worlds, Ancient Texts: The Power of Tradition and the Shock of Discovery*. Cambridge: Harvard University Press, 1992.

Grafton, Anthony, and Jardine. "'Studied for Action': How Gabriel Harvey Read His Livy." *Past and Present* 129 (1990): 30–78.

Grafton, Anthony, et al. "How Revolutionary Was the Print Revolution?" *American Historical Review* 107, no. 1 (February 2002): 84–86.

Greenberg, John L. "Geodesy in Paris in the 1730s and the Paduan Connection." *Historical Studies in the Physical Sciences* 13, no. 2 (1981): 239–60.

———. *The Problem of the Earth's Shape from Newton to Clairaut: The Rise of Mathematical Science in Eighteenth-Century Paris and the Fall of "Normal" Science*. Cambridge: Cambridge University Press, 1995.

Greenblatt, Stephen. *Marvelous Possessions: The Wonder of the New World*. Chicago: University of Chicago Press, 1991.

———. *Renaissance Self-Fashioning: From More to Shakespeare*. Chicago: University of Chicago Press, 1980.

Groult, Martine, ed. *Discours préliminaire des éditeurs de 1751 et articles de l'Encyclopédie*. Paris: Champion, 1999.

Grove, Richard. *Green Imperialism: Colonial Expansion, Tropical Island Edens and the*

Origins of Environmentalism, 1600–1860. Cambridge: Cambridge University Press, 1995.

Gruzinski, Serge. "Colonial Indian Maps in Sixteenth-Century Mexico: An Essay in Mixed Cartography." *Res* 13 (1987): 46–61.

———. *The Conquest of Mexico: The Incorporation of Indian Societies into the Western World, 16th–18th Centuries*. Trans. by Eileen Corrigan. Cambridge: Polity Press, 1993.

———. "Os índios construtores de catedrais. Mestiçagens, trabalho e produção na Cidade do México, 1550–1600." In *O trabalho mestiço. Maneiras de Pensar e Formas de Viver – Séculos XVI a XIX*. Edited by Eduardo França Paiva and Carla Maria Junho Anastasia, 323–39. São Paulo: Annablume, 2002.

———. *La pensée métisse*. Paris: Fayard, 1999. Translated by Deke Dusinberre as *The Mestizo Mind: The Intellectual Dynamics of Colonization and Globalization*. New York: Routledge, 2002.

———. *Les quatres parties du monde. Histoire d'une mondialisation*. Paris: La Martinière, 2004.

Guerreiro, Inácio, ed. *Cartografia e diplomacia no Brasil do século XVIII*. Lisbon: CNCDP, 1997.

Guillén, Julio F. "Don Antonio de Ulloa y el descubrimiento del platino." *Anales de la Asociación Española para el Progreso de las Ciencias* 2 (1939): 413–16.

Guillén Tato, Jorge. *Los tenientes de navio Jorge Juan y Santacilia y Antonio de Ulloa de la Torre-Guiral y la medición del meridiano*. Madrid: Galo Sáez, 1973.

Gutierres Vocanegra, Bernardo. "Inventario y reconocimiento del equipaje de los Académicos Franceses" (1 June 1736). In *Documentos para la historia de la Audiencia de Quito*. Edited by José Rumazo, vol. 6. Madrid: Afrodisio Aguado, 1948–59.

Haase, Wolfgang, and Meyer Reinhold. *The Classical Tradition and the Americas*. Berlin: Walter de Gruyter, 1994.

Habermas, Jürgen. *The Structural Transformation of the Public Sphere: An Inquiry into a Category of Bourgeois Society*. Translated by Thomas Burger. Cambridge: MIT Press, 1989.

Haechler, Jean. *L'Encyclopédie: Les combats et les hommes*. Paris: Les Belles Lettres, 1998.

Hahn, Roger. *The Anatomy of a Scientific Institution: The Paris Academy of Sciences, 1666–1803*. Berkeley: University of California Press, 1971.

Halévi, Ran. *Les loges maçonniques dans la France d'Ancien Régime*. Paris: Librairie Armand Colin, 1984.

Hankins, Thomas L. *Jean d'Alembert: Science and the Enlightenment*. Oxford: Clarendon Press, 1977.

———. *Science and the Enlightenment*. Cambridge: Cambridge University Press, 1985.

Hanks, Lesley. *Buffon avant l'"Histoire naturelle"*. Paris: PUF, 1966.

Hannaway, Owen. "Laboratory Design and the Aim of Science: Andreas Libavius versus Tycho Brahe." *Isis* 77 (1986): 585–610.

Harbsmeier, Michael. "Towards a Prehistory of Ethnography: Early Modern German Travel Writing as Traditions of Knowledge." In *Fieldwork and Footnotes: Studies in the History of European Anthropology*. Edited by Han F. Vermeulen and Arturo Alvarez Roldán. New York: Routledge, 1995.

Harley, J. B. *The New Nature of Maps: Essays in the History of Cartography*. Baltimore: Johns Hopkins University Press, 2001.

———. "Silences and Secrecy. The Hidden Agenda of Cartography in Early Modern Europe." In *The New Nature of Maps*, 83–108. Originally published in *Imago Mundi* 40 (1988): 57–76.

Harris, Max. *Aztecs, Moors, and Christians: Festivals of Reconquest in Mexico and Spain*. Austin: University of Texas Press, 2000.

———. *The Dialogical Theatre: Dramatizations of the Conquest of Mexico and the Question of the Other*. New York: St. Martin's Press, 1993.

Hartog, François. "L'art du récit historique." *Autrement* 150–51 (January 1995).

———. "L'écriture du voyage." In *Le voyage mystique: Michel de Certeau*. Edited by Luce Giard, 123–32. Paris: Recherches de Science Religieuse, 1988.

Hemming, John. *Amazon Frontier: The Defeat of the Brazilian Indians*. London: Macmillan, 1987.

Henkel, Thomas. *Chronique d'un chasseur des âmes: Un jésuite suisse en Amazonie au XVIIIe siècle*. Fribourg: Éditions de l'Hèbe, 1993.

Herzog, Tamar. *Upholding Justice: Society, State, and the Penal System in Quito (1650–1750)*. Ann Arbor: University of Michigan Press, 2004. Originally published as *La administración como un fenónemo social: La justicia penal de la ciudad de Quito (1650–1750)*, Madrid: Centro de Estudios Constitucionales, 1995.

Hesse, Carla. "Enlightenment Epistemology and the Laws of Authorship in Revolutionary France, 1777–1793." *Representations* 30 (Spring 1990): 109–37.

———. "Reading Signatures: Female Authorship and Revolutionary Law in France, 1750–1850." *Eighteenth-Century Studies* 22, no. 3 (Spring 1989): 469–87.

Hill, Ruth. *Sceptres and Sciences in the Spains*. Liverpool: Liverpool University Press, 2000.

Hoare, Michael Rand. *The Quest for the True Figure of the Earth: Ideas and Expeditions in Four Centuries of Geodesy*. Aldershot: Ashgate, 2005.

Hodgen, Margaret T. *Early Anthropology in the Sixteenth and Seventeenth Centuries*. Philadelphia: University of Pennsylvania Press, 1998.

Honigmann, E. A. J. *The Stability of Shakespeare's Texts*. London: Edward Arnold, 1965.

Honigsbaum, Mark. *Valverde's Gold: A True Tale of Greed, Obsession and Grit*. New York: Macmillan, 2004.

Hulme, Peter, and Ludmilla Jordanova, eds. *Enlightenment and Its Shadows*. London: Routledge, 1990.

Hunt, Lynn. *The Family Romance of the French Revolution*. Berkeley: University of California Press, 1992.

Iliffe, Rob. "'Aplatisseur du monde et de Cassini': Maupertuis, Precision Measurement and the Shape of the Earth in the 1730s." *History of Science* 31 (1993): 335–75.

———. "Butter for Parsnips: Authorship, Audience, and the Incomprehensibility of the *Principia*." In *Scientific Authorship*. Edited by Mario Biagioli and Peter Galison, 33–66. New York: Routledge, 2003.

———. "'Ce que Newton connut sans sortir de chez lui': La mesure de la figure de la terre en France, 1700–1750." *Histoire et mesure* 4 (1995): 1–41.

La imprenta en Iberoamérica, 1539–1833. Madrid: Biblioteca Nacional, 1983.

Iversen, Erik. *The Myth of Egypt and Its Hieroglyphs in European Tradition*. 1961. Reprint, Princeton: Princeton University Press, 1993.

Jacob, Christian. *L'empire des cartes. Approche théorique de la cartographie à travers l'histoire*. Paris: Albin Michel, 1992. Translated by Tom Conley as *The Sovereign Map*. Edited by Edward H. Dahl. Chicago: University of Chicago Press, 2006.

Jacob, Margaret C. *Living the Enlightenment: Freemasonry and Politics in Eighteenth-Century Europe*. Oxford: Oxford University Press, 1991.

———. *The Radical Enlightenment: Pantheists, Freemasons and Republicans*. London: Allen and Unwin, 1981.

Jaramillo-Arango, Jaime. "Estudio crítico acerca de los hechos básicos de la historia de la quina." *Revista de la Academía Colombiana de Ciencias Exactas Físicas y Naturales* (Bogotá) 8 (1951): 245–73.

Jardine, Lisa. *Erasmus, Man of Letters*. Princeton: Princeton University Press, 1993.

———. "Strange Specimens." In *Ingenious Pursuits: Building the Scientific Revolution*, 223–72. New York: Anchor Books, 2000.

Jardine, Lisa and Anthony Grafton. "'Studied for Action': How Gabriel Harvey Read His Livy." *Past and Present* 129 (1990): 30–78.

Jardine, Nicholas, James A. Secord, and E. C. Spary, eds. *Cultures of Natural History*. Cambridge: Cambridge University Press, 1996.

Jasanoff, Maya. *Edge of Empire: Lives, Culture, and Conquest in the East, 1750–1850*. New York: Knopf, 2005.

Johns, Adrian. *The Nature of the Book: Print and Knowledge in the Making*. Chicago: University of Chicago Press, 1999.

———. "How to Acknowledge a Revolution." *American Historical Review* 107, no. 1 (February 2002): 106–25.

Johnson, Julie Green, and Susan L. Newbury. *The Book in the Americas: The Role of Books and Printing in the Development of Culture and Society in Colonial Latin America: Catalogue of an Exhibition*. Providence: John Carter Brown Library, 1988.

Jones, Colin. "The Great Chain of Buying: Medical Advertisement, the Bourgeois Public Sphere, and the Origins of the French Revolution." *American Historical Review* 101, no. 1 (February 1996): 13–40.

Jones, Tom B. "The French Expedition to Lapland, 1736–1737." *Terrae Incognitae* 2 (1970): 15–24.

Kafker, Frank A., and Serena L. Kafker. *The Encyclopedists as Individuals: A Biographical Dictionary of the Authors of the Encyclopédie.*" Studies on Voltaire and the Eighteenth Century, 257. Oxford: Voltaire Foundation, 1988.

Kaplan, Steven L. *The Bakers of Paris and the Bread Question, 1700–1775.* Durham: Duke University Press, 1996.

———. *Bread, Politics, and Political Economy in the Reign of Louis XV.* 2 vols. The Hague: Martinus Nijhoff, 1976.

———. "The Famine Plot Persuasion in Eighteenth-Century France." *Transactions of the American Philosophical Society* 72, no. 3 (1982): 1–79.

Katzew, Ilona. *Casta Painting: Images of Race in Eighteenth-Century Mexico.* New Haven: Yale University Press, 2005.

Kellman, Jordan. "Discovery and Enlightenment at Sea: Maritime Exploration and Observation in the Eighteenth-Century French Scientific Community." Ph.D. diss., Princeton University, 1998.

Kellogg, Susan. "Depicting Mestizaje: Gendered Images of Ethnorace in Colonial Mexican Texts." *Journal of Women's History* 12, no. 3 (2000): 69–92.

Kenny, Neil. *Curiosity in Early Modern Europe: Word Histories.* Wiesbaden: Harrassowitz Verlag, 1998.

Knorr-Cetina, Karin D., and Michael Mulkay, eds. *Science Observed: Perspectives on the Social Study of Science.* London: Sage, 1983.

Koller, Armin Hajman. *The Theory of Environment: An Outline of the History of the Idea of Milieu, and Its Present Status.* Menasha: George Banta Publishing, 1918.

Konvitz, Josef W. *Cartography in France, 1660–1848: Science, Engineering, and Statecraft.* Chicago: University of Chicago Press, 1987.

Kosseleck, Reinhard. *Critique and Crisis: Enlightenment and the Pathogensis of Modern Society.* Cambridge: MIT Press, 1988.

Kratz, Guillermo. *El tratado hispano-portugués de límites de 1750 y sus consecuencias.* Rome: IHSI, 1954.

Kriegel, Blandine. *Les académies de l'histoire.* Paris: Presses Universitaires de France, 1988.

Kupperman, Karen Ordall, ed. *America in European Consciousness, 1493–1750.* Chapel Hill: University of North Carolina Press, 1995.

Kury, Lorelai. *Histoire naturelle et voyages scientifiques (1780–1830).* Paris: Harmattan, 2001.

———. "Les instructions de voyage dans les expeditions scientifiques françaises, 1750–1830." *Revue d'histoire des sciences* 51 (1998): 65–91.

Lacombe, Henri, and Pierre Costabel, eds. *La figure de la terre du XVIIIe siècle à l'ère spatiale.* Paris: Académie des Sciences, 1988.

Lacroix, Alfred. "Les de Jussieu et leurs correspondants, leur rôle dans les colonies françaises." *Mémoires de l'Académie Royale des Sciences et de l'Institut de France* 63 (1941): 48–58.

Lafuente, Antonio. "Una ciencia para el Estado: la expedición geodésica hispano-francesa al virreinato del Perú, 1734–1743." *Revista de Indias* 43 (July–December 1983): 549–629.

Lafuente, Antonio, and Antonio J. Delgado. *La geometrización de la tierra: Observaciones y resultados de la expedición geodésica hispano-francesa al virreinato del Perú, 1735–1744*. Madrid: CSIC, 1984.

Lafuente, Antonio, and Antonio Mazuecos. "La academia itinerante: La expedición franco-española al Reino de Quito de 1736." In *Carlos III y la ciencia de la Ilustración*. Edited by José Luis Peset and Antonio Lafuente, 299–312. Madrid: Alianza Universidad, 1989.

———. *Los caballeros del punto fijo. Ciencia, política y aventura en la expedición geodésica hispanofrancesa al virreinato del Perú en el siglo XVIII*. Madrid: Serbal/CSIC, 1987.

Lafuente, Antonio, and Eduardo Estrella. Introduction to *Viaje a la América Meridional por el río de las Amazonas*. Edited by Antonio Lafuente and Eduardo Estrella. Barcelona: Editorial Alta Fulla, 1986.

Lafuente, Antonio, and José Luis Peset. "Las actividades e instituciones científicas en la España Ilustrada." In *Carlos III y la ciencia de la Ilustración*. Edited by José Luis Peset and Antonio Lafuente, 29–80. Madrid: Alianza Universidad, 1989.

———. "La question de la figure de la Terre au XVIIIe siècle" *Revue d'Histoire des Sciences* 37, nos. 3–4 (1984): 235–54.

Laissus, Yves. "Le Jardin du Roi." In *Enseignement et diffusion des sciences en France au XVIIIe siècle*. Edited by René Taton. Paris: Hermann, 1964.

———. *Les naturalistes français en Amérique du Sud*. Paris: Éditions du CTHS, 1995.

———. "Note sur les manuscrits de Joseph de Jussieu (1704–1779) conservés à la Bibliothèque central du Museum d'histoire naturelle." In *89e Congrès des sociétés savants*, 9–16. Paris: Bibliothèque Nationale, 1964.

———. "Les voyageurs naturalistes du Jardin du roi et du Muséum d'histoire naturelle: Essai de portrait robot." In *Revue d'histoire des sciences* 34, nos. 3–4 (July–October 1981): 259–317.

Lara, Dario. "L'amitié de deux hommes de sciences: Charles-Marie de la Condamine et Pedro Vicente Maldonado, et l'origine de l'amitié de deux peuples." In *Colloque International "La Condamine": Paris, 22/23 Novembre 1985*. Mexico: IPGH, 1987.

Larrea, Carlos Manuel. *La Real Audiencia de Quito y su territorio*. Quito: Imprenta del Ministerio de Relaciones Exteriores, 1987.

Larrère, Catherine. *L'invention de l'économie au XVIIIe siècle: Du droit natural à la physiocratie*. Paris: Presses Universitares de France, 1992.

Larson, James L. *Interpreting Nature: The Science of Living Form from Linnaeus to Kant*. Baltimore: Johns Hopkins University Press, 1994.

Latorre, Octavio. *Los mapas del Amazonas e el desarrollo de la cartografía ecuatoriana en el siglo XVIII*. Guayaquil: Museo del Banco Central del Ecuador, 1988.

Latour, Bruno. "Comment redistribuer le Grand Partage?" *Revue de Synthèse* 110 (1983): 202–36.

———. *Science in Action: How to Follow Scientists and Engineers through Society*. Cambridge: Harvard University Press, 1987.

————. "Visualization and Cognition: Thinking with Eyes and Hands." *Knowledge and Society* 6 (1986): 1–40.

Latour, Bruno, and Steve Woolgar. *Laboratory Life: The Social Construction of Scientific Facts.* Princeton: Princeton University Press, 1986.

Laubat, François de Chasseloup. *François Fresneau, seigneur de la Gataudière, père du caoutchouc.* Paris: Plon, 1942.

Lavallé, Bernard. *Al filo de la navaja: Luchas y derivas caciquiles en Latacunga, 1730–1790.* Quito: IFEA, 2002.

Le Hir, Gaston. "L'oeuvre de Joseph de Jussieu (1704–1779) en Amérique méridionale. D'après les manuscrits conservés à la Bibliothèque centrale du Muséum d'histoire naturelle." In *Les naturalistes français en Amérique du Sud.* Edited by Yves Laissus, 121–35. Paris: Éditions du CTHS, 1995.

Le Sueur, Achille Ambroise Anatole. *La Condamine d'après ses papiers inédits.* Paris: A. Picard, 1911.

Leibsohn, Dana. "Colony and Cartography: Shifting Signs on Indigenous Maps of New Spain." In *Reframing the Renaissance: Visual Culture in Europe and Latin America, 1450–1650.* Edited by Claire J. Farago. New Haven: Yale University Press, 1995.

Leite, Serafim. *História da Companhia de Jesus no Brasil.* 10 vols. Lisbon: Livraria Portugalia, 1938–50.

Lenclud, Gérard. "Quand voir, c'est reconnaître. Les récits de voyage et le regard anthropologique." *enquête* 1 (1995): 113–29.

Lenoir, Timothy, ed. *Inscribing Science: Scientific Texts and the Materiality of Communication.* Stanford: Stanford University Press, 1998.

Lestringant, Frank. Introduction to *Le Brésil d'André Thévêt.* Paris: Chandeigne, 1997.

Levy, Darline Gay. *The Ideas and Careers of Simon-Nicolas-Henri Linguet.* Urbana: University of Illinois Press, 1980.

Licoppe, Christian. *La formation de la pratique scientifique: Le discours de l'expérience en France et en Angleterre, 1630–1820.* Paris: La Découverte, 1996.

Linchet, Dominique. "La relation de la rivière des Amazones de M. de Gomberville: La traduction au service du colonialisme, du patriotisme et de la propagande sous Louis XIV." Ph.D. diss., University of North Carolina, 1995.

Lindqvist, Svante. "The Spectacle of Science: An Experiment in 1744 concerning the Aurora Borealis." *Configurations* 1, no. 1 (1993): 57–94.

Livingstone, David N., and Charles W.J. Withers, eds. *Geography and Enlightenment.* Chicago: University of Chicago Press, 1999.

Llana, James. "Natural History and the *Encyclopédie.*" *Journal of the History of Biology* 33 (2000): 1–25.

Loewenstein, Joseph. *The Author's Due: Printing and the Prehistory of Copyright.* Chicago: University of Chicago Press, 2002.

Lombard, Alfred. *L'abbé Du Bos: Un initiateur de la pensée moderne, 1670–1742.* 1913. Reprint, Geneva: Slatkine reprints, 1969.

Long, Pamela. *Openness, Secrecy, Authorship: Technical Arts and the Culture of Knowledge from Antiquity to the Renaissance*. Baltimore: Johns Hopkins University Press, 2001.

Lopez, Michel. "La caractérisation de l'électricité dans la foudre du XVIIIe siècle par Thomas-François Dalibard, un physicien français méconnu." *Actes du XIe colloque Forum Industrie de l'Université du Maine*. Le Mans, 1999.

Loridan, Jules. *Voyages des astronomes français à la recherche de la figure de la terre et de ses dimensions*. Lille: Desclée de Brouwer, 1890.

Lough, John. *The Encyclopédie*. Geneva: Slatkine Reprints, 1989.

Lubar, Steven, and W. David Kingery. *History from Things: Essays on Material Culture*. Washington, D.C.: Smithsonian Institution Press, 1993.

Lynch, John. *Bourbon Spain, 1700–1808*. Oxford: Blackwell, 1989.

Lynn, Michael R. "Divining the Enlightenment: Public Opinion and Popular Science in Old Regime France." *Isis* 92, no. 1 (March 2001): 34–54.

———. "Enlightenment in the Public Sphere: The Musée de Monsieur and Scientific Culture in Late Eighteenth-Century Paris." *Eighteenth-Century Studies* 32, no. 4 (1999): 463–76.

———. *Popular Science and Public Opinion in Eighteenth-Century France*. Manchester: Manchester University Press, 2006.

Macchi, Fernanda. "Imágenes de los Incas en el siglo XVIII." Ph.D. diss., Yale University, 2003.

MacLeod, Roy, ed. "Nature and Empire: Science and the Colonial Enterprise." Special Issue, *Osiris* 15 (2000).

Mallon, Florencia E. "The Promise and Dilemma of Subaltern Studies: Perspectives from Latin American History." *American Historical Review* 99 (5): 1491–1525.

Mansilla, Ronald Escobedo. *Las comunidades indígenas y la economía colonial peruana*. [Bilbao]: Servicio Editorial Universidad del País Vasco, 1997.

Marañón Posadillo, Gregorio. "Consideraciones sobre Feijóo." *La Nueva España* (Oviedo), March 31, 1954.

Marcel, Gabriel. *El geógrafo Tomás López y sus obras; ensayo de biografía y de cartografía*. Madrid: Imprenta del Patronato huérfanos de Administración militar, 1908.

Marcil, Yasmine. "La presse et le compte rendu de recits de voyage scientifique: Le cas de la querelle entre Bouguer et La Condamine." *Sciences et Techniques en Perspective* 3, no. 2 (1999): 285–304.

———. "Récits de voyage et presse périodique au XVIIIe siècle: De l'extrait à la critique." Ph.D. diss., École des Hautes Etudes en Sciences Sociales, Paris, 2000.

Marsak, L. M. *Bernard de Fontenelle: The Idea of Science in the French Enlightenment*. Transactions of the American Philosophical Society, 49, pt. 7. Philadelphia: American Philosophical Society, 1959.

Marshall, P. J., and Glyndwr Williams. *The Great Map of Mankind: Perceptions of New Worlds in the Age of Enlightenment*. Cambridge: Harvard University Press, 1982.

Martínez, Bernabé Bartolomé. "Andrés Marcos Burriel: Un pionero de reformas en investigación y enseñanza," *Revista Complutense de Educación* 2, no. 3 (1991): 481–90

Martínez, José Luis. *El libro en hispanoamérica: Origen y desarrollo.* Madrid: Ediciones Pirámide, 1986.

Mason, Peter. *Infelicities: Representations of the Exotic.* Baltimore: Johns Hopkins University Press, 1998.

Masten, Jeffrey, Peter Stallybrass, and Nancy Vickers, eds. *Language Machines: Technologies of Literary and Cultural Production.* New York: Routledge, 1997.

Maxwell, Kenneth. *Pombal: Paradox of the Enlightenment.* Cambridge: Cambridge University Press, 1995.

Maza, Sarah. "Women, the Bourgeoisie, and the Public Sphere: Response to Daniel Gordon and David Bell." *French Historical Studies* 17, no. 4 (Autumn 1992): 935–50.

———. *Private Lives and Public Affairs: The Causes Célèbres of Prerevolutionary France.* Berkeley: University of California Press, 1993.

Mazzotti, José Antonio. *Coros mestizos del Inca Garcilaso: Resonancias andinas.* Mexico: Fondo de Cultura Económica, 1996.

McClellan, James E., III. *Colonialism and Science: Saint Domingue in the Old Regime.* Baltimore: Johns Hopkins University Press, 1992.

———. *Specialist Control: The Publications Committee of the Académie Royale des Sciences (Paris), 1700–1793.* Philadelphia: American Philosophical Society, 2003.

McConnell, Anita. "La Condamine's Scientific Journey down the River Amazon, 1743–44." *Annals of Science* 48 (1991): 1–19.

McDonald, Christie V. *The Dialogue of Writing: Essays in Eighteenth-Century French Literature.* Waterloo: Wilfrid Laurier University Press, 1984.

McKenzie, D. F. *Bibliography and the Sociology of Texts.* 1986. Reprint edition. Cambridge: Cambridge University Press, 1999.

———. "Printers of the Mind: Some Notes on Bibliographical Theories and Printing-House Practices." *Studies in Bibliography* 22 (1969): 1–76.

Meek, Ronald L. *Social Science and the Ignoble Savage.* Cambridge: Cambridge University Press, 1976.

Melzer, Sara E. "The French *Relation* and Its 'Hidden' Colonial History." In *Companion to the Literatures of Colonial America.* Edited by Susan Castillo and Ivy Schweitzer, 220–40. Oxford: Blackwell Publishers, 2005.

Mercier, Roger. "Les français en Amérique du Sud au XVIIIe siècle: La mission de l'Académie des sciences (1735–45)." *Revue française d'histoire d'outre mer* 56, no. 205 (1969): 327–74.

———. "La théorie des climats des 'Réflexions critiques' à 'L'esprit des lois.'" *Revue d'Histoire Littéraire de la France* 53 (1953): pt. 1, 17–37; and pt. 2, 159–74.

Merino Navarro, José P., and Miguel M. Rodriguez San Vicente. Introduction to Jorge Juan and Antonio de Ulloa, *Relacion histórica del viaje a la América meridional.* Madrid: Fundación Universitaria Española, 1978.

Mestre, Antonio. "La imagen de España en el siglo XVIII: Apologistas, críticos y detractores." *Arbor* 449 (May 1983): 49–73.

Michaud, J. F., and L. G., eds. *Biographie universelle, ancienne et moderne.* Paris: Desplaces, 1854.

Middleton, W. E. Knowles. *A History of the Thermometer and Its Use in Meteorology.* Baltimore: Johns Hopkins University Press, 1966.

Mignolo, Walter. *The Darker Side of the Renaissance: Literacy, Territoriality, and Colonization.* Ann Arbor: University of Michigan Press, 1995.

Miller, David Philip, and Peter Hanns Reill, eds. *Visions of Empire: Voyages, Botany, and Representations of Nature.* Cambridge: Cambridge University Press, 1996.

Millones Figueroa, Luis. *Pedro de Cieza de León y su crónica de Indias. La entrada de los Incas en la historia universal.* Lima: Fondo Editorial de la Pontifica Universidad, 2001.

Monmonier, Mark. *From Squaw Tit to Whorehouse Meadow: How Maps Name, Claim, and Inflame.* Chicago: University of Chicago Press, 2006.

Monteiro, Nuno Gonçalo, ed. *Meu pai e meu senhor, muito do meu coração.* Lisbon: Instituto de Ciências Sociais da Universidade de Lisboa/Quetzal, 2000.

Mornet, Daniel. *Les sciences de la nature en France, au XVIIIe siècle.* 1911. Reprint, Geneva: Slatkine Reprints, 2001.

Morrissey, Robert, et Philippe Roger, eds. *L'Encyclopédie: Du réseau au livre et du livre au réseau.* Paris: Honoré Champion, 2001.

Moureau, François. *De bonne main: La communication manuscrite au XVIIIe siècle.* Paris: Universitas, 1993.

———. "O Brasil das Luzes Francesas." *Estudos Avançados* (São Paulo) 13, no. 36 (1999): 165–82.

———. *Le roman vrai de l'Encyclopédie.* Paris: Découvertes Gallimard, 1990.

Mundy, Barbara. *The Mapping of New Spain: Indigenous Cartography and the Maps of the Relaciones Geográficas.* Chicago: University of Chicago Press, 1996.

Muthu, Sankar. *Enlightenment against Empire.* Princeton: Princeton University Press, 2003.

Myrone, Martin, and Lucy Peltz, eds. *Producing the Past: Aspects of Antiquarian Culture and Practice, 1700–1850.* Aldershot, Hampshire: Ashgate, 1999.

Navarro García, Luis. *Las reformas borbónicas en América, el plan de intendencias y su aplicación.* Seville: Universidad de Sevilla, 1995.

Nicholl, Charles. *The Creature in the Map.* London: Jonathan Cape, 1995.

Nijenhuis, I. J. A. "The Passions of an Enlightened Jew: Isaac de Pinto (1717–1787) as a 'Solliciteur du Bien Public.'" In *The Low Countries and Beyond.* Edited by Robert S. Kirsner, 47–54. Lanham: University Press of America, 1993.

———. *Een Joodse Philosophe: Isaac de Pinto (1717–1787).* Amsterdam: NEHA, 1992.

Nussbaum, Felicity A., ed. *The Global Eighteenth Century*. Baltimore: Johns Hopkins University Press, 2003.

O'Gorman, Edmundo. *La invención de América*. Mexico City: Fondo de Cultura Económica, 1986.

O'Malley, John, S.J. *The Ratio Studiorum: The Official Plan for Jesuit Education*. St. Louis: Institute for Jesuit Sources, 2005.

O'Neal, John C. *The Authority of Experience: Sensationist Theory in the French Enlightenment*. University Park: Pennsylvania State University, 1996.

Ogilvie, Brian W. "The Many Books of Nature: Renaissance Naturalists and Information Overload." *Journal of the History of Ideas* 64, no. 1 (January 2003): 29–40.

Ophir, Adi, and Steven Shapin. "The Place of Knowledge: A Methodological Survey." *Science in Context* 4 (1991): 3–21.

Orgel, Stephen. "What Is a Text?" In *Staging the Renaissance: Essays on Elizabethan and Jacobean Drama*. Edited by David Scott Kastan and Peter Stallybrass. New York: Routledge, 1991.

Ortega, Francisco A. "Trauma and Narrative in Early Modernity: Garcilaso's *Comentarios reales* (1609–1616)." *Modern Language Notes* 118, no. 2 (2003): 393–426.

Ortiz Arellano, Carlos. *Pedro Vicente Maldonado, forjador de la patria ecuatoriana, 1704–1748*. Quito: Casa de la Cultura Ecuatoriana, 2004.

Osborne, Michael A. *Nature, the Exotic, and the Science of French Colonialism*. Bloomington: Indiana University Press, 1994.

Outram, Dorinda. *The Enlightenment*. Cambridge: Cambridge University Press, 1995.

———. "On Being Perseus: New Knowledge, Dislocation, and Enlightenment Exploration." In *Geography and Enlightenment*. Edited by David N. Livingstone and Charles W. J. Withers, 281–94. Chicago: University of Chicago Press, 1999.

Ozanam, Didier. "Un consul de France à Cadix: Pierre-Nicolas Partyet (1716–1729)." In *L'ouvrier, l'Espagne, la Bourgogne et la vie provinciale. Parcours d'un historien (Mélanges offert à Pierre Ponsot)*. Madrid: Casa de Velasquez, 1994.

Ozanam, Didier, ed. *La diplomacia de Fernando VI*. Madrid: CSIC, 1975.

Padrón, Ricardo. *The Spacious Word: Cartography, Literature, and Empire in Early Modern Spain*. Chicago: University of Chicago Press, 2003.

Pagden, Anthony. *European Encounters with the New World*. New Haven: Yale University Press, 1993.

———. *The Fall of Natural Man: The American Indian and the Origins of Comparative Ethnology*. Cambridge: Cambridge University Press, 1982.

———. *Spanish Imperialism and the Political Imagination*. New Haven: Yale University Press, 1990.

Parrish, Susan Scott. *American Curiosity: Cultures of Natural History in the Colonial British Atlantic World*. Chapel Hill: University of North Carolina Press, 2006.

Passeron, Irène. "La forme de la terre est-elle une preuve de la vérité du système newtonien?" In *Terres à découvrir, terres à parcourir.* Edited by Danielle Lecoq and Antoine Chambord, 129–45. Paris: Publications de l'Université de Paris VII-Denis Diderot, 1996.

Paul, Charles B. *Science and Immortality: The Éloges of the Paris Academy of Sciences, 1699–1791.* Berkeley: University of California Press, 1980.

Pedley, Mary Sponberg. *Bel et Utile: Work of the Robert de Vaugondy Family of Mapmakers.* Tring: Map Collector Publications, 1992.

———. *The Commerce of Cartography: Making and Marketing Maps in Eighteenth-Century France and England.* Chicago: University of Chicago Press, 2004.

———. "New Light on an Old Atlas: Documents concerning the Publication of the *Atlas Universel* (1757)." *Imago Mundi* 36 (1984): 48–63.

Pelayo, Francisco, and Miguel-Ángel Puig-Samper. "Las actividades científicas de Joseph de Jussieu en América del Sur." In *Nouveau monde et renouveau de l'histoire naturelle.* Edited by Marie-Cécile Bénassy and Jean-Pierre Clément, 2: 67–84. Paris: Presses de la Sorbonne Nouvelle, 1993.

Perez Magallón, Jesús. *Construyendo la modernidad: La cultura española en el tiempo de los novatores, 1675–1725.* Madrid: CSIC/Anejos de Revista de Literatura, 2002.

Perrier, Georges. *La mission française de l'équateur.* Paris: F. Leve, 1907.

Perrone-Moisés, Leyla. *Le voyage de Gonneville (1503–1505) et la découverte de la Normandie par les Indiens du Brésil.* Paris: Éditions Chandeigne, 1995.

Peset, José Luis, and Antonio Lafuente, eds. *Carlos III y la ciencia de la Ilustración.* Madrid: Alianza Universidad, 1989.

Pillon, Marie-José. "Le voyage de La Condamine en Amazonie au XVIIIe siècle, étude critique de certains aspects." Ph.D. diss., Paris III–Sorbonne la Nouvelle, 1986.

Pimentel, Juan. *Testigos del mundo: Ciencia, literatura y viajes en la Ilustración.* Madrid: Marcial Pons, 2003.

Pinedo, Fernández de, Alberto Gil Novales, and Albert Dérozier. *Centralismo, Ilustracion y agonía del Antiguo Régimen.* Barcelona: Labor, 1980.

Platt, Peter G., ed. *Wonders, Marvels, and Monsters in Early Modern Culture.* Newark: University of Delaware, 1999.

Pomian, Krzysztof. *Collectionneurs, amateurs et curieux. Paris, Venise: XVIe–XVIIIe siècle.* Paris: Gallimard, 1987.

Poole, Deborah. *Vision, Race, and Modernity: A Visual Economy of the Andean Image World.* Princeton: Princeton University Press, 1997.

Popkin, Richard H. "Hume and Isaac de Pinto." *Texas Studies in Literature and Language* 12 (1970): 417–30.

Porter, Roy. "The New Taste for Nature in the Eighteenth Century." *Linnean* 4, no. 1 (1988): 14–30.

Porter, Roy, Simon Schaffer, Jim Bennett, and Olivia Brown. *Science and Profit in 18th-Century London.* Cambridge: Whipple Museum of the History of Science, 1985.

Pratt, Mary Louise. *Imperial Eyes: Travel Writing and Transculturation*. New York: Routledge, 1991.

"El proceso contra Jorge Juan y Antonio de Ulloa en Quito (1737)," *Anuario de estudios americanos* 5 (1948): 747–80.

Puerto Sarmiento, Francisco Javier. *La ilusión quebrada. Botánica, sanidad y política científica en la España Ilustrada*. Madrid: CSIC/Serbal, 1988.

Putnam, Lara. "To Study the Fragments/Whole: Microhistory and the Atlantic World." *Journal of Social History* 39, no. 3 (2006): 615–30.

Quesada, Aurelio Miró. *El Inca Garcilaso*. Lima: Pontificia Universidad Católica del Peru, 1994.

Quintanilla, Joaquín F. *Naturalistas para una corte ilustrada*. Madrid: Ediciones Doce Calles, 1999.

Raber, Karen L. "Warrior Women in the Plays of Cavendish and Killigrew." *SEL: Studies in English Literature* 40, no. 3 (Summer 2000): 413–33.

Raj, Kapil. "Circulation and the Emergence of Modern Mapping: Great Britain and Early Colonial India, 1764–1820." In *Society and Circulation: Mobile People and Itinerant Cultures in South Asia, 1750–1950*. Edited by Claude Markovits, Jacques Pouchepadass, and Sanjay Subrahmanyam. Delhi: Permanent Black, 2003.

Raminelli, Ronald. "Charles La Condamine." In *Dicionário do Brasil Colonial (1500–1808)*. Edited by Ronaldo Vainfas. Rio de Janeiro: Editora Objetiva, 2000.

Ramos Gomez, Luis J. *Epoca, genesis y texto de las "Noticias Secretas de America", de Jorge Juan y Antonio de Ulloa*. Vol. 1, *El viaje a America (1735–1745) de los tenientes de navio Jorge Juan y Antonio de Ulloa, y sus consecuencias literarias*. Madrid: Consejo Superior de Investigaciones Científicas, 1985.

Ravel, Jeffrey. *The Contested Parterre: Public Theater and French Political Culture, 1680–1791*. Ithaca: Cornell University Press, 1999.

Regourd, François. "Sciences et colonisation sous l'Ancien Régime: Le cas de la Guyane et des Antilles françaises, XVIIe–XVIIIe siècles." Ph.D. diss., University of Bordeaux-3, 2000.

Reilly, P. Conor, S.J. *Athanasius Kircher S.J.: Master of a Hundred Arts 1602–1680*. Wiesbaden: Edizioni del Mondo, 1974.

Reinhartz, Dennis. *The Cartographer and the Literati: Herman Moll and His Intellectual Circle*. Lewiston: E. Mellen Press, 1997.

Reisz, Emma. "Curiosity and Rubber in the French Atlantic." *Atlantic Studies* 4 (2007): 5–26.

Revue Générale du Caoutchouc. Organe Officiel du Syndicat du Caoutchouc, 125. Paris, 1936.

Riskin, Jessica. *Science in the Age of Sensibility: The Sentimental Empiricists of the French Enlightenment*. Chicago: University of Chicago, 2002.

Robbins, Louise E. *Elephant Slaves and Pampered Parrots: Exotic Animals in Eighteenth-Century Paris*. Baltimore: Johns Hopkins University Press, 2002.

Roberts, J. A. G. "L'image de la Chine, dans l'Encyclopédie." *Recherches sur Diderot et sur* l'Encyclopédie 22 (April 1997): 87–107.

Robinson, David J. *Mil leguas por América: De Lima a Caracas, 1740–41.* Bogota: Banco de la República, 1992.

Roche, Daniel. *Les républicains des lettres.* Paris: Fayard, 1988.

Rodríguez Castelo, Hernán. *El Padre Samuel Fritz: Diario.* Quito: Academia Ecuatoriana de la Lengua, 1997.

Rodríguez Garza, Francisco Javier, and Lucino Gutiérrez Herrera. *Ilustración española, reformas borbónicas y liberalismo temprano en México.* Azcapotzalco: Universidad Autónoma Metropolitana, 1992.

Román Gutiérrez, José Francisco. *Las reformas borbónicas y el nuevo orden colonial.* Mexico City: Instituto Nacional de Antropología e Historia, 1998.

Rose, Carol M. "Possession as the Origin of Property." *University of Chicago Law Review* 51 (1985): 73–88.

Rose, Mark. "The Author as Proprietor: *Donaldson v. Beckett* and the Genealogy of Modern Authorship." *Representations* 23 (1988): 51–85.

Rosenberg, Daniel. "Early Modern Information Overload." *Journal of the History of Ideas* 64, no. 1 (January 2003): 1–9.

Rosenfeld, Sophia. *A Revolution in Language: The Problem of Signs in Late Eighteenth-Century France.* Stanford: Stanford University Press, 2001.

Rouse, Joseph. "What Are Cultural Studies of Scientific Knowledge?" *Configurations* 1, no. 1 (1993): 57–94.

Rousseau, G. S., and Roy Porter. Introduction to *Exoticism in the Enlightenment.* Edited by G. S. Rousseau and Roy Porter. Manchester: Manchester University Press, 1990.

Rousseau, G. S., and Roy Porter, eds. *Exoticism in the Enlightenment.* Manchester: Manchester University Press, 1990.

Rudel, Yves-Marie. *Duguay-Trouin, corsaire et chef d'escadre (1673–1763).* Paris: Perrin, 1973.

Rumazo, José. *Documentos para la Historia de la Audiencia de Quito.* 8 vols. Madrid: Afrodisio Aguado, 1948–50.

Russo, Alessandra. *El realismo circular: Tierras, espacios y paisajes de la cartografía novohispana, siglos XVI y XVII.* Mexico: Universidad Autónoma de México and Instituto de Investigaciones Estéticas, 2005.

Safier, Neil. "The Confines of the Colony: Boundaries, Ethnographic Landscapes, and Imperial Cartography in Iberoamerica." Edited by James R. Akerman. Chicago: University of Chicago Press, 2009.

———. "'Every day that I travel . . . is a page that I turn': Reading and Observing in Eighteenth-Century Amazonia." *Huntington Library Quarterly* 70, no. 1 (2007): 103–28.

———. "Fruitless Botany: Joseph de Jussieu's South American Odyssey." In *Science and Empire in the Atlantic World.* Edited by James Delbourgo and Nicholas Dew, 203–24. New York: Routledge, 2007.

———. "Subalternidade Tropical? Interrogar o trabalho do índio remador nos caminhos fluviais amazônicos." In *O trabalho mestiço: Maneiras de pensar e formas de viver — séculos XVI a XIX*. Edited by Eduardo França Paiva and Carla Anastasia, 427–42. São Paulo: Annablume/PPGH-UFMG, 2002.

———. "Unveiling the Amazon to European Science and Society: The Reading and Reception of La Condamine's *Relation abrégée*." *Terrae Incognitae* 33 (2001): 33–47.

Said, Edward W. *Orientalism*. New York: Vintage Books, 1994.

Saint-Amand, Pierre. *Diderot: Le labyrinthe de la relation*. Paris: J. Vrin, 1984.

Salomon, Frank. "Yumbo Ñan: La vialidad indígena en el noroccidente de Pichincha y el trasfondo aborigen del camino de Pedro Vicente Maldonado." *Cultura: Revista del Banco Central del Ecuador* 8, no. 24B (January–April 1986): 611–26.

———. *Los Yumbos, Niguas y Tsatchila o "Colorados" durante la Colonia Española: Etnohistoria del Noroccidente de Pichincha, Ecuador*. Quito: Ediciones Abya-Yala, 1997.

Sanchez, Jean-Pierre. *Mythes et légendes de la conquête de l'Amérique*. 2 vols. Rennes: Presses universitaires de Rennes, 1996.

Sánchez-Blanco, Francisco. *La mentalidad ilustrada*. Madrid: Taurus, 1999.

Sarukkai, Sundar. *Translating the World: Science and Language*. Lanham, Md.: University Press of America, 2002.

Schaffer, Simon. "Natural Philosophy and Public Spectacle in the Eighteenth Century." *History of Science* 21, part 1, no. 51 (1983): 1–43.

———. "Self Evidence." In *Questions of Evidence: Proof, Practice, and Persuasion across the Disciplines*. Edited by James Chandler, Arnold I. Davidson, and Harry Harootunian. Chicago: University of Chicago Press, 1993.

Schiebinger, Londa. "Forum Introduction: The European Colonial Science Complex." *Isis* 96 (2005): 52–55.

———. *Plants and Empire: Colonial Bioprospecting in the Atlantic World*. Cambridge: Harvard University Press, 2004.

———. "Prospecting for Drugs." In *Colonial Botany: Science, Commerce, and Politics in the Early Modern World*. Edited by Londa Schiebinger and Claudia Swan. Philadelphia: University of Pennsylvania Press, 2005.

Schiebinger, Londa, and Claudia Swan, eds. *Colonial Botany: Science, Commerce, and Politics in the Early Modern World*. Philadelphia: University of Pennsylvania Press, 2005.

Schiff, Stacy. *A Great Improvisation: Franklin, France, and the Birth of America*. New York: Henry Holt, 2005.

Schmidt, Benjamin. *Innocence Abroad: The Dutch Imagination and the New World, 1570–1670*. Cambridge: Cambridge University Press, 2001.

Schnapper, Antoine. *Le géant, la licorne, la tulipe: Collections françaises au XVIIe siècle*. Paris: Flammarion, 1988.

Schramm, Helmar, Ludger Schwarte, and Jan Lazardzig, eds. *Collection, Laboratory, Theater: Scenes of Knowledge in the Seventeenth Century*. Berlin: Walter de Gruyter, 2005.

Schwartz, Stuart B., ed. *Implicit Understandings: Observing, Reporting, and Reflecting on the Encounters between Europeans and Other Peoples in the Early Modern World*. Cambridge: Cambridge University Press, 1994.

Schwarz, Kathryn. *Tough Love: Amazon Encounters in the English Renaissance*. Durham, N.C.: Duke University Press, 2000.

Secord, Ann. "Science in the Pub: Artisan Botanists in Early Nineteenth-Century Lancashire." *History of Science* 32, no. 97 (1994): 269–315.

Seed, Patricia. *Ceremonies of Possession in Europe's Conquest of the New World, 1492–1640*. Cambridge: Cambridge University Press, 1995.

Seremetakis, C. Nadia, ed. *The Senses Still: Perception and Memory as Material Culture in Modernity*. Boulder: Westview Press, 1994.

Sgard, Jean, ed. *Dictionnaire des Journaux*. Paris: Universitas, 1991.

Shalev, Zur. "Measurer of All Things: John Greaves (1602–1652), the Great Pyramid, and Early Modern Metrology." *Journal of the History of Ideas* 64, no. 4 (October 2002): 555–75.

Shapin, Steven. "The House of Experiment in Seventeenth-Century England." *Isis* 79 (1988): 373–404.

———. "Pump and Circumstance: Robert Boyle's Literary Technology." *Social Studies of Science* 14, no. 4 (November 1984): 481–520.

———. *The Scientific Revolution*. Chicago: University of Chicago Press, 1996.

———. *A Social History of Truth: Civility and Science in Seventeenth-Century England*. Chicago: University of Chicago Press, 1994.

Shapin, Steven, and Simon Schaffer. *Leviathan and the Air-Pump: Hobbes, Boyle, and the Experimental Life*. Princeton: Princeton University Press, 1985.

Sheehan, Jonathan. "From Philology to Fossils: The Biblical Encyclopedia in Early Modern Europe." *Journal of the History of Ideas* 64, no. 1 (January 2003): 41–60.

Shelton, Anthony Alan. "Cabinets of Trangression: Renaissance Collections and the Incorporation of the New World." In *The Cultures of Collecting*. Edited by John Elsner and Roger Cardinal, 177–203. Cambridge: Harvard University Press, 1994.

Showalter, English. *Françoise de Graffigny: Her Life and Works*. Studies on Voltaire and the Eighteenth Century, 11. Oxford: Voltaire Foundation, 2004.

———. "Les *Lettres d'une Péruvienne*: Composition, Publication, Suites." *Archives et Bibliothèques de Belgique* 54 (1983): 14–28.

Simpson, Percy. *Proof-reading in the Sixteenth, Seventeenth and Eighteenth Centuries*. 1935. Reprint, Oxford: Oxford University Press, 1970.

Smith, Anthony. *The Lost Lady of the Amazon: The Story of Isabela Godin and Her Epic Journey*. New York: Carroll and Graf, 2003.

Smith, James R. *From Plane to Spheroid: Determining the Figure of the Earth from 3000 B.C. to the Eighteenth-Century Lapland and Peruvian Survey Expeditions*. Rancho Cordova: Landmark Enterprises, 1986.

Smith, Pamela H., and Paula Findlen, eds. *Merchants and Marvels: Commerce, Science, and Art in Early Modern Europe*. New York: Routledge, 2002.

Smith, Thomas R. "Cruz Cano's Map of South America, Madrid, 1775: Its Creation, Adversities, and Rehabilitation." *Imago Mundi* 20 (1966): 49–78.

Solomon, Julie Robin. "'To Know, To Fly, To Conjure': Situating Baconian Science at the Juncture of Early Modern Modes of Reading." *Renaissance Quarterly* 44 (1991): 513–58.

Soto Arango, Diana, Miguel Ángel Puig-Samper, and Luis Carlos Arboleda, eds. *La Ilustración en América colonial*. Madrid: CSIC/Doce Calles, 1995.

Soto Arango, Diana, Miguel Ángel Puig-Samper, and Maria Dolores González-Ripoll, eds. *Científicos criollos e Ilustración*. Madrid: CSIC/Doce Calles, 1999.

Spary, E. C. "Of Nutmegs and Botanists: The Colonial Cultivation of Botanical Identity." In *Colonial Botany: Science, Commerce, and Politics in the Early Modern World*. Edited by Londa Schiebinger and Claudia Swan. Philadelphia: University of Pennsylvania Press, 2005.

———. *Utopia's Garden: French Natural History from Old Regime to Revolution*. Chicago: University of Chicago Press, 2000.

Stafford, Barbara. *Voyage into Substance: Art, Science, Nature, and the Illustrated Travel Account, 1760–1840*. Cambridge: MIT Press, 1984.

Stagl, Justin. *A History of Curiosity: The Theory of Travel, 1550–1800*. Australia: Harwood Academic Publishers, 1995.

———. "The Methodising of Sixteenth-Century Travel." *History and Anthropology* 4 (1990): 303–38.

Stallybrass, Peter, and Margreta DeGrazia. "The Materiality of the Shakespearean Text." *Shakespeare Quarterly* 44, no. 3. (Autumn 1993): 255–83.

Stallybrass, Peter, Roger Chartier, J. Franklin Mowry, and Heather Wolfe. "Hamlet's Tables and the Technologies of Writing in Renaissance England." *Shakespeare Quarterly* 55, no. 4 (2004): 379–419.

Starobinski, Jean. "Remarques sur *l'Encyclopédie*." *Revue de metaphysique et de morale* 75 (1970): 284–91.

Stenger, Gerhard. "L'horreur des voyages." *Le magazine littéraire* (France) 391 (2000): 30–32.

Stoler, Ann Laura, and Frederick Cooper. "Between Metropole and Colony." In *Tensions of Empire: Colonial Cultures in a Bourgeois World*. Edited by Frederick Cooper and Ann Laura Stoler. Berkeley: University of California Press, 1997.

Stolzenberg, Daniel. "Egyptian Oedipus: Antiquarianism, Oriental Studies and Occult Philosophy in the Work of Athanasius Kircher." Ph.D. diss., Stanford University, 2004.

Stolzenberg, Daniel, ed. *The Great Art of Knowing: The Baroque Encyclopedia of Athanasius Kircher*. Fiesole: Edizioni Cadmo, 2001.

Storey, William K., ed. *Scientific Aspects of European Expansion*. Aldershot: Variorum, 1996.

Stroup, Alice. *A Company of Scientists: Botany, Patronage, and Community at the Seventeenth-Century Parisian Royal Academy of Sciences*. Berkeley: University of California, 1990.

Subrahmanyam, Sanjay. "Connected Histories: Notes toward a Reconfiguration of Early Modern Eurasia." In *Beyond Binary Histories: Re-imagining Eurasia to c. 1830*. Edited by Victor Lieberman, 289–316. Ann Arbor: University of Michigan Press, 1999.

———. *Explorations in Connected Histories: Mughals and Franks*. New York: Oxford University Press, 2005.

Sutcliffe, Adam. "Can a Jew be a Philosophe? Isaac de Pinto, Voltaire, and Jewish Participation in the European Enlightenment." *Jewish Social Studies* 6, no. 3 (2000): 31–51.

Swann, Marjorie. *Curiosities and Texts*. Philadelphia: University of Pennsylvania Press, 2001.

Switzer, Richard. "America in the *Encyclopédie*." *Studies on Voltaire and the Eighteenth Century* 58 (1967): 1481–99.

Taton, René, ed. *Enseignement et diffusion des sciences en France au XVIIIe siècle*. Paris: Hermann, 1964.

Taussig, Michael. *Mimesis and Alterity: A Particular History of the Senses*. New York: Routledge, 1993.

Terán Najar, Rosemarie. *Los proyectos del imperio borbónico en la Real Audiencia*. Quito: Abya-Yala, 1988.

Terrall, Mary. "Heroic Narratives of Quest and Discovery." *Configurations* 6, no. 2 (1998): 223–42.

———. *The Man Who Flattened the Earth: Maupertuis and the Sciences in the Enlightenment*. Chicago: University of Chicago Press, 2002.

———. "Representing the Earth's Shape: The Polemics Surrounding Maupertuis' Expedition to Lapland." *Isis* 83, no. 2 (1992): 218–37.

Thomas, Keith. *Man and the Natural World: A History of the Modern Sensibility*. Oxford: Oxford University Press, 1983.

"Thomas François Dalibard." *Dictionary of Scientific Biography*. Edited by Charles Coulston Gillispie. New York: Scribner's, 1971.

Thorndike, Lynn. "*L'Encyclopédie* and the History of Science." *Isis* 6 (1924): 361–86.

Toribio Medina, José. *Historia de la imprenta en los antiguos dominios españoles de América y Oceania*. Santiago: Fondo Histórico y Bibliográfico José Toribio Medina, 1958.

Torre Revello, José. *El Libro, la imprenta y el periodismo en América durante la Dominación Española*. Buenos Aires: Casa Jacobo Peuser, 1940.

El Tratado de Tordesillas y su época. 3 vols. [Valladolid]: Junta de Castilla y León, 1995.

Triger, Robert-Gustave-Marie. "Le Collège de Crannes et Thomas-François Dalibard, naturaliste et physicien." *Bulletin de la Société d'agriculture, sciences et arts de la Sarthe* 30 (1885): 189–204.

Trouille, Mary. "Eighteenth-Century Amazons of the Pen: Stéphanie de Genlis and Olympe de Gouges." In *Femmes savantes et femmes d'esprit*. Edited by Roland Bonnel and Catherine Rubinger, 341–70. New York: Peter Lang, 1997.

Trouillot, Michel-Rolph. *Silencing the Past: Power and the Production of History.* Boston: Beacon Press, 1995.

Trystram, Florence. *Le procès des étoiles: Récit de la prestigieuse expédition de trois savants français en Amérique du Sud et des mésaventures qui s'ensuivent (1735–1771).* Paris: Seghers, 1979.

Uzcátegui Andrade, Byron. *Los llanganates y la tumba de Atahualpa.* Quito: Ediciones Abya-Yala/Instituto Panamericano de Geografía e Historia, 1992.

Van Damme, Stéphane. *Paris, capitale philosophique: De la Fronde à la Révolution.* Paris: Odile Jacob, 2005.

Van Helden, Albert, and Thomas L. Hankins, eds. *Instruments.* In *Osiris,* second series, no. 9. Chicago: University of Chicago Press, 1994.

Varner, John Grier. *El Inca: The Life and Times of Garcilaso de la Vega.* Austin: University of Texas Press, 1968.

Vergé-Franceschi, Michel. "Duguay-Trouin (1673–1736): Un corsaire, un officier général, un mythe." *Revue historique* 295, no. 2 (1996): 333–52.

Vermeulen, Han F., and Arturo Alvarez Roldán, eds. *Fieldwork and Footnotes: Studies in the History of European Anthropology.* New York: Routledge, 1995.

Villani, Anna Luigia. *Dizionario di concordanze e frequenze della "Relation abrégée" di Charles-Marie de La Condamine.* Florence: Università di Firenze, 1984.

Villavicencio, Manuel de. *Geografía de la República del Ecuador.* New York: Impr. De R. Craighead, 1858.

Vovelle, Michel, ed. *L'homme des lumières.* Paris: Éditions du Seuil, 1996.

Vyverberg, Henry. *Human Nature, Cultural Diversity, and the French Enlightenment.* Oxford: Oxford University Press, 1989.

Wachtel, Nathan. *The Vision of the Vanquished.* New York: Barnes and Noble Imports, 1977.

Wakefield, Celia. *Searching for Isabel Godin.* Berkeley: Creative Arts, 1999.

Warner, Deborah Jean. "What Is a Scientific Instrument, When Did It Become One, and Why?" *British Journal for the History of Science* 23 (1990): 83–93.

Washburn, Douglas Alan. "The Bourbon Reforms: A Social and Economic History of the Audiencia of Quito, 1760–1810." Ph.D. diss., University of Texas at Austin, 1984.

Whitaker, Katie. "The Culture of Curiosity." In *Cultures of Natural History.* Edited by Nicholas Jardine, James A. Secord, and E. C. Spary, 75–90. Cambridge: Cambridge University Press, 1996.

Whitaker, Robert. *The Mapmaker's Wife: A True Tale of Love, Murder, and Survival in the Amazon.* New York: Basic Books, 2004.

White, Hayden. *The Content of the Form: Narrative Discourse and Historical Representation.* Baltimore: Johns Hopkins University Press, 1987.

Whitehead, Neil L. *Histories and Historicities in Amazonia.* Lincoln: University of Nebraska Press, 2003.

———. Introduction to Walter Ralegh, *The Discoverie of the Large, Rich, and Bewtiful Empyre of Guiana.* Norman: University of Oklahoma Press, 1997.

Wijler, Jacob Samuel. *Isaac de Pinto, sa vie et ses oeuvres*. Apeldoorn: C. M. B. Dixon, 1923.

Woodmansee, Martha. "The Genius and the Copyright: Economic and Legal Conditions of the Emergence of the 'Author.'" *Eighteenth-Century Studies* 17 (1984): 425–48.

Woodmansee, Martha, and Peter Jaszi, eds. *The Construction of Authorship: Textual Appropriation in Law and Literature*. Durham, N.C.: Duke University Press, 1994.

Woodward, David. *Maps as Prints in the Italian Renaissance: Makers, Distributors and Consumers*. London: British Library, 1996.

Woodward, David, and J. B. Harley. Introduction to *The History of Cartography*. Chicago: University of Chicago Press, 1987.

Woodward, David, and J. B. Harley, eds. *History of Cartography*. 3 vols. Chicago: University of Chicago, 1987–2007.

Woodward, David, and G. Malcolm Lewis, eds. *The History of Cartography: Cartography in the Traditional African, American, Arctic, and Pacific Societies*. Vol. 2, no. 3. Chicago: University of Chicago Press, 1998.

Woolf, Harry. *The Transit of Venus: A Study of Eighteenth-Century Science*. Princeton: Princeton University Press, 1959.

Yeo, Richard. "Copyright and Public Knowledge." In *Encyclopaedic Visions*, 195–221.

———. "Encyclopaedic Knowledge." In *Books and the Sciences in History*. Edited by Marina Frasca-Spada and Nick Jardine, 207–24. Cambridge: Cambridge University Press, 2000.

———. *Encyclopaedic Visions: Scientific Dictionaries and Enlightenment Culture*. Cambridge: Cambridge University Press, 2001.

———. "Ephraim Chambers's *Cyclopaedia* (1728) and the Tradition of Commonplaces." *Journal of the History of Ideas* 57, no. 1 (1996): 157–75.

———. "Reading Encyclopedias: Science and the Organization of Knowledge in British Dictionaries of Arts and Sciences, 1730–1850." *Isis* 82 (1991): 24–49.

———. "A Solution to the Multitude of Books: Ephraim Chambers's *Cyclopaedia* (1728) as "the Best Book in the Universe." *Journal of the History of Ideas* 64, no. 1 (January 2003): 61–72.

Zamora, Margarita. *Language, Authority, and Indigenous History in the* Comentarios reales de los Incas. Cambridge: Cambridge University Press, 1988.

Zúñiga, Neptalí. *Pedro Vicente Maldonado, un científico de América*. Madrid: Publicaciones Españolas, 1951.

———. *La expedición científica de Francia del siglo XVIII en la Presidencia de Quito*. Quito: Municipio de Quito, 1986.

Index

Page numbers in italics refer to figures.